S0-AGA-183

TEACHER'S EDITION

EARTH'S NATURAL RESOURCES

EarthComm®

EARTH SYSTEM SCIENCE IN THE COMMUNITY

Michael J. Smith Ph.D.
Principal Investigator

John B. Southard Ph.D.
Senior Writer

Ruta Demery
Editor

Emily Crum
Contributing Writer

Developed by the American Geological Institute
Supported by the National Science Foundation and
the American Geological Institute Foundation

Published by
It's About Time Inc., Armonk, NY

It's About Time, Inc.
84 Business Park Drive, Armonk, NY 10504
Phone (914) 273-2233 Fax (914) 273-2227
Toll Free (888) 698-TIME
www.Its-About-Time.com

Publisher
Laurie Kreindler

| **Project Manager** | **Project Coordinator** | **Design** |
| Ruta Demery | Matthew Smith | John Nordland |

| **Editor** | **Contributing Writer** | **Production Manager** |
| Ena de Jong | Emily Crum | Joan Lee |

All student activities in this textbook have been designed to be as safe as possible, and have been reviewed by professionals specifically for that purpose. As well, appropriate warnings concerning potential safety hazards are included where applicable to particular activities. However, responsibility for safety remains with the student, the classroom teacher, the school principals, and the school board.

EarthComm® is a registered trademark of the American Geological Institute. Registered names and trademarks, etc., used in this publication, even without specific indication thereof, are not to be considered unprotected by law.

It's About Time™ is a registered trademark of It's About Time, Inc. Registered names and trademarks, etc., used in this publication, even without specific indication thereof, are not to be considered unprotected by law.

© Copyright 2002: American Geological Institute

All rights reserved. No part of this publication may be reproduced, stored in a retrieval system, or transmitted, in any form or by any means, electronic, mechanical, photocopying, recording, or otherwise, without the prior written permission of the copyright owner.

Care has been taken to trace the ownership of copyright material contained in this publication. The publisher will gladly receive any information that will rectify any reference or credit line in subsequent editions.

Printed and bound in the United States of America

ISBN #1-58591-068-6

2 3 4 5 QC 06 05 04

This project was supported, in part, by the
National Science Foundation (grant no. ESI-9452789)

Opinions expressed are those of the authors and not necessarily those of the National Science Foundation or the donors of the American Geological Institute Foundation.

Student's Edition Illustrations and Photos

R5, R6, R9, R30, R46, R48, R64, R66, R73, R74 (anemometer), R78 Fig. 1–2, R98, R99, R105, R119, R120, R125, R128, R137, R138 (left, right), R153 (bottom), R157, R159, R160, R164 (top, bottom), R167 (top, bottom), R178, R181 Figs. 3, 4a–4b, R182 Figs. 5a–5b, R185, R186, R197, illustrations by Stuart Armstrong

R40, illustration by Stuart Armstrong, source: Energy Information Administration

R23, R58, R59, illustrations by Stuart Armstrong, source: Energy Information Agency

R31, R32, R41 Fig. 2, illustrations by Stuart Armstrong, source: Kentucky Geological Survey

R74, map illustrations by Stuart Armstrong, source: National Renewable Energy Laboratory

R47, illustration by Stuart Armstrong, source: Oak Ridge National Laboratories

R44, R200, illustrations by Stuart Armstrong, source: United States Environmental Protection Agency

R39, R112, R148, R173, illustrations by Stuart Armstrong, source: United States Geological Survey (US Geological Survey)

R126, photo by Atlantic-Richfield Company (ARCO)

R115, R132, R140, photos by ASARCO

R60, photo by The Bakersfield Californian

R10 Fig. 2, photo by Charles Bickford, The Taunton Press

R163, photo by John Buchanan

R4, R16, R25, R35, R43, R53, R62, R72, R88, R96, R111, R118, R127, R136, R146, R156, R169, R177, R184, R196, illustrations by Tomas Bunk

R20, R21, R22, R27, maps by Bureau of Economic Geology, source: Energy Information Agency

R56, R81, maps by Bureau of Economic Geology at the University of Texas at Austin

R123 Fig. 2, illustration by Burmar Technical Corporation

R97 Figs. A, C–E, G–L, R103, R110 Fig. 2, photos by Richard Busch

R166, R175 Fig. 5, R189, R190, R191, R193 Figs. 5–6, photos by California Department of Water Resources

R95, photo by Caitlin Callahan

R102, photo by Corbis

R201, photo by Cornerstone Water Systems

R12 Fig. 5, R70 Fig. 3, R79 Fig. 3, photos by Department of Energy

R19, R187, photos by Digital Stock Royalty Free Images

R29, R59 Fig. 2, R69, R70 Fig, 2, photos by Digital Vision Royalty Free Images

R133, photo by Gold Institute

R139 Fig. 2, photo by George James

R192, photo by Maureen Keller, Bigelow Laboratory for Ocean Sciences

R91, photo by Nathan Lamkey

R124, photo by Michelle Markley

R162, photo by Gene McSweeney

R151, R199, photos by Bruce Molnia

R174 Fig. 4, photo by Martin Muller

R130, photo by Alma Paty

R86, (left, bottom), R87 (bottom), R90, R93, R113, R139 Fig. 1, R144 (left, center), R145 (right), R152, R2, R3, R10 Fig. 3, R13 Fig. 6, R76, R80, R80, R153 (top), R174 Fig. 3, R180 Figs. 1–2, photos by PhotoDisc

R26 A–D, R49, R87 (top), R92, R97 (B, F), R103 Fig. 3, R104, R106 Figs. 6A–D, R107 Figs. 7A–D, R108 Figs. 8–9, R26 A–D, R49, R114 Fig. 2, R164, R174 Fig. 2, photos by Doug Sherman, Geo File Photography

R129, photo by Mark Snyder Photography

R175 Fig. 6, photo by South Florida Water Management District

R15, photo by Tom Sponheim

R123 Fig. 3, photo by George Springston, Vermont Geological Survey

R154, photo by Leslie Taylor

R114 Fig. 3, R122, R165, photos by US Geological Survey

R41 Fig. 3, R131, photos by Peter Warwick, US Geological Survey

R11 Fig. 4, photo by D. R. "Doc" Young Photography

Taking Full Advantage of *EarthComm* Through Professional Development

Implementing a new curriculum is challenging. That is why It's About Time Publishing has partnered with the American Geological Institute, developers of *EarthComm*, to provide a full range of professional development services. The sessions described below were designed to help you deepen your understanding of the content, pedagogy, and assessment strategies outlined in this Teacher's Edition, and adapt the program to suit the needs of your students and your local and state standards and curriculum frameworks.

Professional Development Services Available

Implementation Workshops

Two- to five-day sessions held at your site that prepare you to implement the inquiry, systems, and community-based approach to learning Earth Science featured in *EarthComm*. These workshops can be tailored to serve the needs of your school district, with chapters selected from the modules based on local or state curricula and framework criteria.

Program Overviews

One- to three-day introductory sessions that provide a complete overview of the content and pedagogy of the *EarthComm* program, as well as hands-on experience with activities from specific chapters. Program overviews are designed in consultation with school districts, counties, and SSI organizations.

Regional New-User Summer Institutes

Two- to five-day sessions that are designed to deepen your Earth Science content knowledge, and to prepare you to teach through inquiry. Guidance is provided in the gathering and use of appropriate materials and resources and specific attention is directed to the assessment of student learning.

Leadership Institutes

Six-day summer sessions supported by the American Geological Institute that are designed to prepare current users for professional development leadership and mentoring within their districts or as consultants for It's About Time.

Follow-up Workshops

One- to two-day sessions that provide additional Earth Science content and pedagogy support to teachers using the program. These workshops focus on identifying and solving practical issues and challenges to implementing an inquiry-based program.

Mentoring Visits

One-day visits that can be tailored to your specific needs that include class visits, mentoring users of the program, and in-service sessions.

Please fill in the form below to receive more information about participating in one of these Professional Development Services. The form can be directly faxed to our Professional Development at 914-273-2227. Our department will contact you to discuss further details and fees.

District/School: _____ Phone: _____

Address: _____

Contact Name: _____ Title: _____

E-mail: _____ Fax: _____

School Enrollment: _____ Number of Students Impacted: _____ Grade Level: _____

Have you purchased the following: ❏ Student Editions   ❏ Teacher Editions   ❏ Kits

Briefly explain how you plan to implement or how you are implementing the program in your school.

_____

_____

_____

Table of Contents

EarthComm Team   vii

Acknowledgements   viii

EarthComm: Earth System Science in the Community   xiii

EarthComm Modules and Chapters   xiv

EarthComm: Correlation to the National Science Education Standards   xv

EarthComm "Big Ideas"   xvi

EarthComm Goals and Expectations for Teachers   xviii

EarthComm Goals and Expectations for Students   xix

EarthComm Curriculum Design   xx

Using *EarthComm* Features in Your Classroom   xxii

EarthComm Assessment Opportunities   xxxiii

EarthComm Assessment Tools   xxxiv

Reviewing and Reflecting upon Your Teaching   xxxvi

Using the *EarthComm* Web Site   xxxvii

Managing Collaborative Group Learning   xxxix

Enhancing *Natural Resources* with *EarthView Explorer*   xl

Enhancing *Natural Resources* with GETIT   xliv

Module Overview: *Earth's Natural Resources*   xlv

Chapter 1 Energy Resources and Your Community   1

Chapter Overview   2

Sample Outline   3

National Science Education Standards   5

Key Science Concepts and Skills   7

Equipment List   9

Getting Started   13

Chapter Challenge and Assessment Criteria   15

Assessment Rubric for Chapter Challenge on Energy Resources   16

Activity 1: Exploring Energy Resource Concepts   18

Activity 2: Electricity and Your Community   58

Activity 3: Energy from Coal   94

Activity 4: Coal and Your Community   124

Activity 5: Environmental Impacts and Energy Consumption   154

Activity 6: Petroleum and Your Community   184

Activity 7: Oil and Gas Production   210

Activity 8: Renewable Energy Sources—Solar and Wind   246

End-of-Chapter Assessment   286

Teacher Review   290

Chapter 2 Mineral Resources and Your Community 293

Chapter Overview 294
Sample Outline 295
National Science Education Standards 296
Key Science Concepts and Skills 297
Equipment List 298
Getting Started 301
Chapter Challenge and Assessment Criteria 303
Assessment Rubric for Chapter Challenge on Mineral Resources 304

Activity 1: Materials Used for Beverage Containers in Your Community 306
Activity 2: What are Minerals 332
Activity 3: Where are Mineral Resources Found? 380
Activity 4: How are Minerals Found? 406
Activity 5: What are the Costs and Benefits of Mining Minerals? 434
Activity 6: How are Minerals Turned into Usable Materials? 458

End-of-Chapter Assessment 486
Teacher Review 490

Chapter 3 Water Resources and Your Community 493

Chapter Overview 494
Sample Outline 495
National Science Education Standards 496
Key Science Concepts and Skills 497
Equipment List 498
Getting Started 501
Chapter Challenge and Assessment Criteria 503
Assessment Rubric for Chapter Challenge on Water Resources 504

Activity 1: Sources of Water in the World and in Your Community 506
Activity 2: How Does Your Community Maintain Its Water Supply? 538
Activity 3: Using and Conserving Water 574
Activity 4: Water Supply and Demand: Water Budgets 602
Activity 5: Water Pollution 626
Activity 6: Water Treatment 660

End-of-Chapter Assessment 686
Teacher Review 690
EarthComm Assessments 693

EarthComm Team

EarthComm Project Staff

Michael J. Smith, Principal Investigator
        Director of Education, American Geological Institute
John B. Southard, Senior Writer
        Professor of Geology Emeritus, Massachusetts Institute of Technology
Emily J. Crum, Contributing Writer
        American Geological Institute
Matthew Smith, Project Coordinator
        American Geological Institute
Caitlin N. Callahan, Project Assistant
        American Geological Institute
William S. Houston, Field Test Coordinator
        American Geological Institute
Robert A. Bernoff, Field Test Evaluator
        Professor Emeritus, Penn State University
Do Yong Park, Field Test Evaluator
        University of Iowa
Larry G. Enochs, Pilot Test Evaluator
        Professor of Science Education, Oregon State University

Original *EarthComm* Project Personnel

Charles Groat, United States Geological Survey
Marilyn Suiter, American Geological Institute
Bonnie Brunkhorst, UC San Bernardino
Richard M. Busch, West Chester University
Steven C. Good, West Chester University
John Carpenter, University of South Carolina
Linda Knight, Houston, Texas
Bob Ridky, University of Maryland

National Advisory Board

Harold Pratt, Chair
Jane Crowder, Bellevue, Washington
Don Lewis, Lafayette, California
Arthur Eisenkraft, Bedford Public Schools, New York
Tom Ervin, LeClaire, Iowa
Mary Kay Hemenway, University of Texas at Austin
William Leonard, Clemson University
Wendell Mohling, National Science Teachers Association
Barb Tewksbury, Hamilton College
Laure Wallace, United States Geological Survey

National Science Foundation Program Officers

Gerhard Salinger
Patricia Morse

Acknowledgements

Principal Investigator

Michael Smith is Director of Education at the American Geological Institute in Alexandria, Virginia. Dr. Smith worked as an exploration geologist and hydrogeologist. He began his Earth Science teaching career with Shady Side Academy in Pittsburgh, PA in 1988 and most recently taught Earth Science at the Charter School of Wilmington, DE. He earned a doctorate from the University of Pittsburgh's Cognitive Studies in Education Program and joined the faculty of the University of Delaware School of Education in 1995. Dr. Smith received the Outstanding Earth Science Teacher Award for Pennsylvania from the National Association of Geoscience Teachers in 1991, served as Secretary of the National Earth Science Teachers Association, and is a reviewer for Science Education and The Journal of Research in Science Teaching. He worked on the Delaware Teacher Standards, Delaware Science Assessment, National Board of Teacher Certification, and AAAS Project 2061 Curriculum Evaluation programs.

Senior Writer

Dr. Southard received his undergraduate degree from the Massachusetts Institute of Technology in 1960 and his doctorate in geology from Harvard University in 1966. After a National Science Foundation postdoctoral fellowship at the California Institute of Technology, he joined the faculty at the Massachusetts Institute of Technology, where he is currently Professor of Geology Emeritus. He was awarded the MIT School of Science teaching prize in 1989 and was one of the first cohorts of the MacVicar Fellows at MIT, in recognition of excellence in undergraduate teaching. He has taught numerous undergraduate courses in introductory geology, sedimentary geology, field geology, and environmental Earth Science both at MIT and in Harvard's adult education program. He was editor of the Journal of Sedimentary Petrology from 1992 to 1996, and he continues to do technical editing of scientific books and papers for SEPM, a professional society for sedimentary geology. Dr. Southard received the 2001 Neil Miner Award from the National Association of Geoscience Teachers.

Safety Reviewer Dr. Ed Robeck, Salisbury University, MD.

PRIMARY AND CONTRIBUTING AUTHORS

Earth's Dynamic Geosphere
Daniel J. Bisaccio
Souhegan High School
Amherst, NH

Steve Carlson
Middle School, OR

Warren Fish
Paul Revere School
Los Angeles, CA

Miriam Fuhrman
Carlsbad, CA

Steve Mattox
Grand Valley State University

Keith McKain
Milford Senior High School
Milford, DE

Mary McMillan
Niwot High School
Niwot, CO

Bill Romey
Orleans, MA

Michael Smith
American Geological Institute

Tom Vandewater
Colton, NY

Understanding Your Environment
Geoffrey A. Briggs
Batavia Senior High School
Batavia, NY

Cathey Donald
Auburn High School
Auburn, AL

Richard Duschl
Kings College
London, UK

Fran Hess
Cooperstown High School
Cooperstown, NY

Laurie Martin-Vermilyea
American Geological Institute

Molly Miller
Vanderbilt University

Mary-Russell Roberson
Durham, NC

Charles Savrda
Auburn University

Michael Smith
American Geological Institute

Earth's Fluid Spheres
Chet Bolay
Cape Coral High School
Cape Coral, FL

Steven Dutch
University of Wisconsin

Virginia Jones
Bonneville High School
Idaho Falls, ID

Acknowledgements (continued)

Laurie Martin-Vermilyea
American Geological Institute
Joseph Moran
University of Wisconsin
Mary-Russell Roberson
Durham, NC
Bruce G. Smith
Appleton North High School
Appleton, WI
Michael Smith
American Geological Institute

Earth's Natural Resources
Chuck Bell
Deer Valley High School
Glendale, AZ
Jay Hackett
Colorado Springs, CO
John Kemeny
University of Arizona
John Kounas
Westwood High School
Sloan, IA
Laurie Martin-Vermilyea
American Geological Institute
Mary Poulton
University of Arizona
David Shah
Deer Valley High School
Glendale, AZ
Janine Shigihara
Shelley Junior High School
Shelley, ID
Michael Smith
American Geological Institute

Earth System Evolution
Julie Bartley
University of West Georgia
Lori Borroni-Engle
Taft High School
San Antonio, TX
Richard M. Busch
West Chester University
West Chester, PA
Kathleen Cochrane
Our Lady of Ransom School
Niles, IL
Cathey Donald
Auburn High School, AL
Robert Gasataldo
Colby College
William Leonard
Clemson University
Tim Lutz
West Chester University
Carolyn Collins Petersen
C. Collins Petersen Productions
Groton, MA
Michael Smith
American Geological Institute

Content Reviewers
Gary Beck
BP Exploration
Phil Bennett
University of Texas, Austin

Steve Bergman
Southern Methodist University
Samuel Berkheiser
Pennsylvania Geologic Survey
Arthur Bloom
Cornell University
Craig Bohren
Penn State University
Bruce Bolt
University of California, Berkeley
John Callahan
Appalachian State University
Sandip Chattopadhyay
R.S. Kerr Environmental Research Center
Beth Ellen Clark
Cornell University
Jimmy Diehl
Michigan Technological University
Sue Beske-Diehl
Michigan Technological University
Neil M. Dubrovsky
United States Geological Survey
Frank Ethridge
Colorado State University
Catherine Finley
University of Northern Colorado
Ronald Greeley
Arizona State University
Michelle Hall-Wallace
University of Arizona
Judy Hannah
Colorado State University
Blaine Hanson
Dept. of Land, Air, and Water Resources
James W. Head III
Brown University
Patricia Heiser
Ohio University
John R. Hill
Indiana Geological Survey
Travis Hudson
American Geological Institute
Jackie Huntoon
Michigan Tech. University
Teresa Jordan
Cornell University
Allan Juhas
Lakewood, Colorado
Robert Kay
Cornell University
Chris Keane
American Geological Institute
Bill Kirby
United States Geological Survey
Mark Kirschbaum
United States Geological Survey
Dave Kirtland
United States Geological Survey
Jessica Elzea Kogel
Thiele Kaolin Company
Melinda Laituri
Colorado State University
Martha Leake
Valdosta State University

Donald Lewis
Happy Valley, CA
Steven Losh
Cornell University
Jerry McManus
Woods Hole Oceanographic Institution
Marcus Milling
American Geological Institute
Alexandra Moore
Cornell University
Jack Oliver
Cornell University
Don Pair
University of Dayton
Mauri Pelto
Nicolas College
Bruce Pivetz
ManTech Environmental Research Services Corp.
Stephen Pompea
Pompea & Associates
Peter Ray
Florida State University
William Rose
Michigan Technological Univ.
Lou Solebello
Macon, Gerogia
Robert Stewart
Texas A&M University
Ellen Stofan
NASA
Barbara Sullivan
University of Rhode Island
Carol Tang
Arizona State University
Bob Tilling
United States Geological Survey
Stanley Totten
Hanover College
Scott Tyler
University of Nevada, Reno
Michael Velbel
Michigan State University
Ellen Wohl
Colorado State University
David Wunsch
State Geologist of New Hampshire

Pilot Test Evaluator
Larry Enochs
Oregon State University

Pilot Test Teachers
Rhonda Artho
Dumas High School
Dumas, TX
Mary Jane Bell
Lyons-Decatur Northeast
Lyons, NE
Rebecca Brewster
Plant City High School
Plant City, FL
Terry Clifton
Jackson High School
Jackson, MI

Acknowledgements (continued)

Virginia Cooter
North Greene High School
Greeneville, TN

Monica Davis
North Little Rock High School
North Little Rock, AR

Joseph Drahuschak
Troxell Jr. High School
Allentown, PA

Ron Fabick
Brunswick High School
Brunswick, OH

Virginia Jones
Bonneville High School
Idaho Falls, ID

Troy Lilly
Snyder High School
Snyder, TX

Sherman Lundy
Burlington High School
Burlington, IA

Norma Martof
Fairmont Heights High School
Capitol Heights, MD

Keith McKain
Milford Senior High School
Milford, DE

Mary McMillan
Niwot High School
Niwot, CO

Kristin Michalski
Mukwonago High School
Mukwonago, WI

Dianne Mollica
Bishop Denis J. O'Connell
High School
Arlington, VA

Arden Rauch
Schenectady High School
Schenectady, NY

Laura Reysz
Lawrence Central High School
Indianapolis, IN

Floyd Rogers
Palatine High School
Palatine, IL

Ed Ruszczyk
New Canaan High School
New Canaan, CT

Jane Skinner
Farragut High School
Knoxville, TN

Shelley Snyder
Mount Abraham High School
Bristol, VT

Joy Tanigawa
El Rancho High School
Pico Rivera, CA

Dennis Wilcox
Milwaukee School of Languages
Milwaukee, WI

Kim Willoughby
SE Raleigh High School
Raleigh, NC

Field Test Workshop Staff

Don W. Byerly
University of Tennessee

Derek Geise
University of Nebraska

Michael A. Gibson
University of Tennessee

David C. Gosselin
University of Nebraska

Robert Hartshorn
University of Tennessee

William Kean
University of Wisconsin

Ellen Metzger
San Jose State University

Tracy Posnanski
University of Wisconsin

J. Preston Prather
University of Tennessee

Ed Robeck
Salisbury University

Richard Sedlock
San Jose State University

Bridget Wyatt
San Jose State University

Field Test Evaluators

Bob Bernoff
Dresher, PA

Do Yong Park
University of Iowa

Field Test Teachers

Kerry Adams
Alamosa High School
Alamosa, CO

Jason Ahlberg
Lincoln High
Lincoln, NE

Gregory Bailey
Fulton High School
Knoxville, TN

Mary Jane Bell
Lyons-Decatur Northeast
Lyons, NE

Rod Benson
Helena High
Helena, MT

Sandra Bethel
Greenfield High School
Greenfield, TN

John Cary
Malibu High School
Malibu, CA

Elke Christoffersen
Poland Regional High School
Poland, ME

Tom Clark
Benicia High School
Benicia, CA

Julie Cook
Jefferson City High School
Jefferson City, MO

Virginia Cooter
North Greene High School
Greeneville, TN

Mary Cummane
Perspectives Charter
Chicago, IL

Sharon D'Agosta
Creighton Preparatory
Omaha, NE

Mark Daniels
Kettle Morraine High School
Milwaukee, WI

Beth Droughton
Bloomfield High School
Bloomfield, NJ

Steve Ferris
Lincoln High
Lincoln, NE

Bob Feurer
North Bend Central Public
North Bend, NE

Sue Frack
Lincoln Northeast High
Lincoln, NE

Rebecca Fredrickson
Greendale High School
Greendale, WI

Sally Ghilarducci
Hamilton High School
Milwaukee, WI

Kerin Goedert
Lincoln High School
Ypsilanti, MI

Martin Goldsmith
Menominee Falls High School
Menominee Falls, WI

Randall Hall
Arlington High School
St. Paul, MN

Theresa Harrison
Wichita West High
Wichita, KS

Gilbert Highlander
Red Bank High School
Chattanooga, TN

Jim Hunt
Chattanooga School of Arts
& Sciences
Chattanooga, TN

Patricia Jarzynski
Watertown High School
Watertown, WI

Pam Kasprowicz
Bartlett High School
Bartlett, IL

Caren Kershner
Moffat Consolidated
Moffat, CO

Mary Jane Kirkham
Fulton High School

Ted Koehn
Lincoln East High
Lincoln, NE

Acknowledgements (continued)

Philip Lacey
East Liverpool High School
East Liverpool, OH

Joan Lahm
Scotus Central Catholic
Columbus, NE

Erica Larson
Tipton Community

Michael Laura
Banning High School
Wilmington, CA

Fawn LeMay
Plattsmouth High
Plattsmouth, NE

Christine Lightner
Smethport Area High School
Smethport, PA

Nick Mason
Normandy High School
St. Louis, MO

James Matson
Wichita West High
Wichita, KS

Jeffrey Messer
Western High School
Parma, MI

Dave Miller
Parkview High
Springfield, MO

Rick Nettesheim
Waukesha South
Waukesha, WI

John Niemoth
Niobrara Public
Niobrara, NE

Margaret Olsen
Woodward Academy
College Park, GA

Ronald Ozuna
Roosevelt High School
Los Angeles, CA

Paul Parra
Omaha North High
Omaha, NE

D. Keith Patton
West High
Denver, CO

Phyllis Peck
Fairfield High School
Fairfield, CA

Randy Pelton
Jackson High School
Massillon, OH

Reggie Pettitt
Holderness High School
Holderness, NH

June Rasmussen
Brighton High School
South Brighton, TN

Russ Reese
Kalama High School
Kalama, WA

Janet Ricker
South Greene High School
Greeneville, TN

Wendy Saber
Washington Park High School
Racine, WI

Garry Sampson
Wauwatosa West High School
Tosa, WI

Daniel Sauls
Chuckey-Doak High School
Afton, TN

Todd Shattuck
L.A. Center for Enriched Studies
Los Angeles, CA

Heather Shedd
Tennyson High School
Hayward, CA

Lynn Sironen
North Kingstown High School
North Kingstown, RI

Jane Skinner
Farragut High School
Knoxville, TN

Sarah Smith
Garringer High School
Charlotte, NC

Aaron Spurr
Malcolm Price Laboratory
Cedar Falls, IA

Karen Tiffany
Watertown High School
Watertown, WI

Tom Tyler
Bishop O'Dowd High School
Oakland, CA

Valerie Walter
Freedom High School
Bethlehem, PA

Christopher J. Akin Williams
Milford Mill Academy
Baltimore, MD

Roseanne Williby
Skutt Catholic High School
Omaha, NE

Carmen Woodhall
Canton South High School
Canton, OH

Field Test Coordinator

William Houston
American Geological Institute

Advisory Board

Jane Crowder
Bellevue, WA

Arthur Eisenkraft
Bedford (NY) Public Schools

Tom Ervin
LeClaire, IA

Mary Kay Hemenway
University of Texas at Austin

Bill Leonard
Clemson University

Don Lewis
Lafayette, CA

Wendell Mohling
National Science Teachers
Association

Harold Pratt
Littleton, CO

Barb Tewksbury
Hamilton College

Laure Wallace
USGS

AGI Foundation

Jan van Sant
Executive Director

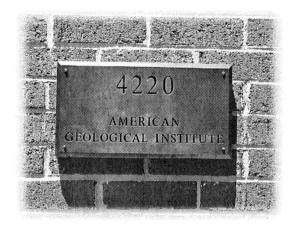

The American Geological Institute and EarthComm

Imagine more than 500,000 Earth scientists worldwide sharing a common voice, and you've just imagined the mission of the American Geological Institute. Our mission is to raise public awareness of the Earth sciences and the role that they play in mankind's use of natural resources, mitigation of natural hazards, and stewardship of the environment. For more than 50 years, AGI has served the scientists and teachers of its Member Societies and hundreds of associated colleges, universities, and corporations by producing Earth science educational materials, *Geotimes*–a geoscience news magazine, GeoRef–a reference database, and government affairs and public awareness programs.

So many important decisions made every day that affect our lives depend upon an understanding of how our Earth works. That's why AGI created *EarthComm*. In your *EarthComm* classroom, you'll discover the wonder and importance of Earth science by studying it where it counts—in your community. As you use the rock record to investigate climate change, do field work in nearby beaches, parks, or streams, explore the evolution and extinction of life, understand where your energy resources come from, or find out how to forecast severe weather, you'll gain a better understanding of how to use your knowledge of Earth science to make wise personal decisions.

We would like to thank the AGI Foundation Members that have been supportive in bringing Earth science to students. These AGI Foundation Members include: Anadarko Petroleum Corp., The Anschutz Foundation, Baker Hughes Foundation, Barrett Resources Corp., Elizabeth and Stephen Bechtel, Jr. Foundation, BPAmoco Foundation, Burlington Resources Foundation, CGG Americas, Inc., ChevronTexaco Corp., Conoco Inc., Consolidated Natural Gas Foundation, Diamond Offshore Co., Dominion Exploration & Production, Inc., EEX Corp., ExxonMobil Foundation, Global Marine Drilling Co., Halliburton Foundation, Inc., Kerr McGee Foundation, Maxus Energy Corp., Noble Drilling Corp., Occidental Petroleum Charitable Foundation, Parker Drilling Co., Phillips Petroleum Co., Santa Fe Snyder Corp., Schlumberger Foundation, Shell Oil Company Foundation, Southwestern Energy Co., Texaco, Inc., Texas Crude Energy, Inc., Unocal Corp. USX Foundation (Marathon Oil Co.).

We at AGI wish you success in your exploration of the Earth System and your Community.

Michael J. Smith
Director of Education, AGI

Marcus E. Milling
Executive Director, AGI

EarthComm: Earth System Science in the Community

Goals of *EarthComm*

Earth System Science in the Community (*EarthComm*) is an NSF-funded curriculum project guided in design and approach by the National Science Education Standards (1996), AGI's *Earth Science Content Guidelines for Grades K-12*, and other major science education curriculum and reform programs. This program builds on the strength of other successful AGI Earth Science education projects such as the *Earth Science Curriculum Project* (known to many as *Investigating the Earth*). *EarthComm* provides a comprehensive secondary-level educational program in the Earth Sciences that includes student learning materials, teacher resources (both materials and teacher-support networks), and assessment tools for a hands-on, inquiry-driven, instructional program and an *EarthComm* web site.

EarthComm covers fewer topics than the traditional Earth Science textbook. It emphasizes important concepts, understandings, and abilities that all students can use to make wise decisions, think critically, and understand and appreciate the Earth system. The goals of the *EarthComm* program are:

- To teach students the principles and practices of Earth Science and to demonstrate the relevance of Earth Science to their life and environment.
- To approach Earth Science through the problem-solving, community-based model in which the teacher plays the role of facilitator.
- To establish an expanded learning environment which incorporates field work, technological access to data, and traditional classroom and laboratory activities.
- To support the development of communities of learners by establishing student teams and by building a greater regional and national community through telecommunication access.
- To utilize local and regional issues and concerns to stimulate problem-solving activities and to foster a sense of Earth stewardship by students in their communities.

Developing *EarthComm*

Hundreds of teachers, scientists, and students helped develop *EarthComm*. In the summer of 1998, six teams of Earth Science educators wrote 122 inquiry-based investigations. Teachers and scientists reviewed draft chapters, which were then revised for pilot testing by 26 teachers in the spring of 1999. Seventeen teachers from the National Earth Science Teachers Association collaborated with project staff to revise *EarthComm* in the summer of 1999. In the 1999-2000 school year, *EarthComm* underwent a national field test with 77 teachers in 27 states. Results of field testing and further content review by more than 40 professional scientists were used to produce the commercial edition of *EarthComm*.

EarthComm Modules and Chapters

I. Earth's Dynamic Geosphere

Volcanoes and Your Community
Plate Tectonics and Your Community
Earthquakes and Your Community

II. Understanding Your Environment

Bedrock Geology and Your Community
River Systems and Your Community
Land Use Planning and Your Community

III. Earth's Fluid Spheres

Oceans and Your Community
Severe Weather and Your Community
Cryosphere and Your Community

IV. Earth's Natural Resources

Energy Resources and Your Community
Mineral Resources and Your Community
Water Resources and Your Community

V. Earth System Evolution

Astronomy and Your Community
Climate Change and Your Community
Changing Life and Your Community

EarthComm: Correlation to the National Science Education Standards

National Science Education Content Standards	EarthComm Modules / Chapters														
	Earth's Dynamic Geosphere			Understanding Your Environment			Earth's Fluid Spheres			Earth's Natural Resources			Earth System Evolution		
	1	2	3	1	2	3	1	2	3	1	2	3	1	2	3
UNIFYING CONCEPTS AND PROCESSES															
System, order and organization	•	•	•	•	•	•	•	•	•	•	•	•	•	•	•
Evidence, models, and explanation	•	•	•	•	•	•	•	•	•	•	•	•	•	•	•
Constancy, change, and measurement	•	•	•	•	•	•	•	•	•	•	•	•	•	•	•
Evolution and equilibrium		•		•	•	•	•	•	•	•	•	•	•	•	•
Form and function						•		•							
SCIENCE AS INQUIRY															
Identify questions and concepts that guide scientific investigations	•	•	•	•	•	•	•	•	•	•	•	•	•	•	•
Design and conduct scientific investigations	•	•	•	•	•	•	•	•	•	•	•	•	•	•	•
Use technology and mathematics to improve investigations	•	•	•	•	•	•	•	•	•	•	•	•	•	•	•
Formulate and revise scientific explanations and models using logic and evidence	•	•	•	•	•	•	•	•	•	•	•	•	•	•	•
Communicate and defend a scientific argument	•	•	•	•	•	•	•	•	•	•	•	•	•	•	•
Understand scientific inquiry	•	•	•	•	•	•	•	•	•	•	•	•	•	•	•
EARTH AND SPACE SCIENCE															
Energy in the Earth system	•	•	•			•	•	•	•	•	•	•	•	•	•
Geochemical cycles	•	•		•	•	•	•	•	•	•	•	•	•	•	
Origin and evolution of the Earth system	•	•	•	•	•	•	•	•	•	•	•	•	•	•	•
Origin and evolution of the universe											•		•		
SCIENCE AND TECHNOLOGY															
Identify a problem or design an opportunity	•				•				•	•		•			
Propose designs and choose between alternative solutions	•				•				•	•		•			
Implement a proposed solution	•				•				•	•		•			
Evaluate the solution and its consequences	•				•				•	•		•			
Communicate the problem, process, and solution	•	•	•	•	•	•	•	•	•	•	•	•	•	•	•
Understand science and technology	•	•	•						•	•	•	•	•	•	
SCIENCE IN PERSONAL AND SOCIAL PERSPECTIVES															
Personal and community health	•		•		•	•		•		•	•	•	•	•	
Population growth						•				•				•	•
Natural Resources	•					•	•			•	•	•		•	
Environmental quality	•			•	•	•	•			•	•	•			
Natural and human-induced hazards	•	•	•		•	•			•	•	•	•	•	•	
Science and technology in local, national, and global challenges	•	•	•			•	•	•	•	•	•	•		•	
HISTORY AND NATURE OF SCIENCE															
Science as a Human Endeavor	•	•	•	•	•	•	•	•	•	•	•	•	•	•	•
Nature of Scientific Knowledge	•	•	•	•	•	•	•	•	•	•	•	•	•	•	•
Historical Perspectives		•	•	•				•		•	•		•	•	•

EarthComm "Big Ideas"

EarthComm curriculum development was guided by 10 fundamental ideas that are emphasized in the five modules and are the primary goals for student learning:

- Earth science literacy empowers us to understand our environment, make wise decisions that affect quality of life, and manage resources, environments, and hazards.

- Earth's dynamic equilibrium system contains subsystems from atoms to planetary spheres. Materials interact among these subsystems due to natural forces and energy that flows from sources inside and outside of the planet. These interactions, changes, forces, and flows tend to occur in offsetting directions and amounts. Materials tend to flow in chains, cycles, and webs that tend toward equilibrium states in which energy is distributed as uniformly as possible. The net result is a state of balanced change or dynamic equilibrium, a condition that appears to have existed for billions of years.

- Change through time produced Earth, the net result of constancy, gradual changes, and episodic changes over human, geological, and astronomical scales of time and space.

- Extraterrestrial influences upon Earth include extraterrestrial energy and materials, and influences due to Earth's position and motion as a subsystem of an evolving solar system, galaxy, and universe.

- The dynamic geosphere includes a rocky exterior upon which ecosystems and human communities developed and a partially molten interior with convection circulation that generates the magnetosphere and drives plate tectonics. It contains resources that sustain life, causes natural hazards that may threaten life, and affects all of Earth's other geospheres.

- Fluid spheres within the Earth system include the hydrosphere, atmosphere, and cryosphere, which interact and flow to produce ever-changing weather, climate, glaciers, seascapes, and water resources that affect human communities, and which shape the land, transfer Earth materials and energy, and change surface environments and ecosystems.

- Dynamic environments and ecosystems are produced by the interaction of all the geospheres at the Earth's surface and include many different environments, ecosystems, and communities that affect one another and change through time.

- Earth resources include the nonrenewable and renewable supplies of energy, mineral, and water resources upon which individuals and communities depend in order to maintain quality of human life, economic prosperity, and requirements for industrialization.

- Natural hazards associated with Earth processes and events include drought, floods, storms, volcanic activity, earthquakes, and climate change and can pose risks to humans, their property, and communities. Earth science is used to study, predict, and mitigate natural hazards so that we can assess risks, plan wisely, and adapt to the effects of natural hazards.

- In order to sustain the presence and quality of human life, humans and communities must understand their dependence on Earth resources and environments, realize how they influence Earth systems, appreciate Earth's carrying capacity, manage and conserve nonrenewable resources and environments, develop alternate sources of energy and materials needed for human sustenance, and invent new technologies.

EarthComm Goals and Expectations for Teachers

- Use motivational teaching methods, interactive technologies, and manipulatives to pique student interest and help all students to understand the practical effects of Earth science and essential concepts and principles that underlie energy within the Earth system, geochemical cycles, and the origin and evolution of the Earth system.

- Facilitate students' understanding of inquiry and ability to inquire scientifically by having students answer questions about local problems and issues, design and conduct investigations, use technology and mathematics, form scientific explanations using logic and evidence, analyze alternative explanations, and communicate and defend scientific arguments.

- Emphasize the connections and relationships between Earth science and other academic disciplines.

- Establish an expanded learning environment for students through fieldwork, technological access to data, laboratory and other classroom activities.

- Nurture communities of science learners by establishing student teams, orchestrating discourse about scientific ideas, building networks of local, regional and national information exchange, and using the services of Earth and space organizations.

- Raise students' awareness of environmental and resource issues and problems in their communities.

EarthComm Goals and Expectations for Students

- Develop knowledge and understanding of practical and essential Earth science concepts and the principles Earth science shares with other disciplines.

- Understand basic principles of Earth system science and think from an Earth system science perspective.

- Develop an understanding of scientific inquiry and abilities needed to conduct scientific inquiry.

- Develop technology-oriented abilities for human enterprises in Earth and space.

- Understand the nature, origin, and distribution of Earth's energy, mineral, and water resources; technologies used to locate, extract, and process these resources; and dependence on these resources to satisfy our wants, needs, and expectations.

- Understand how terrestrial and extraterrestrial processes affect Earth's materials, environments, and organisms, how scientists study these processes on Earth and from space, and how some processes benefit humans while others pose risks.

- Understand how human activities influence Earth's spheres, processes, resources, and environments — factors that affect the size and distribution of human population and Earth's capacity to support life.

- Become aware of career opportunities in the Earth and space sciences, how professions and businesses benefit from technologies used by Earth and space scientists, and how these combined professions and businesses are related to regional economies.

EarthComm Curriculum Design

EarthComm modules are each three chapters connected to a common theme. Every chapter begins with a community-based problem or issue that can only be solved by developing key ideas and understandings in the chapter activities. Activities follow a learning cycle model.

Component of *EarthComm*	What Happens in the Classroom	Stages of 5-E Learning Cycle Model
Chapter Challenge	Students read and discuss a scenario that presents a community-based issue to solve through Earth Science and inquiry. They also explore the criteria and expectations for solving the challenge. Teachers allow students to share their current thinking openly and without closure.	Engage
1. Think about It	Students answer an open-ended question (or two) that sets the context for an activity and provides the teacher with a pre-assessment of their ideas. They briefly discuss their ideas in groups and/or as a class. Teachers allow students to share their ideas openly. They avoid assigning formal labels to concepts or seeking closure.	Engage
2. Investigate	Students collaborate on an inquiry activity that requires hands-on work, literature or web research, or fieldwork. Teachers facilitate and guide student-driven inquiry.	Explore
3. Reflecting upon the Activity and the Challenge	Students read a brief summary of the main ideas explored in the investigation and their relationship to the challenge. Teachers review the main ideas with students and affirm the relevance of the activity to the challenge.	Explain
4. Digging Deeper	Students read text, illustrations, and photographs that explain concepts explored in the investigation. Terms are defined and clarified. Teachers provide further information and clarification of concepts through lecture, slides, videos, or laser disk presentations.	Explain

Component of *EarthComm*	What Happens in the Classroom	Stages of 5-E Learning Cycle Model
5. Understanding and Applying What You Have Learned	Students respond to questions that check their understanding of key principles and concepts (learning goals) for the activity. New, yet familiar situations and scenarios provide contexts for students to apply their developing understandings. Teachers review student responses and use the questions to further probe and hone understanding of key learning goals.	Elaborate
6. Preparing for the Chapter Challenge	Students put their investigative results into the context of the challenge by preparing or organizing their work as it relates to their final product. Teachers review student performance in terms of its consistency with criteria set forth in the expectations for the activity and the challenge.	Elaborate/ Evaluate
7. Inquiring Further	Students are presented with options for deepening their understanding of concepts and skills developed within the activity. Teachers promote and encourage further inquiry.	Elaborate/ Evaluate
Chapter Assessment	Students present their solution to the Chapter Challenge in a variety of formats and consider ways to share their findings beyond the classroom. Teachers use the Assessment Criteria to assess the extent to which student work demonstrates mastery of concepts and skills. They also explore creative ways to share student solutions with the community.	Evaluate
Alternative Assessment	Students respond to a chapter test of essential knowledge and skills targeted throughout the chapter. Teachers score and review the test with students. They help students to understand how to use the results to guide future efforts.	Evaluate

Using *EarthComm* Features in Your Classroom

1. Getting Started

Each *EarthComm* chapter begins with one or more open-ended questions that give teachers the opportunity to explore what their students know about the central concepts of the chapter. Uncovering students' thinking (their prior knowledge) and exposing the diversity of ideas in the classroom are the first steps in the learning cycle. Some teachers prefer to have students record their responses to these questions. They then call for volunteers to offer ideas up for discussion. Other teachers prefer to start with discussion by asking students to volunteer their ideas. In either situation, it is important that teachers encourage the sharing of ideas by not judging responses as "right" or "wrong." It is also important that teachers keep a record of the variety of ideas, which can be displayed in the classroom (on a sheet of easel pad paper or on an overhead transparency) and referred to as students explore the concepts in the chapter. Teachers often find that they can group responses into a few categories and record the number of students who hold each idea.

2. Scenario

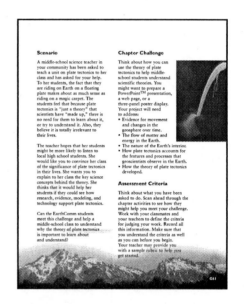

Each *EarthComm* chapter begins with an engaging description of an event or situation in the Earth system that has happened or could actually take place. The scenario (only a paragraph or two in length) sets the stage for the **Chapter Challenge**, which comes next. Many teachers read the scenario aloud to the class as a way to introduce the new chapter. Some teachers expand on the scenario by using videos of actual events, or by inviting persons from the field to present the scenario.

3. Chapter Challenge

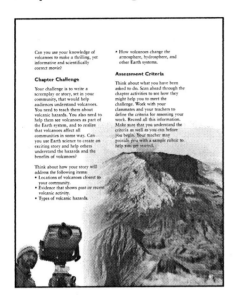

The **Chapter Challenge** is the central core of *EarthComm*. The challenge provides the context for all activities within the chapter. The **Chapter Challenge** provides a ready answer to the question asked all too often by students, "Why am I doing this?" because every activity contributes to solving the central problem set forth in the challenge. It also makes learning relevant to high school students. Each challenge is grounded in the community and designed to make the learning of Earth Science more relevant to the lives of students.

For example, in *Earth's Dynamic Geosphere*, Chapter 1: Volcanoes and Your Community, students are asked to create a screenplay for a thrilling yet informative and scientifically correct movie about volcanoes. Students are naturally intrigued by the dramatic effects and forces of natural hazards, and can easily relate to films they have seen that focus on natural disasters (*Volcano, Dante's Peak, Deep Impact,* and so on).

But unless they live in a volcanic region, it is unlikely that they have contemplated how volcanoes might affect their lives. Writing a story that is set in their community makes students think more deeply about the causes and effects of volcanism, and how volcanoes impact all communities because we live within a set of interconnected systems on Earth. All challenges require that students demonstrate solid understanding of Earth Science concepts and principles.

Another important element of the **Chapter Challenge** is that it provides opportunities for students with diverse interests and abilities to express their understanding in different ways. Challenges are completed in various ways. All involve writing to one extent or another, but some feature oral presentations, teaching, designing brochures, constructing models, creating web sites, or preparing formal presentations. Students who express themselves artistically will shine in some challenges, while those who enjoy designing and constructing will take a leading role in others.

Challenges are flexible enough to engage students at all levels of high school. Classes ranging from 9th grade integrated science to grades 11-12 honors, studying Earth Science, tested the challenges. Teachers establish different expectations for the students they teach, but the challenge is consistent.

4. Assessment Criteria

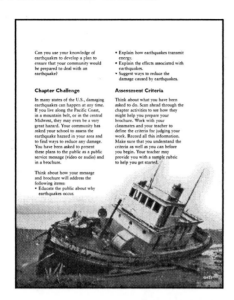

Can you use your knowledge of earthquakes to develop a plan to ensure that your community would be prepared to deal with an earthquake?

- Explain how earthquakes transmit energy.
- Explain the effects associated with earthquakes.
- Suggest ways to reduce the damage caused by earthquakes.

Chapter Challenge

In many states of the U.S., damaging earthquakes can happen at any time. If you live along the Pacific Coast, in a mountain belt, or in the central Midwest, they may even be a very great hazard. Your community has asked your school to assess the earthquake hazard in your area and to find ways to reduce any damage. You have been asked to present these plans to the public as a public service message (video or audio) and in a brochure.

Think about how your message and brochure will address the following items:
- Educate the public about why earthquakes occur.

Assessment Criteria

Think about what you have been asked to do. Scan ahead through the chapter activities to see how they might help you prepare your brochure. Work with your classmates and your teacher to define the criteria for judging your work. Record all this information. Make sure that you understand the criteria as well as you can before you begin. Your teacher may provide you with a sample rubric to help you get started.

The completion of the challenge (the final report or project) serves as the primary source of summative assessment information. Traditional assessment strategies often give too much attention to the memorization of terms or the recall of information. As a result, they often fall short of providing information about students' ability to think and reason critically and apply information that they have learned. In *EarthComm*, the solutions students provide to **Chapter Challenges** provide information used to assess thinking, reasoning, and problem-solving skills that are essential to lifelong learning.

Assessment is one of the key areas that teachers need to be familiar with and understand when trying to envision implementing *EarthComm*. In any curriculum model, the mode of instruction and the mode of assessment are connected. In the best scheme, instruction and assessment are aligned in both content and process. However, to the extent that one becomes an impediment to reform the other, they can also be uncoupled. *EarthComm* uses multiple assessment formats. Some are non-traditional and are consistent with reform movements in science education that *EarthComm* is designed to promote. **Project-based assessment,** for example, is built into every *EarthComm* **Chapter Challenge.** At the same time, the developers acknowledge the need to support teachers whose classroom context does not allow them to depart completely from traditional assessment formats, such as paper and pencil tests.

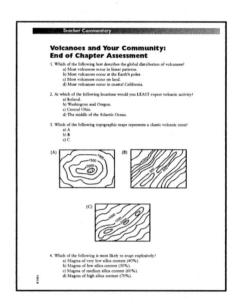

Teacher Commentary

Volcanoes and Your Community: End of Chapter Assessment

1. Which of the following best describes the global distribution of volcanoes?
   a) Most volcanoes occur in linear patterns.
   b) Most volcanoes occur at the Earth's poles.
   c) Most volcanoes occur on land.
   d) Most volcanoes occur in coastal California.

2. At which of the following locations would you LEAST expect volcanic activity?
   a) Iceland.
   b) Washington and Oregon.
   c) Central Ohio.
   d) The middle of the Atlantic Ocean.

3. Which of the following topographic maps represents a classic volcanic cone?
   a) A
   b) B
   c) C

4. Which of the following is most likely to erupt explosively?
   a) Magma of very low silica content (40%).
   b) Magma of low silica content (50%).
   c) Magma of medium silica content (60%).
   d) Magma of high silica content (70%).

An assessment instrument can imply but not determine its own best use. This means that *EarthComm* teachers can inadvertently assess chapter reports in ways that work against integrative thinking, a focus on important ideas, flexibility in approach, and consistency between assessment and the inferences made from that assessment.

All expectations should be communicated to students. Discussing the grading criteria and creating a general rubric are critical to student success. Better still, teachers can engage students in modifying and/or creating the criteria that will be used to assess their performance. Start by sharing the sample rubric with students and holding a class discussion. Questions that can be used to focus the discussion include: Why are these criteria included? Which activities will help you to meet these expectations? How much is required? What does an "A" presentation or report look like? The criteria should be revisited throughout the completion of the chapter, but for now students will have a clearer understanding of the challenge and the expectations they should set for themselves.

Teacher Commentary

Assessment Rubric for Chapter Report on Volcanoes

Meets the standard of excellence. **5**	*Significant* information is presented about all of the following: • Locations of volcanoes closest to your community • Evidence of past or recent volcanic activity • Volcanic hazards • How volcanoes change Earth systems All the information is accurate and appropriate The writing is clear and interesting
Approaches the standard of excellence. **4**	*Significant* information is presented about most of the following: • Locations of volcanoes closest to your community • Evidence of past or recent volcanic activity • Volcanic hazards • How volcanoes change Earth systems All the information is accurate and appropriate The writing is clear and interesting
Meets an acceptable standard. **3**	*Significant* information is presented about most of the following: • Locations of volcanoes closest to your community • Evidence of past or recent volcanic activity • Volcanic hazards • How volcanoes change Earth systems Most of the information is accurate and appropriate The writing is clear and interesting
Below acceptable standard and requires remedial help. **2**	*Limited* information is presented about the following: • Locations of volcanoes closest to your community • Evidence of past or recent volcanic activity • Volcanic hazards • How volcanoes change Earth systems Most of the information is accurate and appropriate Generally, the writing does not hold the reader's attention
Basic level that requires remedial help or demonstrates a lack of effort. **1**	*Limited* information is presented about the following: • Locations of volcanoes closest to your community • Evidence of past or recent volcanic activity • Volcanic hazards • How volcanoes change Earth systems Little of the information is accurate and appropriate The writing is difficult to follow

10

By the conclusion of the discussion of Assessment Criteria, students should have a clear "road map" of the structure of the chapter. They should have a sense as to why it is important to complete each activity in order to successfully meet the **Chapter Challenge**. They should be able to describe how each activity contributes toward the long-term goal.

5. Goals

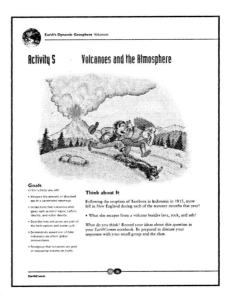

At the beginning of each activity students are provided with a list of goals that they should be able to achieve by completing the activity. Throughout this Teacher's Edition, we point out where each goal is addressed within each activity and provide some suggestions for assessing the goal. In most cases the goals are addressed directly by the hands-on investigation, as well as through reading the text or working on the **Chapter Challenge**. Pointing out the goals at the start of the activity reminds students about the expectations for learning. It is often helpful to point out how specific goals relate to the **Chapter Challenge**. For example, one element of the **Chapter Challenge** in *Earth's Dynamic Geosphere*, Chapter 1, Volcanoes and Your Community, is for students to address how volcanoes change the atmosphere, hydrosphere, and other Earth systems. One of the goals of Activity 5 in that chapter (Volcanoes and the Atmosphere) is to "describe how volcanoes are part of the hydrosphere and water cycle." When introducing the activity to students, it helps to point out how this particular goal contributes to the **Chapter Challenge**. It also serves to remind students "why we are doing this."

6. Think about It

One of the most fundamental principles derived from many years of research on student learning is that:

"Students come to the classroom with preconceptions about how the world works. If their initial understanding is not engaged, they may fail to grasp the new concepts and information that are taught, or they may learn them for the purposes of a test but revert to their preconceptions outside the classroom." (*How People Learn: Bridging Research and Practice*, National Research Council, 1999, P. 10.)

This principle has been illustrated through the Private Universe series of videotapes that show Harvard graduates responding to basic science questions in much the same way that fourth grade students do. Although the videotapes revealed that the Harvard graduates used a more sophisticated vocabulary, the majority held onto the same naïve incorrect conceptions of elementary school students. Research on learning suggests that the belief systems of students who are not confronted with what they believe and adequately shown why they should give up that belief system remain intact. Real learning requires confronting one's beliefs and testing them in light of competing explanations.

Drawing out and working with students' preconceptions is important for learners. In *EarthComm*, **Think about It** is used to ascertain students' prior knowledge about the key concept or Earth Science processes or events explored in each activity. Students verbalize what they think about the age of the Earth, the causes of volcanoes, or the way that the landscape changes over time before they embark on an activity designed to challenge and test these beliefs. A brief discussion about the diversity of beliefs in the classroom makes students consider how their ideas compare to others and the evidence that supports their view of volcanoes, earthquakes, or seasons.

The **Think about It** question is not a conclusion, but a lead into inquiry. It is not designed to produce the correct answer or a debate about the features of the question, or to bring closure. The activity that follows will provide that discussion as the results of inquiry are analyzed. Students are encouraged to record their ideas in words and/or drawings to ensure that they have considered their prior knowledge. After students discuss their ideas in pairs or in small groups, teachers activate a class discussion. A discussion with fellow students prior to class discussion may encourage students to exchange ideas without the fear of personally giving a "wrong answer." Teachers sometimes have students exchange papers and volunteer responses that they find interesting.

The "humorous illustration" above each **Think about It** section was designed to stimulate student thinking. In our field test edition of *EarthComm*, we used photographs of events and processes to stimulate thinking. However, we came to realize that illustrations would provide greater flexibility to stimulate students to begin to make the specific kinds of connections emphasized in each activity. For example, the first activity in *Earth's Dynamic Geosphere* Chapter 1 is titled "Where are the Volcanoes?" The **Think about It** question asks students "Can volcanoes form anywhere on Earth? Why or why not?" While some might argue that a photograph of a volcanic eruption would be most appropriate here, the drawing of a volcanic eruption occurring in a backyard barbeque used in the activity stimulates further thinking: "Could a volcano occur in my backyard?" Most students have experienced a summer cookout, but few have experienced a volcanic eruption. The context of the drawing is more relevant to students and makes it easier to stimulate student thinking about phenomena that they have never experienced.

7. Investigate

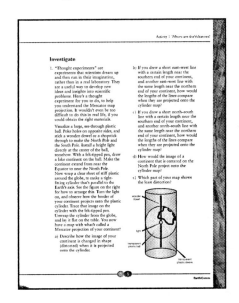

EarthComm is a hands-on, minds-on curriculum. In designing *EarthComm*, we were guided by the belief that doing Earth Science is essential to learning Earth Science. Testing of *EarthComm* activities by teachers across America provided critical testimonial about the importance of the activities to student learning. In small groups and as a class, students take part in doing hands-on experiments, participating in field work, or searching for answers using the Internet and reference materials. Blackline Masters are included in the Teacher's Edition for any maps or illustrations that are essential for students to complete the activity.

Each part of an *EarthComm* activity, as well as the sequence of activities within a chapter, moves from concrete to abstract. Hands-on activities provide the basis for exploring student beliefs about how the world works and to manipulate variables that affect the outcomes of experiments, models, or simulations. Later in each activity, formal labels are applied to concepts by introducing terminology used to describe the processes that students have explored through hands-on activity. This flow from concrete (hands-on) to abstract (formal explanations) is progressive – students begin to develop their own explanations for phenomena by responding to questions within the **Investigate** section.

Each activity has instructions for each part of the investigation. The community focus of *EarthComm* makes investigating the world more relevant to students. Have any volcanoes occurred in my state in the past? Have we ever experienced a major earthquake? What mineral resources exist in my community? Activities were designed with regard to the cost of materials and equipment needed. Many resources can be readily obtained in the community (local rock or soil samples, for example) or brought in to school by students (plastic two-liter soda bottles). Materials kits are available for purchase, but you will also need to obtain some resources from outside suppliers, such as topographic and geologic maps of your community, state, or region. The *EarthComm* web site will direct you to sources where you can gather such materials.

Most **Investigate** activities will require between one and two class periods. The variety of school schedules and student needs makes it difficult to predict exactly how much time your class will need. For example, if students need to construct a graph for part of an investigation, and the students have never been exposed to graphing, then this may require additional introductory lessons on the construction and interpretation of graphs. The most challenging aspect of

EarthComm for teachers to "master" is that the **Investigate** section of each activity has been designed to be student-driven. Students learn more when they have to struggle to "figure things out" and work in collaborative groups to solve problems as a team. Teachers will have to resist the temptation to provide the answers to students when they get "stuck" or hung up on part of a problem. Eventually, students learn that while they can call upon their teacher for assistance, the teacher is not going to "show them the answer." Field testing of *EarthComm* revealed that teachers who were most successful in getting their students to solve problems as a team were patient with this process and steadfast in their determination to act as facilitators of learning during the **Investigate** portion of activities. As one teacher noted, "My response to questions during the investigation was like a mantra, 'What do you think you need to do to solve this?' My students eventually realized that although I was there to provide guidance, they weren't going to get the solution out of me."

Another concern that many teachers have when examining *EarthComm* for the first time is that their students do not have the background knowledge to do the investigations. They want to deliver a lecture about the phenomena before allowing students to do the investigation. Such an approach is common to many traditional programs and is inconsistent with the pedagogical theory used to design *EarthComm*. The appropriate place for delivering a lecture or reading text in *EarthComm* is following the investigation, not preceding it. For example, suppose a group of students has been asked to interpret a map. The traditional approach to science education is for the teacher to give a lecture or assign a reading, "How to Interpret Maps," then give students practice reading maps. *EarthComm* teachers recognize that while students may lack some specific skills (reading latitude and longitude, for example), within a group of four students, it is not uncommon for at least one of the students to have a vital skill or piece of knowledge that is required to solve a problem. The one or two students who have been exposed to (or better yet, understood) latitude and longitude have the opportunity to shine within the group by contributing that vital piece of information or demonstrating a skill. That's how scientific research teams work – specialists bring expertise to the group, and by working together, the group achieves something that no one could achieve working alone. The **Investigate** section of *EarthComm* is modeled in the spirit of the scientific research team.

8. Reflecting on the Activity and the Challenge

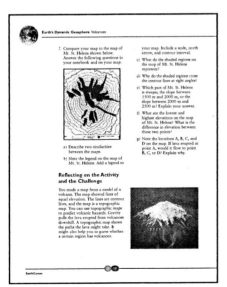

Each activity contributes to the solution of the **Chapter Challenge**. This feature gives students a brief summary of the activity. It helps students to relate the activity that they just completed to the "big picture." Teachers also find this section useful for students who were absent for an investigation. In situations where students cannot make up the investigation (after school or during off-hours), teachers can use this section to provide an overview of what was missed. Although reading about the main point of an activity is a poor substitute to actually doing it, teachers find that the overview helps them deal with the reality of student absences and the hectic pace of school schedules.

9. Digging Deeper

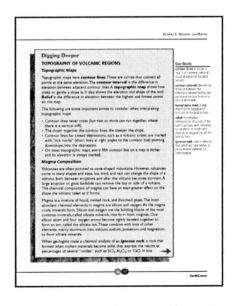

This section provides text, illustrations, data tables, and photographs that give students greater insight into the concepts explored in the activity. Words that may be new or unfamiliar to students are defined and explained (so-called **Geo Words**). These are words that geologists use when discussing the concepts presented. This is not the same thing as stating that **Geo Words** are "important words," or "words to be memorized." Teachers use their own judgment about selecting the **Geo Words** that are most important for their students to learn. Teachers typically use discretion and consider their state and local guidelines for science content understanding when assigning importance to particular vocabulary, which in most cases is very likely to be a small subset of all the **Geo Words** introduced in each chapter.

Teachers often assign **Check Your Understanding** questions as homework to guide students to think about the major ideas in the text. Teachers can also select questions to use as quizzes, rephrasing the questions into multiple choice or "true/false" formats. This provides assessment information about student understanding and as a "motivational tool" to ensure that students complete the reading assignment and comprehend the main ideas.

This is the stage of the activity that is most appropriate for teachers to explain concepts to students in whole-class lectures or discussions. References to Blackline Masters are available throughout the Teacher's Edition. They refer to illustrations from the textbook that teachers may photocopy and distribute to students or make overhead transparencies for lectures or presentations.

10. Understanding and Applying What You Have Learned

Questions in this feature ask students to use the key principles and concepts introduced in the activity. Students are sometimes presented with new situations in which they are asked to apply what they have learned. The questions in this section typically require higher-order thinking and reasoning skills than **Check Your Understanding**. Teachers can assign these questions as homework, or have students complete them in groups during class. Assigning them as homework economizes time available in class, but has the drawback of making it difficult for students to collectively revisit the understanding that they developed as they worked through the concepts as a group during the investigation. A third alternative is of course to assign the work individually in class. When students work through application problems in class, teachers have the opportunity to interact with students at a critical juncture in their learning – when they may be just on the verge of "getting it."

11. Preparing for the Chapter Challenge

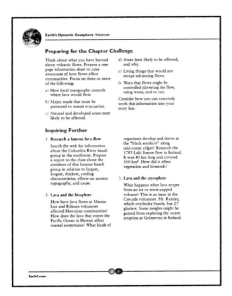

This feature suggests ways in which students can organize their work and get ready for the challenge. It prompts students to combine the results of their inquiry as they work through the chapter. Another one of the important principles of learning used to guide the selection of content in *EarthComm* is that:

"To develop competence in an area of inquiry, students must (a) have a deep foundation of factual knowledge, (b) understand facts and ideas in the context of a conceptual framework, and (c) organize knowledge in ways that facilitate retrieval and application." *(How People Learn: Bridging Research and Practice,* National Research Council, 1999, P. 12.)

This phase of an activity (**Preparing for the Chapter Challenge**) is an important metacognitive tool that makes students examine what they have learned in the activity and then think critically about

the usefulness of the results of their inquiry. The process of synthesizing what they have learned in order to solve the **Chapter Challenge** forces students to take stock of their learning and evaluate whether or not they really understand "how it fits into the big picture." It is important for teachers to guide students through this process with questions such as: "What part of your work best helps you to solve the challenge? How does what you learned help you to solve the challenge? How does this assignment relate to the criteria that we established for your chapter report? Are you making the best possible use of the evidence you have gathered?"

12. Inquiring Further

This feature provides lots of suggestions for helping students to deepen their understanding of the concepts and skills developed in the activity. It also gives students the opportunity to relate what they have learned to the Earth system. Teachers should review the suggestions and consider how the time available in class, the specific resources available to their students and/or in the school, and the needs and abilities of their students. Some of these suggestions make for excellent "do at home" investigations or Internet and library-based research projects. Some teachers assign **Inquiring Further** as "extra credit" projects. Some of the suggested activities in **Inquiring Further** may have particular relevance to your community. In such cases, make every attempt to integrate the activity into your instruction.

The most common complaint teachers make about Internet-based research stems from a concern about the limited amount of time students have available at school computers. The *EarthComm* web site has been designed to help students focus their research. By providing specific links helpful to each **Inquiring Further** activity on the site, students will gain access to useful information from stable web sites without spending time searching for information.

(Reference: *How People Learn: Bridging Research and Practice* (1999) Suzanne Donovan, John Bransford, and James Pellegrino, editors. National Academy Press, Washington, DC. 78 pages. The report is also available online at http://www.nap.edu.

EarthComm Assessment Opportunities

In keeping with the discussion of assessment outlined in the *National Science Education Standards* (NSES), teachers must be careful while developing the specific expectations for each chapter. Four issues are of particular importance in that they may present somewhat new considerations for teachers and students:

1. Integrative Thinking

The *National Science Education Standards* (NSES) state: "Assessments must be consistent with the decisions they are designed to inform." This means that as a prerequisite to establishing expectations, teachers should consider the use of assessment information. In *EarthComm*, students must be able to articulate the connection between Earth Science concepts and their own community. This means that they have to integrate traditional Earth Science content with knowledge of their surroundings. It is likely that this kind of integration will be new to students, and that they will require some practice at accomplishing it. Assessment in one chapter can inform how the next chapter is approached so that the ability to apply Earth Science concepts to local situations is enhanced on an ongoing basis.

2. Importance

An explicit focus of NSES is to promote a shift to deeper instruction on a smaller set of core science concepts and principles. Assessment can support or undermine that intent. It can support it by raising the priority of in-depth treatment of concepts, such as students evaluating the relevance of core concepts to their communities. Assessment can undermine a deep treatment of concepts by encouraging students to parrot back large bodies of knowledge-level facts that are not related to any specific context in particular. In short, by focusing on a few concepts and principles, deemed to be of particularly fundamental importance, assessment can help to overcome a bias toward superficial learning. For example, assessment of terminology that emphasizes deeper understanding of science is that which focuses on the use of terminology as a tool for communicating important ideas. Knowledge of terminology is not an end in itself. Teachers must be watchful that the focus remains on terminology in use, rather than on rote recall of definitions. This is an area that some students will find unusual if their prior science instruction has led them to rely largely on memorization skills for success.

3. Flexibility

Students differ in many ways. Assessment that calls on students to give thoughtful responses must allow for those differences. Some students will find the open-ended character of the *EarthComm* chapter reports disquieting. They may ask many questions to try to find out exactly what the finished product should look like. Teachers will have to give a consistent and repeated message to those students, expressed in many different ways, that the ambiguity inherent in the open-ended character of the assessments is an opportunity for students to show what they know in a way that makes sense to them. This also allows for the assessments to be adapted to students with differing abilities and proficiencies.

4. Consistency

While the chapter reports are intended to be flexible, they are also intended to be consistent with the manner in which instruction takes place, and the kinds of inferences that are going to be made about students' learning on the basis of them. The *EarthComm* design is such that students have the opportunity to learn new material in a way that places it in context. Consistent with that, the chapter reports also call for the new material to be expressed in context. Traditional tests are less likely to allow this kind of expression, and are more likely to be inconsistent with the manner of teaching that *EarthComm* is designed to promote. Likewise, in that *EarthComm* is meant to help students relate Earth Science to their community, teachers will be using the chapter reports as the basis for inferences regarding the students' abilities to do that. The design of the chapter reports is intended to facilitate such inferences.

EarthComm Assessment Tools

The series of evaluation sheets and scoring rubrics provided in the back of this Teacher's Edition should be available to students before they begin their first investigation. Consider photocopying a set of the sheets for each student to include in his or her *EarthComm* notebook. The purpose of distributing the evaluation sheets is to help students become familiar with the criteria and expectations for their work. If students have a complete set of the evaluation sheets, you can refer to the relevant evaluation sheet at the appropriate point within an *EarthComm* lesson.

Think about It Evaluation Sheet

This sheet will help students to learn the basic expectations for the warm-up activity. **Think about It** is intended to reveal student conceptions about the phenomena or processes explored in the activity. It is not intended to produce closure and so your assessment of student responses should not be driven by a concern for correctness. Instead, the evaluation sheet emphasizes that you want to see evidence of prior knowledge and that students should communicate their thinking clearly. It is unlikely that you will be able to apply this assessment every time students complete a warm-up activity (there are only so many hours in a teacher's day), yet in order to ensure that students value the importance of committing their initial conceptions to paper and taking the warm-up seriously, you can use this evaluation sheet as a spot check on the quality of their work.

Investigate Notebook-Entry Evaluation Sheet

This evaluation sheet is designed to allow the students to get a sense of the expectations for *EarthComm* notebook entries. When assessing student investigations, keep in mind that the **Investigate** section of an *EarthComm* activity equates to the exploration phase of the 5E learning-cycle model where students explore their conceptions of phenomena through hands-on activity. This evaluation sheet provides a variety of criteria for you to select from when students will be in a better position to ensure that the quality of their work meets the highest possible standards and expectations. Encourage students to internalize the criteria by making the criteria part of your "assessment conversations" with them as you circulate around the classroom.

For example, while students are working, you can ask them criteria-driven questions such as: "Is your work thorough and complete? Are all of you participating in the activity? Do you each have a role to play in solving this problem?" and so on.

EarthComm Notebook-Entry Checklist

The *EarthComm* Notebook-Entry Checklist provides a quick summary of important processes, concepts, and skills that you might wish to assess during and after an investigation. You can add further criteria specific to your classroom needs or a particular investigation. The checklist provides a quick guide for student self-assessment and also provides you with an opportunity to quickly score student work.

Check Your Understanding Notebook-Entry Evaluation Sheet

This evaluation sheet is used to help you evaluate the extent to which students understand the key concepts explored in the activity and explained in the **Digging Deeper** reading section. The two criteria used in the sheet include "Reflects an Understanding of Key Concepts" and "Clarity of Expression."

Student Presentation Evaluation Form

This evaluation form provides three simple yet powerful criteria to help your students prepare their presentations. In order to prepare properly, students must know that you have expectations for the quality of the ideas they present, their ability to answer questions during the presentation, and their overall comprehension of the material. When students work in groups and present the results of their inquiry, they often divide up the work, with some members preparing the presentation and others delivering the presentation. Students need to know that any member of the group can be called upon to demonstrate their understanding of the material during a presentation. This evaluation form will help you to make this clear to students.

Student Evaluation of Group Participation

One of the challenges to assessing students who work in collaborative teams is assessing group participation. Students need to know that each group member must pull his or her weight. As a component of a complete assessment system, especially in a collaborative learning environment, it is often helpful to engage students in conducting a self-assessment of their participation in a group. Knowing that their contributions to the group will be evaluated provides an additional motivational tool to keep students constructively engaged. This evaluation form provides students with an opportunity to assess group participation. In no case should the results of this evaluation be used as the sole source of assessment data. Rather, it is better to assign a weight to the results of this evaluation and factor it in with other sources of assessment data.

Student Ratings and Self Evaluation

This form provides an alternative to the student evaluations of group participation. You might alternate the use of these forms between chapters. The two-page form is flexible in that you can assign values as you see fit for your classroom and it allows students to provide extra comments to explain the ratings.

Reviewing and Reflecting upon Your Teaching

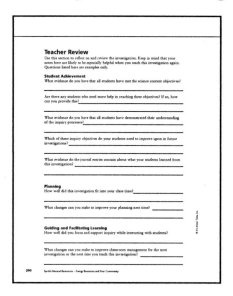

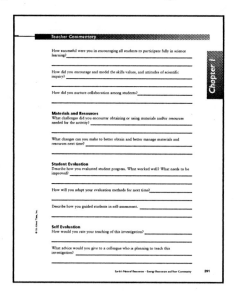

Reviewing and Reflecting upon Your Teaching provides an important opportunity for professional growth. A master copy of a two-page Teacher Review form is included at the back of each chapter. At the back of this Teacher's Edition are various Evaluation sheets for both students and teachers. They will help you to reflect upon your teaching for each investigation. We suggest that you try to answer each question at the completion of each investigation, then go back to the relevant section of this Teacher's Edition and write specific comments in the margins that will help you the next time you teach the investigation. For example, if you found that you were able to make substitutions to the list of materials needed, write a note about those changes in the margin of that page of this guide.

Using the *EarthComm* Web Site

http://www.agiweb.org/earthcomm
The *EarthComm* web site has been designed for teachers and students.

- Each *EarthComm* chapter has its own web page that has been designed specifically for the content addressed within that chapter.

- Chapter web sites are broken down by activity and also contain a section with links to relevant resources that are useful for the chapter.

- Each activity is divided into background information, materials and supplies needed, completing the investigation, and suggested links for completing the **Inquiring Further** portion of each activity.

Additionally, *EarthComm* users can consult the *EarthComm* website to find the latest errata, downloadable images from the Student Edition, and alternative suggestions for **Investigation** materials and setups.

Enhancing Teacher Content Knowledge

Each *EarthComm* chapter has a specific web page that will help teachers to gather further background information about the major topics covered in each activity. Example from Volcanoes and Your Community—Activity 1: Where are the Volcanoes?

To learn more about this topic, visit the following web sites:
1. **Volcanoes Beneath the Sea:**
Volcano World — "Submarine Volcanoes"
Reviews the basics of plate tectonics and examines closely submarine volcanoes at divergent and convergent boundaries and hot spots. The site has good images of underwater lava flows as well as images of the organisms that live near these submarine volcanoes. http://volcano.und.nodak.edu/vwdocs/Submarine/

Obtaining Resources

The community focus of *EarthComm* will require teachers to obtain local or regional maps, rocks, and data. The *EarthComm* web site helps teachers to find these materials.

Completing the Investigation

This portion of each activity provides suggestions for using Internet data in the classroom.

Example from Volcanoes and Your Community — Activity 4: Volcanic Hazards: Airborne Debris.

Using the Internet during the investigation:
To simulate the eruption of one of the three volcanoes, go to the Volcanic Ash Forecast Transport and Dispersion (http://www.arl.noaa.gov/ready/vaftadmenu.html) web site and click on "Run VAFTAD Model."

Inquiring Further

This section of each activity provides suggested web sites that will helps students to complete the **Inquiring Further** portion of *EarthComm* activities. This will help students to make the most of what is often limited time available to conduct Internet-based research during the school day.

Example from Volcanoes and Your Community — Activity 3: Volcanic Hazards: Flows.

To complete the **Inquiring Further** section of this activity:

Investigate how lava flows affect the biosphere:
• USGS Hawaiian Volcano Observatory (http://hvo.wr.usgs.gov/) has information about Hawaiian volcanoes
• University of Hawaii, Hawaii Center for Volcanology (http://www.soest.hawaii.edu/GG/hcv.html) has general and specific information about the Hawaiian shield volcanoes and links to other volcano sites
• VolcanoWorld (http://volcano.und.nodak.edu/vw.html)
• NOAA - VENTS Program (http://www.pmel.noaa.gov/vents/) has information about black smokers and activity along mid-ocean ridges
• RIDGE Program (http://ridge.oce.orst.edu/) has information about activity along mid-ocean ridges and links to further information.

Resources

The web page for each *EarthComm* chapter provides a list of relevant web sites, maps, videos, books, and magazines. Specific links to sources of these materials are often provided.

Managing Collaborative Group Learning

Working in small collaborative groups is seen as an important part of scientific inquiry, and is reinforced by the *National Science Education Standards* and *Benchmarks for Science Literacy*. Scientists, and others, frequently work in teams to investigate things and solve problems. However, there are times when it is important to work alone. You may have students who are more comfortable working on their own. Traditionally, the competitive nature of school curricula has emphasized individual effort through grading, "honors" classes, and so on. Many parents will have been through this experience themselves as students and will be looking for comparisons between their children's performance and other students as a result. Managing collaborative groups may therefore present some initial problems, especially if you have not organized your class in this way before and the idea is new for your students. Below are some key points to keep in mind as you develop a group approach.

- Arrange your classroom furniture into small group areas.
- Explain to students ahead of time how and why they are going to work in groups.
- Stress the responsibility each group member has to the others in the group.
- Choose student groups carefully to ensure each group has a balance of ability, special talents, gender, ethnicity, and so on.
- Make it clear that groups are not fixed for all time and that their composition will change from time to time.
- Promote the idea of fair work-sharing within groups, where everyone is contributing.
- Help students see the benefits of learning with and from each other.
- Ensure that there are some opportunities for students to work alone.
- Provide students with a copy of any rubrics that address group work and discuss the rubrics with them.

There are assessment rubrics provided in this Teacher's Edition to help you and your students manage and evaluate student collaboration (Student Evaluation of Group Participation, page 700; Student Ratings and Self Evaluation, pages 701 and 702).

Enhancing *EarthComm's Earth's Natural Resources* with *EarthView Explorer*

There are many similarities between *EarthComm Earth's Natural Resources* and *EarthView Explorer* that will be extremely useful to those using both sets of materials in their Earth Science classes. These similarities include both pedagogy and content. The organization of the materials in *EarthComm Earth's Natural Resources* into chapters and activities is similar to *EarthView's* unit and activity organization. Both sets of materials are inquiry-based. Activities begin with focus questions. There is a strong Science, Technology, and Society (STS) relevancy, and both sets of materials encourage metacognitive thinking by students.

The **Chapter Challenge** at the beginning of each chapter of *Earth's Natural Resources* is similar to the processes and major data products table in the *EarthView* Teacher's Manual (Table 1). The **Scenario** introduces what in *EarthView* are the Science, Technology and Society issues. Students can relate their existing understanding of the Earth Science topics to their experiences and begin to comprehend how the topic relates to their lives. In *EarthView*, maps of population help students visualize how phenomena may or may not affect people. In *EarthComm* the **Assessment Criteria** are presented to students before they begin their activities so that they can plan their study at a high level. In *EarthView*, this function is performed by the **Basecamp**, in which students' can keep a journal of the purpose and results of each of their activities. These notes are intended to help students to construct a more extensive familiarity with the material.

Most importantly, the *EarthComm* activities share a similar inquiry-based approach to the *EarthView* activities. *EarthComm* has several different approaches to inquiry. It uses hands-on experimentation, charts, and data analyses. *EarthView* focuses on the latter two, but allows for additional in-depth quantitative inquiry.

Following is a correlation between *EarthComm Earth's Natural Resources* and *EarthView Explorer*. For example, specific data investigations in *EarthView* can be used to dig deeper into various topics presented in *Earth's Natural Resources*. Of the three chapters, the matchup to *EarthView* is best for Chapter 1: Energy Resources...and Your Community. However, there are useful matchups in Chapter 3: Water Resources...and Your Community, as well. The good content matchup arises from the fact that both sets of materials are derived from current Earth Science research topics, and thus by necessity share many common elements.

Enhancing *EarthComm's Earth's Natural Resources* with *EarthView Explorer*

General: As for the other *EarthComm* titles, there are similarities between *Earth's Natural Resources* (ENR) and *EarthView Explorer* (EView) that will be useful to teachers using both sets of materials in their Earth Science classes. However, the correlations between these two sets of materials are not as clear-cut. *EarthView Explorer* does not focus on natural resources per se as *Earth's Natural Resources* does. However, several useful correlations between the materials are apparent in cases where the resources covered in the *EarthComm* materials play a role in the processes covered in *EarthView Explorer*. These match-ups are listed in the table below along with usage notes. The usage notes should be considered carefully before classroom use because the context of the material in the two educational titles can be significantly different and cause confusion unless the transition is made properly.

Earth's Natural Resources (ENR) Content	Page no.	Matching EarthView (EView) Content	Usage Notes
CHAPTER 1: ENERGY RESOURCES			
General: Two important processes addressed by *Eview* are relevant to the activities and content investigated in chapter one of *ENR*; namely the flow of solar energy and heat and the role of fossil fuel burning on the level of carbon dioxide in the atmosphere.			
ACTIVITY 1 Exploring Energy Resource Concepts			
Investigate Parts A and C; also **Digging Deeper**	R5	Climate unit – all activities	This activity of *ENR* investigates heat transfer mechanisms and some of the fundamental relationships between energy, power, and work. The transmission of solar energy to the Earth is the core of *EView's* average temperature model, and therefore this *EView* unit can be a useful extension of the treatment of heat transfer in the **Digging Deeper** section. In particular, the diagrams of energy flow that are in each of the Information sections in the Climate activities could be used to identify the different forms of heat transfer involved in each interaction. The exercises with energy units in *ENR* could be investigated further by the more conceptual treatment of energy in the *EView's* Climate section.
ACTIVITY 2 Electricity and Your Community			
Investigate Parts A and B; also **Digging Deeper**	R16	Atmosphere unit – Greenhouse gases Climate unit – Atmosphere	The concept of atmospheric carbon dioxide increases that is introduced in the *EView* activity on Greenhouse gases can be correlated to these *ENR* Investigate sections. The temporal changes in atmospheric carbon dioxide over the last several hundred years, shown in slide 5 of the Greenhouse gases activity, can be related to the recent electric generation figures for fossil fuels presented in the table on p. R17. Likewise, the variation in fossil fuel burning among countries (as demonstrated in the *ENR* data) leads directly to the idea that the amount of carbon dioxide produced by each country can vary widely. This is the major variable in the *EView* Atmosphere activity.
ACTIVITY 3 Energy from Coal			
Digging Deeper	R29	Biosphere unit – Land plants and Land biomes	The process of photosynthesis is introduced in *EView* Land plants. The maps available in this activity and Land biomes can be used to reinforce the idea of localized distributions of coal deposits.

Enhancing *EarthComm's Earth's Natural Resources* with *EarthView Explorer*

Earth's Natural Resources (*ENR*) Content	Page no.	Matching EarthView (*EView*) Content	Usage Notes
CHAPTER 1: ENERGY RESOURCES (continued)			
ACTIVITY 4 Coal and Your Community			
Investigate Parts A and C	R36	Climate unit – Atmosphere	The question of what activities in students' lives involve the burning of fossil fuels relates to these *ENR* sections, which investigate exploration of coal sources, coal production, consumption, and use.
ACTIVITY 5 Enviromental Impacts and Energy Consumption			
Digging Deeper	R45	Atmosphere unit – Greenhouse gases Climate unit – Atmosphere	The figure of the global carbon cycle on p. R46 links directly to a similar depiction in the *EView* Greenhouse gases slide 4. The *EView* Greenhouse gases activity and Atmosphere activity relate to the data on the changes in carbon dioxide in the *ENR* section.
ACTIVITY 6 Petroleum and Your Community			
Investigate Parts A and C	R54	Climate unit – Atmosphere	The question of what activities in students' lives involve the burning of fossil fuels relates to this *ENR* activity.
ACTIVITY 7 Oil and Gas Production			
Investigate Parts A and **Digging Deeper**	R72	Atmosphere unit – Climate zones	The concept of consistent, regional differences in temperature is introduced in *EView's* Climate zone activity and can be related to the solar radiation diagrams on p. R74. The Information slides in the Climate zones activity also relate closely to the idea of changing seasonal insolation in the *ENR* **Digging Deeper** section.
CHAPTER 3: WATER RESOURCES			
General: Various aspects of the water cycle addressed by the *EView* Hydrosphere unit correlate well with this *ENR* chapter.			
ACTIVITY 1 Sources of Water in the World			
Investigate Parts A–D and **Digging Deeper**	R146	Hydrosphere unit – Rivers and Rain; Ocean salinity Climate unit – Hydrosphere Biosphere unit – Land biomes	Much of this *ENR* activity relates to the global water cycle. This is a core process for the *EView* Hydrosphere unit. This *ENR* chapter starts off by investigating the entire cycle in (Investigate Part C of this activity), while the *EView* Hydrosphere unit reserves this for the final activity. Other than this sequence difference, the materials correlate to each other well. In particular, Part A of the *ENR* materials explain in greater detail the breakdown among water reservoirs introduced in the *EView* Information section in Ocean Salinity. The outline of the water cycle in **Investigate** Part C incorporates the ideas of atmospheric transfer presented in both the Hydrosphere – Ocean salinity and Climate – Atmosphere activities. The collection of rainwater suggested in **Investigate** Part D can be supplemented with the precipitation and river discharge information available to students in the *EView* Hydrosphere – Rivers and Rain activity. The concepts of ground water and transpiration in the *ENR* **Digging Deeper** section are not addressed in the *EView* Biosphere unit, but these concepts could be applied in the *EView* Biosphere activity on Land biomes that includes precipitation data.

Enhancing *EarthComm's Earth's Natural Resources* with *EarthView Explorer*

Earth's Natural Resources (*ENR*) Content	Page no.	Matching EarthView (*EView*) Content	Usage Notes
CHAPTER 3: WATER RESOURCES (continued)			
ACTIVITY 2 How Does Your Community Maintain Its Water Supply?			
Investigate Parts A and C and **Digging Deeper**	R156	Hydrosphere – Rivers and Rain; Ocean salinity	The consideration of local surface water flow and ground water supply in the *ENR* materials are more detailed than the global water cycle in the *EView* Hydrosphere unit. A logical progression could be made to lead students from the local (*ENR*) to the global (*EView*) or vice-versa.
ACTIVITY 4 Water Supply and Demand: Water Budgets			
Investigate Parts A–B and **Digging Deeper**	R177	Hydrosphere – Rivers and Rain	The *EView* Rivers and Rain activity contains data on river seasonality of discharge that could be used as the basis for comparing the season rainfall data compiled by students in Part A of this *ENR* investigation. Also, the discharge data itself could be used as the input to the construction of water budgets in Part B. Some of the *EView* Rivers and Rain material is correlated to the *ENR* **Digging Deeper** section on rivers and ground water. This *ENR* section extends the information presented in *EView*.
ACTIVITY 5 Water Pollution			
Digging Deeper	R189	Biosphere – Ocean plants	The association between nutrients and phytoplankton growth is introduced in the *EView* Ocean plants activity, and this association can be related to the *ENR* **Digging Deeper** discussion of excess sewage derived nutrients and the growth of harmful algal blooms.

GETIT™ Geoscience Education Through Interactive Technology for Grades 6-12

Earthquakes, volcanoes, hurricanes, and plate tectonics are all subjects that deal with energy transfer at or below the Earth's surface. The GETIT CD-ROM uses these events to teach the fundamentals of the Earth's dynamism. GETIT contains 63 interactive—inquiry-based—activities that closely simulate real-life science practice. Students work with real data and are encouraged to make their own discoveries—often learning from their mistakes. They use an electronic notebook to answer questions and record ideas, and teachers can monitor their progress using the integrated class-management module. The Teacher's Guide includes Assessments, Evaluation Criteria, Scientific Content, Graphs, Diagrams and Blackline Masters. GETIT conforms to the National Science Education Standards and the American Association for the Advancement of Science benchmarks for Earth Science.

Enhancing *EarthComm's Earth's Natural Resources* with **GETIT**

Earth's Natural Resources				
Chapter	**Activity**	**Page**	**Goals**	**GETIT Activity**
1. Energy Resources	1. Exploring Energy Resource Concepts	R4	Learn about the Second Law of Thermodynamics	• Science Showtime episode: The Big Idea
		R6	Kinetic energy	• Conservation of Energy
		R9	**Digging Deeper**	• Temperature and Heat
		R10	**Digging Deeper**	• See the light
2. Mineral Resources	2. What are minerals?	R110	**Inquiring Further**	• Periodic table (associations button)
	4. How are minerals formed?	R123	Geophysicist	• Seismic wave properties
3. Water Resources	1. Sources of Water	R152	Evaporation and precipitation	• Now you see it, now you don't • Science Showtime episode: Shake your molecules

Module Overview: *Earth's Natural Resources*

A typical day would not be possible without mineral resources to make the products we use, water resources to sustain our lives, keep clean, and cook our food, or energy resources to drive around and to provide our electricity. All these resources exist in abundance on our planet, but none will last forever without careful planning and management. While water is a renewable resource, if it is polluted it is not usable. Mineral resources are nonrenewable. Even given enough time, some kinds of mineral deposits can no longer form because conditions on the Earth are not favorable for their formation at this time. Most of the energy resources we use are nonrenewable. We can find ways to use renewable sources such as the Sun and wind but on a large scale we are still dependent on coal, oil, and gas. This module explores how we use these resources, where they come from, how they exist in our community, and how our use of resources affects the environment in our community. It includes numerous inquiries into aspects of resources that affect our use of them every day.

Themes

Through their inquiry in this module, students develop understandings of the complex Earth systems associated with Earth's natural resources. The major themes addressed include the following portions of the National Science Education Standards for Grades 9-12:

- Humans use resources in the environment in order to maintain and improve their existence.
- Earth does not have infinite resources. Increasing population places severe stress on the natural processes that renew some resources, and it depletes those resources that cannot be renewed.
- People make decisions and conduct activities every day that affect their environment and Earth systems.
- The Earth system contains a fixed amount of each stable element. Each element can exist in several different chemical reservoirs and moves among reservoirs as part of geochemical cycles.

Energy Resources... and Your Community

In this chapter, students are challenged to evaluate energy consumption and usage in the community relative to a hypothetical population growth of 20% and to suggest realistic alternatives to avoid an energy-supply shortage. Students are asked to write a report, aimed at a general audience, that explains the different types of energy resources, how they work, and how they are formed, discovered, and processed. Students begin the chapter with a review of energy concepts, including heat transfer processes and the conversion of mechanical energy into heat. Students determine which energy resources are used most for electricity generation throughout the world, the United States, and within their own state. Students learn how coal is formed, where it is distributed in the United States, how coal deposits are discovered and mined, how coal is used as a source of energy, and the environmental impacts of using coal as an energy resource. Students analyze the dependence of today's society on petroleum. They learn how petroleum and natural gas are formed, how oil and gas deposits are discovered, and how oil and gas are extracted from a reservoir. Finally, students explore renewable energy sources, focusing on the potential for solar and wind energy as sources of power generation.

Mineral Resources... and Your Community

The **Chapter Challenge** for Mineral Resources and Your Community is for students to help a local company use mineral resources from the community to design a beverage container. Students are asked to prepare a report that explains what mineral resources are, how they are used and extracted, and what impact their use has on the environment, all relative to the proposed beverage container. Students begin the chapter by surveying the materials that are currently being used for beverage containers in their community. Students learn to identify minerals. They learn how mineral resources are found, extracted, and processed for use, and explore the environmental impacts of these activities. By the end of the chapter, students are better prepared to understand what makes minerals a valuable resource and how the use of mineral resources impacts the other Earth systems.

Water Resources... and Your Community

The **Chapter Challenge** for Water Resources and Your Community is for students to prepare a report that will help their community leaders to determine how the water resource needs of the community would change with the development of an industrial park, mini-mall, and residential area in the community. Students use an Earth systems approach to investigate the hydrologic cycle and global biogeochemical cycles (nitrogen cycle) within the context of community water resources and development. Students identify the human and natural factors which determine the "income and expenditure" of water resources. They determine how to measure domestic usage and obtain information on the quantity of water used by industry and agriculture. These activities help them to identify methods used to conserve water. They determine how rainfall, temperature, and other natural factors affect proper management and usage. They investigate how water resources are vulnerable to pollution by both human use and natural cycles or processes, and make models of water treatment processes. By the end of the chapter, students are better prepared to understand potential water quality problems and efficiently manage this precious and vital resource.

1

Energy Resources ...and Your Community

EARTH'S NATURAL RESOURCES
CHAPTER I

ENERGY RESOURCES... AND YOUR COMMUNITY

Chapter Overview

In **Chapter 1, Energy Resources and Your Community**, students are challenged to evaluate energy consumption and use in the community relative to a hypothetical population growth of 20%, and to suggest realistic alternatives to avoid an energy-supply shortage. Students are asked to write a report, aimed at a general audience, that explains what the different types of energy resources are, how they work, and how they are formed, discovered, and processed.

Chapter 1 begins with a review of energy concepts, including heat transfer processes and the conversion of mechanical energy into heat. Students determine which energy resources are used most for electricity generation throughout the world, in the United States, and in their own state. They learn how coal is formed, where coal is distributed in the United States, how coal deposits are discovered and mined, how coal is used as a source of energy, and what the environmental impacts of using coal as an energy resource are. Students analyze the dependence of today's society on petroleum. They learn how petroleum and natural gas are formed, how oil and gas deposits are discovered, and how oil and gas are extracted from reservoirs. Finally, students explore renewable energy sources, focusing on the potential for solar and wind energy as sources of power generation in their community.

Chapter Goals for Students

- Understand how energy resources are tied to other Earth systems.
- Participate in scientific inquiry and construct logical conclusions based on evidence.
- Recognize that energy resources are indispensable natural resources whose use and impact on the environment need to be carefully monitored.
- Appreciate the value of Earth science information in improving the quality of lives, globally and within the community.

Chapter Timeline

Chapter 1 takes about four weeks to complete, assuming one 45-minute period per day, five days per week. Adjust this guide to suit your school schedule and standards. Build flexibility into your schedule by manipulating homework and class activities to meet your students' needs.

A sample outline for presenting the chapter is shown below. It assumes that the teacher assigns homework at least three nights a week and assigns **Understanding and Applying What You Have Learned** and **Preparing for the Chapter Challenge** as group work to be completed during class. This outline also assumes that **Inquiring Further** sections are reserved as additional, out-of-class activities. This is only a sample, not a suggested or recommended method of working through the chapter; adjust your daily and weekly plans to meet the needs of your students and your school.

Day	Activity	Homework
1	**Getting Started; Scenario; Chapter Challenge; Assessment Criteria**	
2	**Activity 1 – Investigate, Parts A and B**	**Investigate, Part C; Digging Deeper; Check Your Understanding**
3	**Activity 1 – Review; Understanding and Applying; Preparing for the Chapter Challenge**	
4	**Activity 2 – Investigate**	**Digging Deeper; Check Your Understanding**
5	**Activity 2 – Review; Understanding and Applying; Preparing for the Chapter Challenge**	
6	**Activity 3 – Investigate**	**Digging Deeper; Check Your Understanding**
7	**Activity 3 – Review; Understanding and Applying; Preparing for the Chapter Challenge**	
8	**Activity 4 – Investigate, Parts A and B**	**Investigate, Part C; Digging Deeper; Check Your Understanding**
9	**Activity 4 – Review; Understanding and Applying; Preparing for the Chapter Challenge**	
10	**Activity 5 – Investigate**	**Digging Deeper; Check Your Understanding**

Day	Activity	Homework
11	**Activity 5 – Review; Understanding and Applying; Preparing for the Chapter Challenge**	
12	**Activity 6 – Investigate, Parts A, B, and C**	**Digging Deeper; Check Your Understanding**
13	**Activity 6 – Review; Understanding and Applying; Preparing for the Chapter Challenge**	
14	**Activity 7 – Investigate, Parts A and B**	
15	**Activity 7 – Investigate, Parts C and D**	**Digging Deeper; Check Your Understanding**
16	**Activity 7 – Review, Understanding and Applying; Preparing for the Chapter Challenge**	
17	**Activity 8 – Investigate, Parts A and B**	**Digging Deeper, Check Your Understanding**
18	**Activity 8 – Review, Understanding and Applying; Preparing for the Chapter Challenge**	
19	**Complete Chapter Report**	**Finalize Chapter Report**
20	**Present Chapter Report**	

National Science Education Standards

Analyzing energy use and consumption in the community relative to a hypothetical population increase sets the stage for the **Chapter Challenge**. Through a series of activities, students begin to develop the content understandings outlined below.

CONTENT STANDARDS

Unifying Concepts and Processes
- Systems, order, and organization
- Evidence, models, and explanation
- Constancy, change, and measurement
- Evolution and equilibrium

Science as Inquiry
- Identify questions and concepts that guide scientific investigations
- Design and conduct scientific investigations
- Use technology and mathematics to improve investigations
- Formulate and revise scientific explanations and models using logic and evidence
- Communicate and defend a scientific argument
- Understand scientific inquiry

Earth and Space Science
- Energy in the Earth system
- Geochemical cycles
- Origin and evolution of the Earth system

Science and Technology
- Identify a problem or design an opportunity
- Propose designs and choose between alternative solutions
- Implement a proposed solution
- Evaluate the solution and its consequences
- Communicate the problem, process, and solution
- Understand science and technology

Science in Personal and Social Perspectives

- Personal and community health
- Population growth
- Natural resources
- Environmental quality
- Natural and human-induced hazards
- Science and technology in local, national, and global challenges

History and Nature of Science

- Science as a human endeavor
- Nature of scientific knowledge
- Historical perspectives

Key Science Concepts and Skills

Activities Summaries	Earth Science Principles
Activity 1: Exploring Energy Resource Concepts Students cycle through a series of stations to investigate three processes of heat transfer: conduction, radiation, and convection. They answer questions based on a thought experiment to understand the conversion of mechanical energy into heat. Students learn about the Second Law of Thermodynamics and investigate how it is related to the generation of electricity.	• Heat transfer processes: convection, radiation, and conduction • Kinetic and potential energy • Power and work • Second Law of Thermodynamics
Activity 2: Electricity and Your Community Students compare the use of energy resources for electricity generation in the United States to those used in other countries. Students then identify the energy sources that are most commonly used for electricity generation in the United States and in their state. This helps them think about what current means of producing power their community relies on most, and which power sources are least important.	• Energy resources for electricity generation • Electric power versus electric energy • Methods of generating electric power • Renewable energy source
Activity 3: Energy from Coal Students examine coal samples to understand the physical properties of different types of coal. Students look at a map that shows the distribution of coal resources in the United States. They determine whether or not coal deposits are found in their community today, or if they could be found in their community in the future.	• Physical properties of coal • Distribution of coal deposits in the United States • Formation of coal • Fossil fuels
Activity 4: Coal and Your Community Students investigate the production and consumption of coal in the United States They use data on trends in coal production and consumption to extrapolate into the future. Students correlate a series of well logs to understand how geologists explore for coal. Finally, they look at possible methods to conserve coal resources.	• Trends in coal production and consumption • Coal exploration and mining • Well log correlation • Energy conservation
Activity 5: Environmental Impacts and Energy Consumption Students examine a map that shows the acidity of rainfall across the United States, and correlate the pattern of rainwater pH to the distribution of coal-producing regions. Students complete an experiment to understand how different types of rocks can neutralize the acidity of rainwater. They consider how this relates to environmental impacts of acid rain.	• Acid rain • Carbon cycle • CO_2 concentrations and the greenhouse effect • Advantages and disadvantages of energy resources

Key Science Concepts and Skills

Activities Summaries	Earth Science Principles
Activity 6: Petroleum and Your Community Students investigate oil production, imports, and consumption in the United States to recognize the dependence of today's society on oil as a resource. They use data on trends in oil production and consumption to extrapolate into the future. Students look at a map that shows the distribution of oil and gas deposits in the United States to determine whether oil and gas are found, refined, and/or distributed near their community. Finally, they go online to investigate production, consumption, and distribution of oil and natural gas in their state.	• Nature and origin of petroleum and natural gas • Production and consumption of oil and natural gas
Activity 7: Oil and Gas Production Students design investigations to explore porosity and permeability of rock bodies and to consider how these factors affect the volume and rate of production in oil and gas fields. Students use data to produce a cross section of a petroleum reservoir. They estimate the likelihood of finding oil and gas in various locations.	• Porosity and permeability • Petroleum recovery • Petroleum reserves
Activity 8: Renewable Energy Sources— Solar and Wind Students investigate the use of solar energy by constructing a solar water heater and determining its maximum energy output. They investigate the use of wind energy by constructing an anemometer to measure wind speeds and calculating how much power can be generated by wind.	• Forms of solar energy • Solar energy for heat and electricity • Wind power • Energy conservation

Equipment List for Chapter One:

Materials needed for each group per activity.

Activity I Part A

- Lamp with a 100-W light bulb
- Metal cup
- Styrofoam® cup
- Two alcohol thermometers
- Two solar cookers:
    - Large cardboard box (into which a pizza box will fit)
    - Two cardboard pizza boxes
    - Newspaper
    - Cardboard pieces
    - Aluminum foil
    - Glue
    - Black construction paper
    - Two thermometers (that will fit inside pizza boxes)
    - Plastic wrap
    - Straws

Activity I Part B

No additional materials needed.

Activity I Part C

- Calculator
- Steel ball or billiards ball
- Balance scale

Activity 2 Part A

No additional materials needed.

Activity 2 Part B

- Internet access (or copies of your state electricity profile)*

Activity 3 Part A

- Samples of coal: anthracite, bituminous, lignite, peat
- Bunsen burner (for teacher demonstration)
- magnifying glass or hand lens

Activity 3 Part B

- Geologic map of your state or community*
- Internet access (or printouts of your state's Energy, Consumption, Prices, and Expenditures profiles)*

Activity 4 Part A

- Graph paper

Activity 4 Part B

- Copies of cross section on page R39 (see **Blackline Master 4.1 Cross Section of Core Holes**)

Activity 4 Part C

- Calculator
- Internet access (optional)*

Activity 5

(Used for teacher demonstration; not needed for each group)

- Crushed limestone
- Crushed granite (or clean quartz sand to model sandstone)
- Two 1-L soda bottles
- Two 500-mL beakers
- Dilute sulfuric acid solution (mixed with distilled water to a pH of 4.3 to 4.5)
- pH indicator solution
- Water test kit, pH meter, or pH paper
- Samples of crushed rock from your community
- Ring stands to hold the 1-L soda bottles

Activity 6 Part A

- Graph paper
- Calculator

*The *EarthComm* web site provides suggestions for obtaining these resources.

Activity 6 Part B

No additional materials needed.

Activity 6 Part C

- Internet access (or copies of your state's petroleum and natural gas production, consumption, and distribution data from the Energy Information Agency)*

Activity 7 Part A

- Two 500-mL (16 oz) clear plastic soda bottles with bottoms removed
- Fine cheesecloth
- Electric tape
- Two stands with clamps to hold plastic bottles
- Sand
- Aquarium gravel
- 50-mL graduated cylinder
- 200-mL vegetable oil
- Water
- Calculator

Activity 7 Part B

No additional materials needed

Activity 7 Part C

- Four 500-mL (16 oz) clear plastic soda bottles with bottoms removed
- Fine cheesecloth
- Electric tape
- Four stands with clamps to hold plastic bottles
- Four 500-mL glass beakers
- Coarse sand
- Fine sand
- Silt/clay

- Aquarium gravel
- 50-mL graduated cylinder
- Water

Activity 7 Part D

- 500-mL clear plastic soda bottle
- Vegetable oil
- Protractor
- Graph paper

Activity 8 Part A

- 15 feet of flexible plastic tubing
- Cardboard box (or lid) trimmed to height of 1 in. (photocopy paper box top, or large pizza box). The inside surface should be painted black or lined with black construction paper.
- Clear plastic wrap or sheet (or sheet of thin clear Plexiglas® cut to box size)
- Ruler
- Tape
- Glass or plastic container (2-L minimum)
- Styrofoam® cooler
- Two thermometers
- Siphoning bulb
- Adjustable tubing clamp
- Stopwatch or watch with second hand

Activity 8 Part B

- Five paper cups
- Three wooden dowels, each about 10 in. long
- Modeling clay
- Piece of wood, approximately 10 in. x 10 in. – to serve as a base
- Calculator
- Graph paper

*The *EarthComm* web site provides suggestions for obtaining these resources.

NOTES

1

Energy Resources ...and Your Community

Getting Started

As the population of a community increases, so does the demand for energy resources for transportation, electricity, and heating fuels. Ensuring a supply of energy to meet the growing needs of a community requires careful analysis and planning.

Think about all of the ways that you have used energy resources in your day so far, starting with when you got up in the morning until you arrived at this class.

- What was the source of energy for each of the activities?

What do you think? Sketch out some of your ideas about energy resources on paper. First, make a list of all the ways that energy resources have been used in your day so far, starting with the moment you woke up (the alarm clock, for instance) and ending with your present classroom. There should be at least five items on your list. Next, make a column for the energy resource responsible for the activity (coal burned to produce electricity for the alarm clock, for instance). Make another column for the source of this energy resource (coal deposit in New Mexico, for example). Finally, make a column for energy alternatives for each activity. For instance, one energy alternative to taking the bus to school is to ride your bike to school. An energy alternative to taking a ten-minute hot shower is to take a five-minute hot shower. The alternatives do not necessarily have to reduce the energy required. They could have other benefits, such as reducing an environmental hazard. Be prepared to discuss your table with your small group and the class.

R2

Getting Started

Uncovering student conceptions about Energy Resources and the Earth System

Use **Getting Started** to elicit students' ideas about the main topic. The goal of **Getting Started** is not to seek closure (i.e., the right answer) but to provide you (the teacher) with information about the students' starting point and about the diversity of ideas and beliefs in the classroom. By the end of the chapter, students will have developed more detailed and accurate understandings about energy resources and how they fit into the Earth system.

Have students answer the opening question (How have you used energy resources today, and where did the energy come from?). Suggest that they make a simple table to record their ideas. An example is provided below. Ask students to work independently or in pairs, and to exchange papers with others.

Activity	Energy Resource	Source of Energy	Energy Alternative
1. Woke up (used alarm clock)	Coal – used to produce electricity	Coal deposits	Using a windup clock (no electricity used)
2.			
3.			
4.			
5.			

The purpose of this activity is to get students to think about the many ways that they use energy resources in their daily lives. Students will readily identify how they use energy, and they will have ideas about where the energy comes from and how to accomplish the same activity in another way. Students are most likely to struggle with the source of the energy. That's okay. Remind students that it is okay to speculate about where the energy comes from because the purpose of the activity is to share their thinking (not find "right answers") and to identify opportunities to investigate what they might not know or might be unsure about. In time, they will investigate the sources of energy upon which they have come to depend. You might hold a brief class discussion, recording their ideas on an overhead transparency (use the same table format shown in the example). Avoid labeling answers as right or wrong. Accept all responses, and encourage clarity of expression and detail.

In time, *EarthComm* students will understand that energy resources are a component of the geosphere, and that their use and extraction impacts the atmosphere, biosphere, cryosphere, and hydrosphere. Changes in the other Earth systems can cause changes in the mode of operation of the geosphere; likewise, changes within the geosphere can affect the other Earth systems. Students will come to appreciate the importance of understanding these changes, because they can affect the availability of valuable energy resources that we depend on, as well as the environment in which we live.

Scenario

Community leaders often depend upon experts to outline a 10-year plan that addresses the impact of an increase in population on energy consumption and supply. Your community has called upon your *EarthComm* classroom to evaluate energy consumption and use in your community. Community leaders need realistic solutions or alternatives for energy use in your community, to help prevent possible energy shortages while maintaining the quality of the environment.

Chapter Challenge

Your challenge is to produce a report that is written for a general audience. In the report, you must critically analyze energy use in your community on the assumption of a population growth of 20%, and to provide realistic solutions to avoid an energy-supply shortage. Your report needs to help community members understand the origin, production, and consumption (use) of energy resources from an Earth System perspective. You need to address the following points:

- What are the current energy uses and consumption rates in your community?

- How will energy needs change if the population of your community increases by 20%?

- Will your community be able to meet these needs, and at what costs?

- What solutions or alternatives exist that may help to avoid a potential energy crisis?

Your report should explain at least one example for each of the following questions:

- What is the difference between renewable and nonrenewable energy resources?

- How are energy resources used to do work?

- How are energy resources formed, discovered, and produced?

- How much energy conservation is possible in your community? What steps can be taken?

- Does the production and consumption of energy affect the environment?

Assessment Criteria

Think about what you have been asked to do. Scan ahead through the chapter activities to see how they might help you to meet the challenge. Work with your classmates and your teacher to define the criteria for assessing your work. Record all of this information. Make sure that you understand the criteria as well as you can before you begin. Your teacher may provide you with a sample rubric to help you get started.

R3

Chapter Challenge and Assessment Criteria

Read (or have a student read) the **Scenario** and **Chapter Challenge** aloud to the class. Allow students to discuss what they have been asked to do. Have students meet in teams to begin brainstorming what they would like to include in their **Chapter Challenge** reports. Request a brief summary in their own words of what they have been asked to do, and a description of attributes of a high-quality report.

Alternatively, lead a class discussion about the challenge and the expectations. Review the titles of the activities in the Table of Contents. To remind students that the content of the activities corresponds to the content expected for the Chapter Report, ask them to explain how the title of each activity relates to the expectations for the **Chapter Challenge**. Familiarize students with the structure of each activity. When you come to the section titled **Preparing for the Chapter Challenge**, point out that each activity contributes to the challenge in some way.

Guiding questions for discussion include:

- What do the activities have to do with the expectations of the challenge?
- What have you been asked to do?
- What should a good final report contain?

A sample rubric for assessing the **Chapter Challenge** is shown on the following pages. You can copy and distribute the rubric as is, or you can use it as a baseline for developing scoring guidelines and expectations that suit your needs. For example, you might wish to ensure that core concepts and abilities derived from your local or state science frameworks also appear on the rubric. You might also wish to modify the format of the rubric to make it more consistent with your evaluation system. However you decide to evaluate the Chapter Report, keep in mind that all expectations should be communicated to students and that the expectations should be outlined at the start of their work. Please review **Assessment Criteria** (pages xxii to xxiv of this Teacher's Edition) for a more detailed explanation of the assessment system developed for the *EarthComm* program.

Assessment Rubric for Chapter Challenge on Energy Resources

Meets the standard of excellence. **5**	_Significant_ information is presented about <u>all</u> of the following: • The current energy uses and consumption rates in your community. • How energy needs will change if the population of your community increases by 20%. • Whether your community will be able to meet these needs and at what costs. • Solutions or alternatives that may help to avoid a potential energy crisis. • One example that illustrates the difference between renewable and nonrenewable energy resources. • One example that illustrates how energy resources are used to do work. • One example that illustrates how energy resources are formed, discovered, and produced. • One example that illustrates how energy conservation is possible in your community. • One example that illustrates how the production and consumption of energy affects the environment. _All_ of the information is accurate and appropriate. The writing is clear and interesting.
Approaches the standard of excellence. **4**	_Significant_ information is presented about <u>most</u> of the following: • The current energy uses and consumption rates in your community. • How energy needs will change if the population of your community increases by 20%. • Whether your community will be able to meet these needs and at what costs. • Solutions or alternatives that may help to avoid a potential energy crisis. • One example that illustrates the difference between renewable and nonrenewable energy resources. • One example that illustrates how energy resources are used to do work. • One example that illustrates how energy resources are formed, discovered, and produced. • One example that illustrates how energy conservation is possible in your community. • One example that illustrates how the production and consumption of energy affects the environment. _All_ of the information is accurate and appropriate. The writing is clear and interesting.
Meets an acceptable standard. **3**	_Significant_ information is presented about <u>most</u> of the following: • The current energy uses and consumption rates in your community. • How energy needs will change if the population of your community increases by 20%. • Whether your community will be able to meet these needs and at what costs. • Solutions or alternatives that may help to avoid a potential energy crisis. • One example that illustrates the difference between renewable and nonrenewable energy resources. • One example that illustrates how energy resources are used to do work. • One example that illustrates how energy resources are formed, discovered, and produced. • One example that illustrates how energy conservation is possible in your community. • One example that illustrates how the production and consumption of energy affects the environment. _Most_ of the information is accurate and appropriate. The writing is clear and interesting

Assessment Rubric for Chapter Challenge on Energy Resources

Below acceptable standard and requires remedial help. <div align="center">**2**</div>	<u>*Limited*</u> information is presented about the following: • The current energy uses and consumption rates in your community. • How energy needs will change if the population of your community increases by 20%. • Whether your community will be able to meet these needs and at what costs. • Solutions or alternatives that may help to avoid a potential energy crisis. • One example that illustrates the difference between renewable and nonrenewable energy resources. • One example that illustrates how energy resources are used to do work. • One example that illustrates how energy resources are formed, discovered, and produced. • One example that illustrates how energy conservation is possible in your community. • One example that illustrates how the production and consumption of energy affects the environment. <u>Most</u> of the information is accurate and appropriate. Generally, the writing does not hold the reader's attention.
Basic level that requires remedial help or demonstrates a lack of effort. <div align="center">**1**</div>	<u>*Limited*</u> information is presented about the following: • The current energy uses and consumption rates in your community. • How energy needs will change if the population of your community increases by 20%. • Whether your community will be able to meet these needs and at what costs. • Solutions or alternatives that may help to avoid a potential energy crisis. • One example that illustrates the difference between renewable and nonrenewable energy resources. • One example that illustrates how energy resources are used to do work. • One example that illustrates how energy resources are formed, discovered, and produced. • One example that illustrates how energy conservation is possible in your community. • One example that illustrates how the production and consumption of energy affects the environment. <u>*Little*</u> of the information is accurate and appropriate. The writing is difficult to follow.

ACTIVITY 1—EXPLORING ENERGY RESOURCE CONCEPTS

Background Information

Energy, Heat, Work, and Power

It's not easy to make a simple statement of the meaning and significance of the concept of energy. The term energy is often used in a loose or vague way and does not reveal the concept of energy as it is used in the science of physics. In a loose sense, energy has to do with the motions of matter, or the potential for motions of matter. Think first about the concept of kinetic energy. Newton's Second Law of Motion states that the force exerted on a body is equal to the rate of change of momentum of the body—momentum being the product of the mass of the body times its velocity. One of the elementary things that's done in the science of mechanics is to recast this law into a different form that states that the work done on the body is equal to the change in kinetic energy. Work is defined here as the product of the component of the force in the direction of motion. Kinetic energy is defined as one-half the product of the mass and the velocity.

Another form of energy, called potential energy, reflects the work done by the force that depends on the initial and final positions of the body, and not on the details of the path followed or how the speed of the body changes along its path. Gravity is such a force. This leads to the definition of the gravitational potential energy as equal to mgh, where m is the mass of the body, g is the acceleration of gravity, and h is the difference in elevation between the final position and the initial position of the body. These matters are discussed in a simple way in all good elementary physics textbooks. The sum of the kinetic energy and the potential energy of a body, called the mechanical energy, is a convenient way of expressing the part of the total energy of the body that involves its motion and the part that involves its position in a gravitational field.

A body can have other forms of energy as well. The heat energy of a body reflects the energy of motion of the constituent atoms and molecules of which the body is composed. The chemical energy or biochemical energy of a body reflects the potential for increase in mechanical energy and/or heat energy by chemical changes, like combustion or decomposition undergone by the body. The radioactive energy of a body reflects the potential for certain kinds of atoms to generate energy of heat and motion as they change spontaneously and unpredictably into other kinds of atoms, by a process called radioactive decay.

Power is the term used for the rate at which work is done by a force. Work carries no element of time: the work done by a force on a body as the body moves from one place to another depends only upon how the force varies along the path of the body, not on how fast the body traverses its path. Power, on the other hand, depends not only upon the force but also upon the speed of the body as it traverses its path.

Heat Transfer

Heat, a form of energy, can be moved or transferred from place to place by conduction, convection, or radiation. All matter consists of atoms and/or molecules (molecules being small aggregations of tightly bonded atoms). At all temperatures above absolute zero, the atoms and molecules of matter are in a constant state of random motion. In gases, most of the particles at any given time are in free flight, and they collide only occasionally with one another or with the solid walls of a container. In solids, the particles are locked into a rigid structure but undergo small and extremely rapid vibrations around their equilibrium positions in the structure. In liquids, the motions of the particles are less easy to describe as they weave among adjacent particles.

The heat energy of a body of matter is a measure of the energy of motion of the constituent atoms and molecules. Any effect that increases the energy of motion of the atoms and molecules results in an increase in the heat energy of the body. The temperature of a body, on the other hand, is a measure of the speed of motion of the constituent atoms and molecules (which is related to, but different from, the energy of motion of the particles): the faster they move, the higher the temperature you sense when you are in contact with the body.

If two bodies at different temperatures are placed in contact with one another, the faster-moving particles in the higher-temperature body interact with the slower-moving particles in the lower-temperature body. The effect is a speeding up of the particles in the cooler body and a slowing down of the particles in the warmer body. The result is that the heat energy of the cooler body increases, and the heat energy of the warmer body decreases. It is this process that we call heat conduction. For heat to flow from one body to another by conduction, the bodies must be in direct contact.

Heat conduction is an inherently slow and inefficient process. Heat can be moved bodily from one place to another much more efficiently by the large-scale motion of bodies with different heat energy from place to place. Convection is such a process. Convection is the transfer of heat from one place to another by the actual motion of the material. If the material is moved by a machine like a blower, a pump, a fan, or waving arms, the process is called forced convection; if the material is moved because of differences of density in the presence of a gravitational field, the process is called free convection. Large-scale free convection is an extremely important process of heat transfer in the atmosphere, the oceans, and the Earth's mantle.

All matter radiates energy in the form of electromagnetic waves. The waves are generated by the motions of the atoms and molecules at and near the surface of the body. The waves propagate through space at the speed of light. Wavelength is a function of the temperature of the radiating body: the higher the temperature of the body, the shorter the wavelength of the radiation emitted. The energy of the radiation is a very steeply increasing function of the temperature of the radiating material. The physical laws governing such matters can be found in any good physics textbook on electromagnetism.

The change in the heat energy of a body by electromagnetic radiation depends upon the balance between the radiant energy received by the body and the energy radiated by the body. If the energy received is greater than the energy radiated, the temperature and the heat energy of the body increases. When you stand in the sunlight or near a hot stove, you receive more energy than you radiate; when you stand outdoors on a clear night, the surfaces of your body or your clothing are cooled because you are radiating more heat to space than you are receiving from space.

Thermodynamics

Thermodynamics is the branch of physics that deals with the nature of heat and its conversion to other forms of energy (mechanical, chemical, and electrical). It grew out of efforts to build more efficient devices to extract useful work from expanding hot gases, like steam. The total energy that a body of matter contains as a result of the positions, motions, and chemical nature of its atoms and molecules is called its internal energy. Bodies can change their internal energy by absorbing or giving off heat, or by doing work on their surroundings or having work done on them by their surroundings. The First Law of Thermodynamics states that the change in internal energy of a body is equal to the heat absorbed from the environment minus the work done on the environment. This first law can thus be viewed as a statement of energy bookkeeping.

Thermodynamics is more, however, than just energy bookkeeping. Energy transfer is not entirely a two-way street. It turns out that it is impossible to convert all of the heat extracted from a given body with given internal energy into useful mechanical work. This is the implication of what is called the Second Law of Thermodynamics. One traditional way of stating the Second Law of Thermodynamics (there are others) is that it is impossible to make a transformation whose only final result is to transfer heat from a body at a given temperature to another body at a higher temperature. The implication is that spontaneous flow of heat from a hotter body to a colder body is reversible only upon some expenditure of mechanical (or other nonthermal) energy. You can see from this that the practical consequence of the second law is that a heat engine, like a steam engine or an internal-combustion engine, cannot be 100% efficient: the expenditure of heat energy or chemical energy must always be greater — and is usually much greater — than the gain in mechanical energy.

More Information – on the Web

Visit the *EarthComm* web site www.agiweb.org/earthcomm to access a variety of links to web sites that will help you deepen your understanding of content and prepare you to teach this activity. Many of the sites also contain images that you can download and make into overheads for incorporation into class discussions.

Goals and Assessment

Clarify that the goals indicate what students should understand and be able to do as a result of the activity. Make sure students understand that Chapter Assessments are based upon these goals.

Goal	Location in Activity	Assessment Opportunity
Investigate heat transfer by the processes of conduction, convection, and radiation.	**Investigate** Part A **Digging Deeper; Check Your Understanding** Question 2 **Understanding and Applying What You Have Learned** Questions 1 – 3	Observations at each station are accurate, all hypotheses and observations are recorded in notebook. Answers to questions are reasonable and closely match responses given in Teacher's Edition.
Investigate the conversion of mechanical energy into heat.	**Investigate** Part B **Investigate** Part C **Digging Deeper; Check Your Understanding** Questions 1 and 3	Answers to questions are reasonable and are based on graph in text.
Learn about the Second Law of Thermodynamics and how it relates to the generation of electricity.	**Investigate** Part C **Digging Deeper; Check Your Understanding** Questions 4 and 5 **Understanding and Applying What You Have Learned** Questions 4 and 6	Calculations are correct, work is shown, answers to questions are accurate. Answers to questions are reasonable and closely match responses given in Teacher's Edition.

Earth's Natural Resources Energy Resources

Activity 1   Exploring Energy Resource Concepts

Goals

In this activity you will:

- Investigate heat transfer by the processes of conduction, convection, and radiation.

- Investigate the conversion of mechanical energy into heat.

- Learn about the Second Law of Thermodynamics and how it relates to the generation of electricity.

Think about It

A car moving along a mountain road has energy. It has energy due to its motion (kinetic energy), energy due to its position in a gravity field (potential energy), and energy stored as fuel in its gas tank (chemical energy).

- Classify each item below as having kinetic energy, potential energy, or chemical energy:
  a) a rock balanced at the edge of a cliff;
  b) a piece of coal;
  c) a landslide;
  d) a roller-coaster car;
  e) a diver on a 10-m platform;
  f) a car battery;
  g) tides.

What do you think? Record your ideas in your *EarthComm* notebook. Be prepared to discuss your responses with your small group and the class.

Activity Overview

Students begin the investigation by visiting a series of stations that allow them to investigate heat transfer by the processes of conduction, convection, and radiation. Then they complete a thought experiment in which they model how the mechanical energy of a lump of modeling clay that is tossed into the air is converted into heat. Students complete a series of calculations to become familiar with the units and terms used to discuss energy. These calculations also introduce the concept of the Second Law of Thermodynamics. **Digging Deeper** defines the heat transfer processes of conduction, convection, and radiation, as well as different forms of energy (kinetic, potential, chemical, and mechanical). The reading also explains the Second Law of Thermodynamics and its relationship to the generation of electricity.

Preparation and Materials Needed

Part A

Assemble all of the materials before class. Keep in mind that students will be working at stations, and therefore it is not necessary to collect a set of materials for each student group. For the third station, you will need two solar cookers. You can either purchase these or assemble your own by following the instructions below. (Use the illustration on page R5 of the student text as a guide.) One method is to construct the solar cookers using small pizza boxes, and then insert one into a large pizza box to construct the insulated solar cooker.

Assembling a Solar Cooker inside an Insulated Box
a) Line the bottom of a large box with insulating materials like crumpled newspapers.
b) Set a cardboard pizza box inside the big box (this will serve as the oven).
c) Fill the space between the sides of the two boxes with crumpled newspaper.
d) Fold cardboard pieces over the insulated space between boxes, and secure tightly.
e) Line the sides of the inner box with aluminum foil adhered with nontoxic glue.
f) Line the bottom of the inner box with black construction paper.
g) On the top cover of the small pizza box, draw a square 3 cm from all the sides.
h) Cut along three of the lines, but leave the fourth line near the hinge of the box uncut.
i) Carefully unfold the flap.
j) Wrap a piece of aluminum foil around the flap, smooth wrinkles, and secure with glue.
k) Place a thermometer inside the pizza box.
l) Stretch plastic wrap tightly over the inside top of the box. Smooth the plastic and secure it around the sides with glue or tape so that no air can escape.
m) Use straws or another device to prop open the flap and allow aluminum lining to reflect the maximum sunlight.

Assembling a Standard Solar Cooker
To assemble the standard solar cooker, follow Steps (e) to (m) on the previous page.

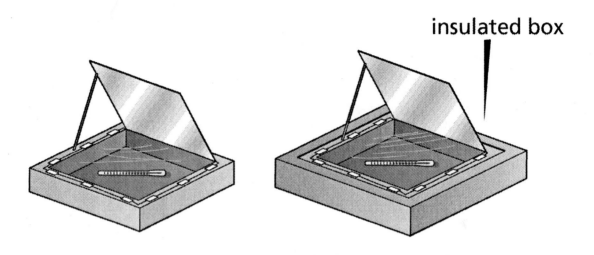

insulated box

Parts B,C

Although no special preparation is required for **Parts B** and **C** of this investigation, it is recommended that you complete the activities before class.

Materials
Part A
- Lamp with a 100-W light bulb
- Metal cup
- Styrofoam cup
- Two alcohol thermometers
- Two solar cookers:
  - Large cardboard box (into which a pizza box will fit)
  - Two cardboard pizza boxes
  - Newspaper
  - Cardboard pieces
  - Aluminum foil
  - Glue
  - Black construction paper
  - Two thermometers (that will fit inside pizza boxes)
  - Plastic wrap
  - Straws

Part B
No additional materials needed.

Part C
- Calculator
- Steel ball or billiards ball
- Balance scale

*The *EarthComm* website provides suggestions for obtaining these resources.

Think about It

Student Conceptions

Students will likely have different levels of understanding of kinetic, potential, and chemical energy. They should be able to use the example of the car moving along a mountain road to help them to classify the given items as having kinetic, potential, or chemical energy. If students are having trouble with this exercise, you may want to review the different forms of energy as a class.

Answer for the Teacher Only

- Kinetic energy is defined as the form of energy associated with motion of a body of matter (landslide, roller-coaster car, tides).
- Potential energy is the mechanical energy associated with position in a gravity field (rock balanced at the edge of a cliff, diver on a 10-m platform).
- Chemical energy is the energy that is stored in a chemical compound and released during chemical reactions, including combustion (piece of coal, car battery).

Assessment Tool

Think about It Evaluation Sheet
Use this evaluation sheet to help students understand and internalize the basic expectations for the warm-up activity.

Investigate

Energy can neither be created nor destroyed (except in nuclear reactions), but it can be changed from one form into another. The following activities will help you to explore basic concepts that govern the use of energy.

Part A: Heat Transfer
Station 1

1. Put your hand close to a 100-W light bulb and notice the heating that occurs in your hand. This is similar to the heat generated from direct sunlight.

   a) Describe what happens to the temperature of your hand as you move it slowly toward and away from the bulb.

   b) Hold a piece of paper between your hand and the light bulb. Describe and explain the change in temperature of your hand.

   c) Compare and explain the temperature difference of your hand when you hold it above the light bulb versus holding it near the side of the bulb.

    Be careful not to touch the hot bulb.

Station 2

1. Which cup will keep the water hot for a longer amount of time, a metal cup or a Styrofoam® cup? Why?

   a) Write down your hypothesis.

2. Design an experiment to test your hypothesis.

   a) Record your experimental design in your notebook.

   b) With the approval of your teacher, carry out your experiment, and record your observations.

3. Five minutes after you fill the cup, place your hand around each of the cups.

   a) Which one feels hotter? Why?

    Be sure your teacher approves your design before you begin. The water should not be hot enough to scald. Wipe up spills immediately. Use alcohol thermometers only.

Station 3

1. Set up two solar cookers as shown in the diagrams below. One is a standard solar cooker and the other is an identical solar cooker inside an insulated box.

   a) What differences do you expect in the temperature inside the two solar cookers over time? Write down your hypothesis in your notebook.

insulated box

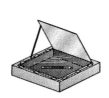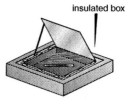

2. Design an investigation to test your hypothesis. Your design should include a plan to measure the temperature in each solar cooker and to record data every minute for at least 25 minutes.

   a) Set up a table to record your data.

R 5

EarthComm

Investigate

Assessment Tool

EarthComm Notebook-Entry Checklist
Use this checklist as a quick guide for student self-assessment and/or an
opportunity to quickly score student work. Add further criteria specific to your
classroom needs or to this particular investigation.

Investigate Notebook-Entry Evaluation Sheet
Point out the criteria listed on this evaluation sheet that are relevant to this
particular investigation. Encourage students to internalize the criteria by making
them part of your assessment conversations as you circulate around the classroom.
For example, while students are working, ask them criteria-driven questions like:
• Is your work thorough and complete?
• Are all of you participating in the activity?
• Do you each have a role to play in solving this problem? And so on.

Part A: Heat Transfer

Station I
1. Remind students not to touch the hot light bulb.
   a) As students move their hands toward the light bulb, they should notice that the
   temperature of their hand increases, and as they move their hands away from
   the light bulb, the temperature decreases. Our sensation of body temperature is
   a rather subjective matter. What the students sense is the relative intensity of
   heating of the skin surface by the radiant energy from the bulb. Measuring the
   actual skin temperature would necessitate putting a sophisticated temperature
   sensor on the surface of the hand.

   b) Students should compare the temperature of their hand—i.e., the intensity of
   radiant heating, as explained in **Step 1 (a)** above—with and without a piece of
   paper held between their hand and the bulb. The piece of paper should NOT
   touch the light bulb. Students should find that their hand feels warmer without
   the paper in front of the light bulb.

   c) Students should find that the temperature of their hand is higher when they
   hold it above the bulb then when they hold it near the side. This is because, in
   addition to receiving the radiant heat from the bulb, the hand is also warmed
   by exposure to the buoyant uprise of air that is heated, by conduction, next to
   the hot surface of the bulb.

Station 2

1. a) Students' hypotheses will vary. Sample student hypothesis: "If heat passes through metal more quickly than through Styrofoam, then water in a Styrofoam cup will stay hotter for a longer time than will water in a metal cup." Encourage students to draw upon their own experiences, reminding them that a hypothesis is not a guess but a testable statement based on prior knowledge.

2. a) Students' experiments may vary slightly, but in general will involve the use of thermometers to record the temperature of hot water in a metal cup and the temperature of hot water in a Styrofoam cup over a period of time.

    b) Be sure to check students' experimental designs before they carry out their experiments.

3. a) Five minutes after the cup is filled with hot water, the metal cup should feel hotter than the Styrofoam cup. The reason for this is that metal is a good conductor of heat whereas Styrofoam is a good thermal insulator.

Station 3

1. a) Students' hypotheses will vary. Most likely, students will think that the temperature inside the solar cooker that is in an insulated box will be greater than the temperature inside the standard solar cooker because the insulation will slow the escape of heat from the solar cooker.

2. a) Students should record the temperature inside each of the solar cookers every minute for 25 minutes.

NOTES

Earth's Natural Resources Energy Resources

3. Place a thermometer in each solar cooker and close the lids. You will want to be able to read the thermometer without blocking the path of solar energy and without opening the boxes.

   a) Record and graph the data.

4. Use the evidence that you have collected to answer the following questions:

   a) How did your results compare with your hypothesis?

   b) What heating mechanism causes the cookers to heat up in the first place?

   c) What are the different heat transfer mechanisms that are taking place in the cookers? Use diagrams to record your ideas in your notebook.

   d) What mechanism keeps the heat from escaping?

   e) What improvements could be made to the cooker if you had to do it over again?

 Be careful when you touch items after they have been in the solar cooker. They will be hot.

Part B: Kinetic Energy, Potential Energy, and Heat

1. The following is a thought experiment. The graph shows the path of a small lump of modeling clay that is thrown into the air.

   a) Copy the graph onto a sheet of graph paper.

2. Imagine that you had thrown the clay into the air so that it landed on a tabletop. In your group, discuss and record your ideas about the following:

   a) How does the kinetic energy of the piece of clay change over time? When is it highest? When is it lowest?

   b) How does the potential energy of the lump of clay change over time? When is it highest? When is it lowest?

   c) How was kinetic energy transformed into potential energy? When did this happen?

   d) How was kinetic energy transformed into heat? When did this happen?

   e) Find a way to represent the changes in these three forms of energy over time. Record your ideas on the sheet of graph paper that shows the path of the modeling clay.

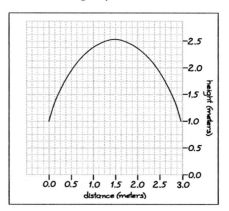

3. a) Students should record the data every minute for at least 25 minutes. They can then graph the data.

4. a) Answers will vary depending upon initial hypothesis.

   b) Solar radiation causes the cooker to heat up.

   c) Radiant energy enters the box through the transparent top and heats the opaque solid inner surfaces of the box; the air inside the box that is in contact with the solid surface is heated by conduction from the hot surfaces; the heated air circulates inside the box by convection.

Heat Transfer Mechanisms at Work in a Solar Cooker

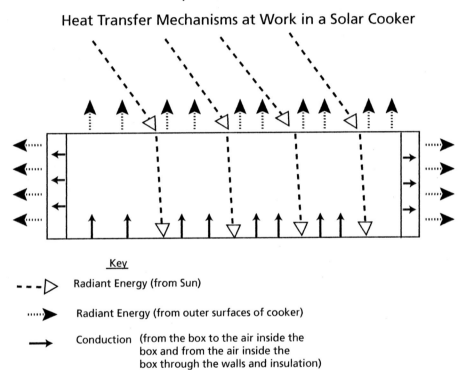

Key

– – ▷ Radiant Energy (from Sun)

⋯⋯▶ Radiant Energy (from outer surfaces of cooker)

⟶ Conduction (from the box to the air inside the box and from the air inside the box through the walls and insulation)

   d) The temperature builds up in the cooker for two reasons: the space is closed, thereby keeping the heated air from rising convectively out of the cooker; and the walls are insulated, to impede heat conduction from inside the box to the ambient air outside the box. A subtle but important point is relevant here as well: once the steady state is reached, the temperature in the box remains constant but higher than outside the box. There eventually has to be a balance between heat gained by the cooker from the incoming solar radiation and heat lost from the cooker by outgoing radiation and by heat conduction through the walls of the cooker. All matter at temperatures above absolute zero radiates energy; the hotter the body, the greater the radiation, and the shorter the wavelength of the radiation. If the box becomes sufficiently warm inside, your students with sensitive skin should be able to feel the radiant heat by putting their faces or the backs of their hands close to the surface of the box.

e) Answers will vary. Two suggestions:
  • Use two layers of transparent material for the top of the cooker, with a thin air space between. This reduces heat loss through the top by providing an insulating air layer, without reducing the transparency very much. Air is a relatively poor conductor of heat.
  • Add more insulation to the walls and bottom of the cooker, to cut down heat loss by conduction.

Part B: Kinetic Energy, Potential Energy, and Heat

1. Have your students copy the graph on page R6 into their notebooks, or provide them with a copy of the graph that they can then insert into their notebooks using **Blackline Master Energy Resources 1.1, Graph of Distance vs Height**

2. a) The kinetic energy of the lump of clay decreases as it is thrown upward, reaches a minimum at the top of its trajectory arc, and increases as it travels downward. If the lump was thrown straight up rather than at an angle, the ball would come to a stop at the instant it reaches its maximum elevation—at that time, its kinetic energy would be zero.

   b) The potential energy of the lump of clay increases as it is thrown upward, reaches a maximum at the top of its trajectory arc, and decreases as it travels downward.

   c) Kinetic energy was transformed into potential energy as the lump of clay rose.

   d) Kinetic energy was transformed into heat when the lump of clay hit the table. The organized motion of the solid body was converted into increased random motions of the constituent atoms of the clay as the clay was deformed upon impact. We sense those atomic motions as the temperature of the clay. In this case, the kinetic energy of the lump was not great enough for the students to notice the very slight increase in the temperature of the clay caused by the impact. A more common mechanism (one that students will likely be familiar with) that facilitates the transformation of kinetic energy into heat is frictional resistance.

   e) Answers will vary. The best way to do this would be to actually plot the values of kinetic energy and potential energy as a function of height, to go along with the vertical axis of the graph shown on page R6 of the student text. A complete job is beyond the reach of your students, because they would need to compute the velocity as a function of height, and then use definition of kinetic energy as $1/2 \, mv^2$, where m is the mass of the lump and v is its velocity. The potential energy is equal to mgh, where g is the acceleration of gravity (9.8 m/s^2) and h is the height of the lump above some arbitrary datum, like the tabletop. They might instead draw a qualitative graph that shows kinetic energy decreasing with height and potential energy increasing with height. Other students might have the idea of drawing a graph of kinetic energy and potential energy on the vertical axis of a graph that has time on the horizontal axis. Again, without further application of the principles of mechanics this would have to be only qualitative.

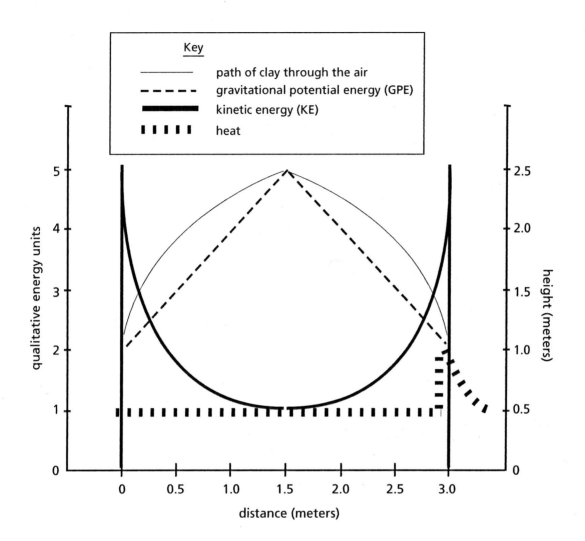

Blackline Master Energy Resources 1.1
Graph of Distance vs Height

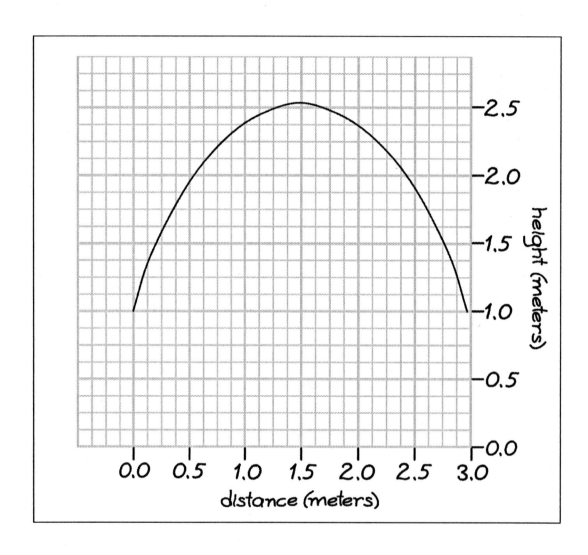

Blackline Master Energy Resources 1.2
Energy Conversion Table

Energy Conversion Table
Heat
1 kcal (kilocalorie) = the heat needed to raise the temperature of one kilogram of water from 14.5°C to 15.5°C
1 Btu (British thermal unit) = the heat needed to raise the temperature of one pound of water from 60°F to 61°F
1 kcal = 1000 cal = 3.968 Btu
Force, mass, and velocity
1 kg = 0.069 slug
acceleration of gravity (g) = 9.8 m/s^2 = 32 ft/s^2
1 N (newton) = 1 J/m (joule per meter) = 0.225 pounds
1 m/s = 3.28 fps (feet per second) = 2.24 mph (miles per hour)
Energy and work (the mechanical equivalent of heat)
1 kcal = 1000 cal = 4184 J (joules)
1 Btu = 252 cal = 777.9 ft-lb = 1055 J
1 kWh = 3,600,000 J = 3413 Btu
1 quad (Q) = 10^{15} Btu
Power (the rate at which work is done)
1 W (watt) = 1 J/s (joules per second)
1 hp (horsepower) = 550 ft-lb/s = 746 W

Part C: Energy Units and Conversions

1. Look at the conversion table. (In this activity you will record all your data in metric units. The table gives both metric and English equivalents to all the units that you will be using in this activity. Refer to this table whenever necessary.)

a) Begin a concept map to show how the units are interconnected. Complete the concept map as you work through this part of the activity.

Energy Conversion Table
Heat
1 kcal (kilocalorie) = the heat needed to raise the temperature of one kilogram of water from 14.5°C to 15.5°C 1 Btu (British thermal unit) = the heat needed to raise the temperature of one pound of water from 60°F to 61°F 1 kcal = 1000 cal = 3.968 Btu
Force, mass, and velocity
1 kg = 0.069 slug acceleration of gravity (g) = 9.8 m/s^2 = 32 ft/s^2 1 N (newton) = 1 J/m (joule per meter) = 0.225 pounds 1 m/s = 3.28 fps (feet per second) = 2.24 mph (miles per hour)
Energy and work (the mechanical equivalent of heat)
1 kcal = 1000 cal = 4184 J (joules) 1 Btu = 252 cal = 777.9 ft-lb = 1055 J 1 kWh = 3,600,000 J = 3413 Btu 1 quad (Q) = 10^{15} Btu
Power (the rate at which work is done)
1 W (watt) = 1 J/s (joules per second) 1 hp (horsepower) = 550 ft-lb/s = 746 W

2. Do you think that you can produce power equal to that of a 100-W light bulb? Obtain and weigh a steel ball.

a) Record the weight of the ball in newtons. As shown by the conversion tables, a newton is a unit of force. The weight of the ball is the same as the force exerted on the ball by the pull of gravity. Show your work in your *EarthComm* notebook.

3. Work is defined as the product of a force times the distance through which the force acts. The work needed to lift the steel ball a certain vertical distance is the force (weight of the ball, in newtons) times the vertical distance, or

$$W = F \cdot d,$$

where W is work in joules (J), F is force in newtons (N), and d is the height it is raised in meters (m).

Part C: Energy Units and Conversions

Teaching Tip

Make copies of the Energy Conversion Table on page R7 of the student text using **Blackline Master Energy Resources 1.2, Energy Conversion Table** and distribute these to your students. They can insert them into their notebooks and use them as a handy reference throughout the chapter.

1. a)

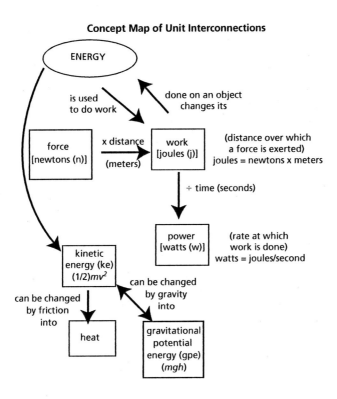

Concept Map of Unit Interconnections

2. a) Students should find the weight of the steel ball in pounds. Each student should independently weigh the ball using a scale. If the weights recorded by each student are not the same, they should average their results. To determine the weight of the ball in newtons, the students will need to complete a simple conversion. For example, if the ball is found to weigh 5 pounds, the following conversion can be set up:

$$5 \text{ lb. } \times \ (1 \text{ N} / 0.225 \text{ lb.}) \ = \ 22.22 \text{ N}$$

Remind students that a newton is a unit of force and that the weight of the ball is the same as the force exerted on the ball by the pull of gravity. The weight of the ball will vary.

3. a) Answers will vary depending upon the weight of the steel ball. If the ball weighs 5 pounds, or 22.22 N, the work needed to lift the ball to a height of 2 m is equal to 22.22 N × 2 m, or 44.44 J.

Earth's Natural Resources Energy Resources

a) In order for an object to obtain kinetic energy, work must be done on it. Calculate the work necessary to lift the steel ball to a height of 2 m.

4. Power is the rate at which work is done. The power you produce when you lift the ball is equal to the work divided by the time it took to lift the ball. If you lift the ball a number of times in a certain time period, the average power you produce is equal to the work of each lift, times the number of lifts, divided by the total time it took to do all of the lifting. Remember that the work is measured in joules and the time is measured in seconds.

$$P = W/t$$

where P is power in watts (W), W is work in joules (J), and t is time in seconds (s).

a) Calculate the power produced by lifting the ball 10 times in one minute. Note from the table that the unit for power is the watt. One watt = one joule per second.

5. In your group, discuss what a person would have to do to produce as much power as a 100-W bulb. Examples include running (how far?) or climbing stairs (how high?). Do this as a 'thought experiment', one that you will describe (with calculations), but not conduct.

a) Record your thought experiment. Show your calculations.

6. The energy it took to produce the power to the ball came from chemical energy. In this case, the chemical energy was energy stored in the food you ate for breakfast. Assume that your body was 100% efficient (all of the stored energy is converted into kinetic energy).

a) Calculate the number of times you could lift the ball to equal a 200 calorie candy bar (use the table and remember that one food calorie = 1000 kilocalories).

b) In nature, no energy change is 100% efficient. Some energy is lost to the environment. In the case of lifting the ball, what form does the lost energy take?

7. As a class discuss the question of whether you think people can produce as much power as a 100-W light bulb.

a) Record the results of your discussion.

Reflecting on the Activity and the Challenge

In Part A of this activity you looked at different ways that heat transfer occurs. Part B helped you to understand the concepts of potential and kinetic energy. In Part C you explored concepts of work, power, and units of energy. You also completed calculations to determine whether or not the exertion required to lift a ball can be equivalent to the power produced by a 100-W light bulb. These activities will help you think about how energy is transformed into a form that you can use. It will also help you think about ways to conserve energy resources so that your community can meet its growing energy needs.

4. **a)** Students should calculate the power produced by lifting the ball to a height
of 2 m 10 times in one minute. Answers will vary depending upon the weight of
the steel ball but will be equal to the value obtained in **Step 3 (a)** times
10 (because the ball is lifted 10 times), divided by 60 seconds. For example,
if the ball weighs 5 pounds, the work required to lift it 10 times is 444.4 J
(44.44 J x 10). The power produced by lifting the 5-pound ball 10 times in
one minute is therefore 7.41 W (444.4 J/60 s).

5. **a)** Results of students' thought experiments will vary depending upon which
activities they choose to discuss. They should show all of their calculations.
Running up stairs is a straightforward example: doing work increases the
potential energy of the student in the gravitational field. Multiply the weight
of the student, in newtons, times the vertical distance ascended, in meters,
divided by the time in seconds it took to run up the stairs. Your students should
easily be able to simulate the energy output of a 100-W light bulb, although
perhaps only in short bursts of intense muscular activity using their largest
muscles, as in a fast run up a flight of stairs.

6. **a)** Answers will vary depending on earlier calculations. Students will first need to
calculate the energy, in joules, needed to lift the ball once, and then convert that
to kilocalories using the table on page R7 of their textbooks. Then, divide the
energy content of the candy bar, in kilocalories, by the result, to obtain the
number of lifts needed to equal the energy content of the candy bar. The candy
bar has 200 food calories (Calories). Note that the calorie used to measure the
energy content of food is actually 1000 calories or 1 kilocalorie. Thus, the
energy content of the candy bar is 200 kilocalories.

   b) Inefficiency in muscular movements; generation of waste heat in the muscles.

7. **a)** Students should refer to their responses to **Step 5** as they proceed through
the discussion.

Reflecting on the Activity and the Challenge

Discuss with your students how each part of the investigation illustrates the
transformation of energy into a form that can be used. Ask them to consider how
their understanding of the various forms of energy will help them think of ways to
conserve energy resources in the community. Remind your students that they must
address this issue in their final reports.

Digging Deeper

HEAT AND ENERGY CONVERSIONS

Heat Transfer

Heat is really the **kinetic energy** of moving molecules. **Temperature** is a measure of this motion. The term **heat transfer** refers to the tendency for heat to move from hotter places to colder places. Many of the important aspects of heat transfer (see *Figure 1*) that you observed with the solar cooker had to do with heat conduction, which is one of the processes of heat transfer. All matter consists of atoms. At temperatures above **absolute zero** (about –273°C, the coldest anything can be!), the atoms vibrate. You sense those vibrations as the temperature of the material. The stronger the vibration, the hotter the material. When a hotter material is in contact with a colder material, collisions between adjacent vibrating atoms in the two materials cause the energy of the vibrations to even out, cooling the hot material and warming the cold material.

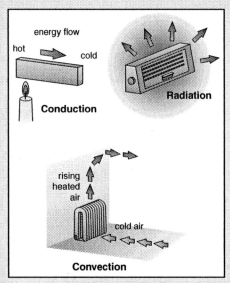

Figure I Three types of heat transfer.

Geo Words

heat: kinetic energy of atoms or molecules associated with the temperature of a body of material.

kinetic energy: a form of energy associated with motion of a body of matter.

temperature: a measure of the energy of vibrations of the atoms or molecules of a body of matter.

heat transfer: the movement of heat from one region to another.

absolute zero: the temperature at which all vibrations of the atoms and molecules of matter cease; the lowest possible temperature.

conduction: a process of heat transfer by which the more vigorous vibrations of relatively hot matter are transferred to adjacent relatively cold matter, thus tending to even out the difference in temperature between the two regions of matter.

thermal insulator: a material that impedes or slows heat transfer.

Conduction is the type of heat transfer you experience when you take a hot bath, when you heat a piece of metal, or when the air cools a cup of hot coffee left on top of a table. For instance, when you put a metal pot on the stove, only the bottom of the pot is in contact with the burner, yet the heat flows through the entire pot all the way to the handle. Materials differ greatly in how well they conduct heat. In **thermal insulators**, like Styrofoam, crumpled paper, or a down jacket, the heat flows slowly. Thermal insulators like these contain a large amount of trapped air. Air is a poor conductor because the air molecules are not in constant contact. Metals, on the other hand, are very good conductors of heat. Heat conduction is very important in your community. Keeping your home warm in the winter

Digging Deeper

As students read the **Digging Deeper** section, the relevance of the concepts investigated in **Activity 1** will become clearer to them. Assign the reading for homework, along with the questions in **Check Your Understanding** if desired.

Assessment Opportunity

Reword or restructure the questions in **Check Your Understanding** for a brief quiz. Use the quiz (or a class discussion of the questions in the textbook) to assess your students' understanding of the main ideas in the reading and the activity.

Assessment Tool

Check Your Understanding Notebook-Entry Evaluation Sheet
Use this sheet to evaluate the extent to which students understand the key concepts explored in **Activity 1** and explained in **Digging Deeper**, and to evaluate the students' clarity of expression.

Teaching Tip

Use **Blackline Master Energy Resources 1.3, Types of Heat Transfer** to make an overhead of *Figure 1* on page R9. Incorporate the overhead into a class discussion about the different types of heat transfer. Encourage students to think about what types of heat transfer they explored at each of the stations in **Part A** of the investigation. Also ask students to think about where in their everyday lives they encounter these heat transfer mechanisms, and why an understanding of these mechanisms is important for the rest of **Chapter 1**.

Earth's Natural Resources Energy Resources

Geo Words

convection: motion of a fluid caused by density differences from place to place in the fluid.

convection cell: a pattern of motion in a fluid in which the fluid moves in a pattern of a closed circulation.

electromagnetic radiation: the movement of energy, at the speed of light, in the form of electromagnetic waves.

would be very difficult (and expensive) without the insulating properties of the walls and the roof. Improving the insulation of your home by using insulating materials like those shown in *Figure 2* can greatly reduce the amount of energy needed to heat or cool your home.

Figure 2 Thermal insulation helps to keep your home warm. It conserves energy needed for space heating.

Another form of heat transfer is **convection**, which is important in liquids and gases. When a liquid or a gas is heated, its density decreases. That causes it to rise above its denser surroundings. In a room heated with a wood stove or a steam or hot water radiator, for instance, a natural circulation pattern is developed. The hot air from the stove rises towards the ceiling and cooler air travels down the walls and across the floor towards the stove. That kind of circulation is called a **convection cell**. Heat convection is also very important to your community. Many of the features of weather, such as sea breezes and thunderstorms, are caused by convection. Also, the way that you heat or cool your home depends strongly on heat convection.

Figure 3 The Sun emits electromagnetic radiation that warms the surface of the Earth.

A third form of heat transfer is **electromagnetic radiation**. Everything emits electromagnetic radiation. Examples of electromagnetic radiation are radio and television waves, visible light, ultraviolet light, and X-rays. Hotter materials emit more energy of electromagnetic radiation than colder materials. The warmth you feel from a hot fire, the Sun, or a light bulb is due to electromagnetic radiation traveling (at the speed of light!) from the hot object to you.

 R 10

Teaching Tip

Thermal insulation minimizes heat transfer from one material to another by reducing conduction, convection, and/or radiation effects. Loosely packed fiberglass strands are often used as insulation between the outer and inner walls of a house, as shown in *Figure 2* on page R10 of the student text. Placed between the walls of a house, the fiberglass material, which is a poor conductor of heat, acts to slow down the transfer of heat by conduction and convection and helps the home to retain heat. Ask students to consider why this would be important from the standpoint of energy conservation.

The standard way to represent how well a material will insulate is by indicating its R-value, which is a measure of the resistance of the material to heat flow. The R-value is defined as the reciprocal of the amount of heat energy per area of material per degree difference between the outside and inside. The units of measurement for R-value are:

$$(ft^2 \times hours \times degrees\ F)/\ Btu$$

R-values of insulation for the home are usually in the range of R-10 up to R-30. Listed below are the R-values of some of the materials that help to insulate houses:

Hardwood siding (1 in. thick)	0.91
Wood shingles (lapped)	0.87
Brick (4 in. thick)	4.00
Concrete block (filled cores)	1.93
Fiberglass batting (3.5 in. thick)	10.90
Fiberglass batting (6 in. thick)	18.80
Fiberglass board (1 in. thick)	4.35
Cellulose fiber (1 in. thick)	3.70
Flat glass (0.125 in thick)	0.89
Insulating glass (0.25 in space)	1.54
Air space (3.5 in. thick)	1.01
Free stagnant air layer	0.17
Drywall (0.5 in. thick)	0.45
Sheathing (0.5 in. thick)	1.32

Radiation is important to the community for many reasons. Solar radiation causes things in the community to be heated. Solar radiation heated the solar cooker. It also heats someone standing in the sunshine on a cold winter day, or a parked car in the Sun in the summer with all its windows closed. If a building is designed appropriately, the heat from the Sun can substitute for heat from other energy resources for space heating and hot water. Using insulation or light-colored reflective materials reduces solar heating in warmer months when heat is not desired.

Energy, Work, and Power

In the investigation, you dealt with four forms of energy: energy of motion, called kinetic energy; energy of position, called **potential energy**; energy stored in the chemical bonds of a substance, called **chemical energy**, and heat. Kinetic energy and potential energy together are called **mechanical energy**. You know that objects in motion have energy, because of what they can do to you when they hit you. The energy of motion is called kinetic energy. The more mass the body has, and the faster it is moving, the more kinetic energy it has. When you threw (or imagined throwing) the lump of modeling clay up in the air, you gave it kinetic energy. The kinetic energy was gradually converted to potential energy. When the lump reached its highest point, its kinetic energy was at a minimum. On the way down, the lump regained its kinetic energy. When it hit the table, all of its kinetic energy was changed to heat. The change in temperature was so small that you would need a very sensitive thermometer to measure it. That's an example of how kinetic energy is changed to heat energy by **friction**. When you rub your hands together to keep them warm, you are converting kinetic energy to heat by friction. Of course, you are always resupplying your hands with kinetic energy by the action of your arm muscles.

Figure 4 A coal-powered train is an example of how chemical energy stored in coal is converted into heat energy that in turn is converted to mechanical energy.

Geo Words

potential energy: mechanical energy associated with position in a gravity field; matter farther away from the center of the Earth has higher potential energy.

chemical energy: energy stored in a chemical compound, which can be released during chemical reactions, including combustion.

mechanical energy: the sum of the kinetic energy and the potential energy of a body of matter.

friction: the force exerted by a body of matter when it slides past another body of matter.

R 11

Teaching Tip

Discuss with students the energy resource concepts that are illustrated by the coal-powered train shown in *Figure 4* on page R11. This should help them understand why it is important to have a firm grasp of these concepts as they proceed through the rest of **Chapter 1**.

Teaching Tip

Energy generated by burning coal in a combustion chamber heats water, which produces steam that turns a turbine to generate electricity. Not all of the constituents of coal are combustible. *Figure 5* on page R12 is a close-up of two ash storage silos that are located behind a coal-feed conveyor. The white silo contains fly ash, which is the particulate matter that is carried in the gas stream that goes up the flue. The red brick silo contains bed ash, which consists mainly of sand, silt, and clay particles that were originally deposited in small concentrations along with the plant material that formed the coal. Students will learn more about how ash content is related to the heat content of a coal sample in **Activity 3**.

Earth's Natural Resources Energy Resources

Geo Words

work: the product of the force exerted on a body and the distance the body moves in the direction of that force; work is equivalent to a change in the mechanical energy of the body.

force: a push or pull exerted on a body of matter.

power: that time rate at which work is done on a body or at which energy is produced or consumed.

watt: a unit of power.

horsepower: a unit of power.

biomass: the total mass of living matter in the form of one or more kinds of organisms present in a particular habitat.

In physics, the term **work** has a very specific meaning. Work is equal to the **force** you exert on some object multiplied by the distance you move the object in the direction of the force. The importance of work is that it causes a change in the mechanical energy (kinetic and/or potential) of the object. When you threw the lump of modeling clay up in the air, your hand did the work. It exerted an upward force on the clay for a certain distance to give it its kinetic energy.

Power is the term used for the rate at which work is done or at which energy is produced or used. Think once more about the now-famous lump of modeling clay. You could have given it its upward kinetic energy by swinging your arm upward slowly for a long distance, generating low power but for a long time. Or, you could have swung your arm upward fast over only a short distance, generating high power but for only a short time. Whenever your muscles move your own body or some other object, you are generating power. The **watt** is the unit of power that is commonly used to describe the power of electrical devices. **Horsepower** is the unit of power that is often used to describe the power of other mechanical devices.

Converting Heat into Mechanical Energy

Figure 5 Coal is fed by a conveyor into a combustion chamber, where it is burned.

You have explored the idea that mechanical energy always tends to be converted into heat by friction. Nothing on Earth is completely frictionless, although some things, like air-hockey pucks, involve very little friction. Only in the emptiness of outer space can bodies move without friction. But how about energy conversion in the opposite direction: from heat to mechanical energy?

The conversion of heat into mechanical energy is central to most of the processes for producing electricity from energy resources. These resources include coal, natural gas, petroleum, sunlight, **biomass**, and nuclear energy. In these processes, water is heated to produce steam. When water boils (at atmospheric pressure) it undergoes about a thousand-fold increase in volume. The pressure of the steam exerts a force that does work to increase the kinetic energy of a turbine. The steam pressure is used to turn a turbine that generates electricity.

 R 12

Blackline Master Energy Resources 1.3
Types of Heat Transfer

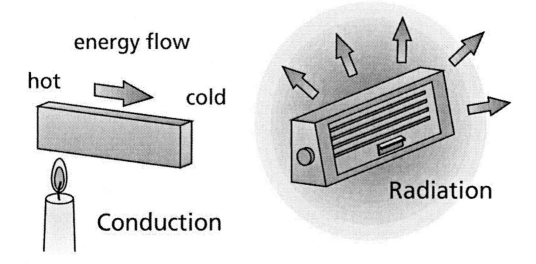

energy flow

hot → cold

Conduction

Radiation

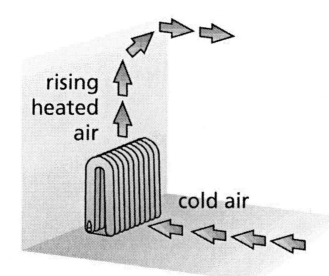

rising heated air

cold air

Convection

Activity 1   Exploring Energy Resource Concepts

The **Second Law of Thermodynamics** states that you can never completely convert heat into mechanical energy. In fact, in converting any form of energy into another, there is always a decrease in the amount of "useful" energy. Stated in general terms, the **efficiency** of a machine or process is the ratio of the desired output (work or energy) to the input:

$$\% \text{ Efficiency} = \frac{\text{useful energy or work out}}{\text{energy or work in}} \times 100$$

Electrical power plants have efficiencies of about 30%. An efficiency of 33% means that for every three trainloads of coal that are burned to produce electricity, the chemical heat energy from only one of those trainloads is converted to electricity.

Some methods for generating electricity are not based on the conversion of heat to mechanical energy. Hydropower and wind power are examples. In hydropower, the mechanical energy of the falling water is converted directly to the mechanical energy of the rotating turbine. The efficiency of hydropower is only about 80% rather than 100%, however, because of friction and the incomplete use of available mechanical energy. Similarly, the wind already has mechanical energy. The efficiency of wind power is no greater than about 60%, mainly because some of the wind goes around the turbine without adding to its rotation. Actual efficiencies of most wind turbines range from 30% to 40% (The windmills shown in *Figure 6* have an efficiency of only 16%.) By comparison, the efficiency of a normal automobile engine is about 22%.

Figure 6 The efficiency of these windmills is only about 16%. Modern wind turbines have efficiencies between 30% and 40%.

Geo Words

Second Law of Thermodynamics: the law that heat cannot be completely converted into a more useful form of energy.

thermodynamics: a branch of physics that deals with the relationships and transformations of energy.

efficiency: the ratio of the useful energy obtained from a machine or device to the energy supplied to it during the same time period.

Check Your Understanding

1. What are some of the methods for generating energy that are based on the conversion of heat to mechanical energy?

2. Describe the three processes of heat transfer.

3. In your own words define mechanical energy.

4. Why can't the efficiency of a device be more than 100%?

5. Why is the efficiency of a device always less than 100%?

Teaching Tip

Figure 6 on page R13 is a photograph of windmills used to generate energy. Students will explore the uses of wind power further in **Activity 8**. The windmills that are used as sources of power vary greatly in size and shape. You can visit the *EarthComm* web site to view additional images of windmills.

Check Your Understanding

1. Energy resources like coal, natural gas, petroleum, sunlight, biomass, and nuclear energy can be used to boil water to generate steam pressure that turns a turbine that generates electricity.

2. • Conduction is a process of heat transfer by which the more vigorous vibrations of relatively hot matter are transferred to adjacent relatively cold matter, thus tending to even out the difference in temperature between the two regions of the matter.

   • Convection is the motion of a fluid (liquid or gas) caused by density differences from place to place in the fluid.

   • Electromagnetic radiation is the movement of energy, at the speed of light, in the form of electromagnetic waves.

3. Mechanical energy is the sum of the kinetic energy and the potential energy of a body of matter. Students should put the definition in their own words.

4. The efficiency of a device cannot be more than 100% because it is not possible to generate more energy than is put into a system.

5. The efficiency of a device is always less than 100% because in converting any form of energy into another form, there is always a decrease in the amount of useful energy. Friction degrades mechanical energy into heat energy, and no machine or device is perfectly frictionless.

Earth's Natural Resources Energy Resources

Understanding and Applying What You Have Learned

1. a) Explain how all of the different parts of the solar cooker work in terms of different heat transfer mechanisms.
   b) How would you adapt your solar cooker to make it more effective and efficient?

2. Describe how a one-liter and two-liter container of water in the same oven differ in their heat and their temperature.

3. Describe how you think heat is transferred in the following situations:

   a) A cold room becomes warm after turning on a hot-water radiator.
   b) Your hand is heated as you grasp the handle of a heated pan on the stove.
   c) The bottom of a pan is heated when placed on an electric burner.
   d) A cold room becomes warm after window drapes are opened on a sunny day.

4. If the energy input of a system is 2500 cal and the energy output is 500 cal, what is the efficiency of the system?

5. A 300 hp engine is equivalent to how many foot-pounds? In your own words state what this means.

6. When you drive in a car, energy is not lost, even though gasoline is being used up. Use what you have learned in this activity to explain what happens to this energy.

Preparing for the Chapter Challenge

Your **Chapter Challenge** is to help community members think about how they will meet their growing energy needs. Draft an introduction to your report. Use what you have learned in this activity to explain how energy resources are used to do work. Help people to understand how energy is converted into heat when mechanical energy is produced from convenience in their everyday lives. You might also begin to think about steps that community members might take to improve their energy efficiency.

Inquiring Further

1. **Perpetual-motion machines**

   The United States Patent Office receives many applications for perpetual-motion machines. All the applications are turned down.

   What is a perpetual-motion machine, and why can no one get a patent for one?

Understanding and Applying What You Have Learned

1. a) Students explored this question in **Part A, Station 3, Step 4 (c)** of the investigation. This question will help them to reaffirm their understanding of the heat transfer mechanisms at work.

    b) Answers will vary. This issue was addressed in part in **Part A, Station 3, Step 4 (e)** of the investigation. You may want to give your students the opportunity to build their own—more effective and more efficient—solar cookers. They could do this as an extension activity and share their results with the class. Encourage students to give their solar cookers a try and see if they can cook a hot dog.

2. Students should assume that the containers are made of the same material and that they have been left in the oven for equal times. They will need to understand that heat and temperature are not the same thing: temperature is a measure of the average speed of motion or vibration of the atoms and molecules of the material, whereas heat, a form of energy, is the average kinetic energy of motion of the atoms and molecules. It's the temperature of a body, not its heat content, that we sense as hotness or coldness. The temperature of the two containers will be exactly the same after they have reached equilibrium with the environment of the interior of the oven. The heat content of the 2-L container, however, will be twice that of the 1-L container.

3. a) A room becomes warm after turning on a radiator through the process of convection, whereby air in contact with the hot radiator is heated by conduction, rises buoyantly toward the ceiling, is cooled by contact with the cold walls and ceiling, and sinks back toward the floor, to complete the circulation. The entire system is called a convection cell.

    b) Heat flows by conduction from the hot handle to your cooler hand.

    c) The cooler metal bottom of the pan is heated by conduction where it is in contact with the hot metal of the burner element. Heat then flows to other parts of the pan, and into the content of the pan, also by conduction.

    d) The radiant energy of the sunlight heats the walls, floor, and objects in the room by radiation. The air in the room is then warmed by contact with the heated materials. The sunlight does not heat the air directly.

4. Students should use the equation given on page R13 of the student text to calculate the efficiency of the system. They should find that a system with an energy input of 2500 cal and an energy output of 500 cal is 20% efficient:

$$(500 \text{ cal}/2500 \text{ cal}) \times 100 = 20\%$$

5. A 300 hp engine is equivalent to how many foot-pounds per second?

A 300-hp engine is equivalent to 165,000 ft-lb/s:

300 hp x 550 ft-lb/s per 1 hp = 165,000 ft-lb/s
Students should use the conversion factor given on page R7 of the student text: 1 hp = 550 ft-lb/s. This means, simply, that the power output of the engine can be expressed equivalently in either horsepower or foot-pounds per second, by use of different systems of units.

6. By the principle of conservation of energy, all of the chemical energy released by combustion of the gasoline must go somewhere. Some of it is converted into the kinetic energy of movement of the car. Some is dissipated into heat by wind friction as the car moves through the air; some is dissipated by tire friction with the road; and some is dissipated by friction in the drive train. Also, by the Second Law of Thermodynamics, heat can never be completely converted to mechanical energy. Much of the chemical energy of the gasoline is wasted as heat rather than being used to power the moving parts of the car.

Preparing for the Chapter Challenge

This section gives students an opportunity to apply what they have learned to the **Chapter Challenge.** They can work on this as a homework assignment, or during class time within groups. **Activity 1** was designed as an introduction to show students how the energy found in natural resources can be transformed into a form that can be used. The information that students produce here can be incorporated as an introduction to their Chapter Report. Refer students to the rubric that you produced at the start of the chapter and make sure that they understand how completing this section will help them to complete the **Chapter Challenge** (see **Assessment Rubric for Chapter Report on Energy Resources** on pages 16 and 17).

Inquiring Further

1. Perpetual-motion machines
Perpetual-motion machines are often called free energy machines because, if such a machine were attainable, it would generate more power than it consumes. This would violate the First and Second Laws of Thermodynamics, which state that the energy of an isolated system is constant and that heat cannot be entirely converted into a more useful form of energy. Recorded attempts to achieve perpetual motion go back at least 15 centuries, and continue today. If a perpetual-motion machine could be created, it would be extremely cost-efficient (imagine running the electricity of an entire community on an AAA battery!). The large number of proposals for perpetual-motion machines has forced the United States Patent Office to issue the following statement:

"The views of the Patent Office are in accord with those scientists who have investigated the subject and are to the effect that such devices are physical impossibilities. The position of the Office can only be rebutted by a working model...The Office hesitates to accept fees from applicants who believe they have discovered Perpetual Motion, and deems it only fair to give such applicants a word of warning that fees cannot be recovered after the case has been considered by the Examiner."

2. **Improving efficiencies of electricity generation**

   Innovative methods for power generation are now being developed to improve the efficiency of generating electricity from energy resources. What are some new methods for generating electricity from coal, natural gas, or oil that have improved efficiencies? Visit the *EarthComm* web site to help you find this information.

3. **History of science**

   Research the work of James Prescott Joule. A Scottish physicist, Joule conducted a famous experiment to observe the conversion of mechanical energy to heat energy. How did the experiments help Joule to conclude that heat is a form of energy?

4. **Solar cooking applications**

   In your investigation, you explored a model of one kind of solar food cooker. Research:

   • How people are using solar cookers and reducing the consumption of wood and fossil fuels for cooking food.

   • Where are solar cookers most commonly used?

   • Are they a suitable energy alternative for your community?

   • How does the use of a solar cooker reduce the effect on the biosphere?

2. Improving efficiencies of electricity generation

Students can visit the *EarthComm* web site for information about improving efficiencies of electricity generation.

Improvements in efficiency of power generation are often dependent on raising the temperature of steam at the entrance of the turbines, and making use of waste heat from those turbines. In recent years there have been a number of new technologies developed (particularly for coal-fired plants) that have both improved the efficiency of power generation and greatly reduced the emissions of pollutants like nitrogen oxides and sulfur dioxide. Additionally, high-temperature superconducting materials hold great promise for increasing the efficiency of power distribution and storage, thus further improving the efficiency in delivery of electricity to the consumer. A few of the newer power generation technologies are discussed briefly below.

Supercritical pulverized fuel (SCPF) technology and ultra super critical (USC) technologies— These technologies raise temperature and steam pressures in thermal power stations to improve generating efficiency greatly.

Integrated coal gasification combined cycle (IGCC) power generation— IGCC systems are highly efficient. They gasify and burn coal gas to drive a gas turbine. Waste heat runs a steam turbine. Boosting gas turbine temperature and pressure improves efficiency even further.

Pressurized fluidized-bed combustion (PFBC) — fluidized-bed combustion (FBC) technology minimizes NO_x compounds. Desulfurization during combustion is within the boilers. This eliminates the need for flue gas desulfurization units, hence facilities are more compact. With PFBC technology, the fluidized bed combusts under pressure. The resulting hot, pressurized gases drive a gas turbine. Gas and steam turbines combine to generate electricity at greater efficiency.

Integrated coal gasification fuel cell (IGFC) technology — IGFC could deliver a thermal efficiency of around 60%, compared with about 40% for conventional coal-fired generating systems. IGFC could thus become as important as liquefied natural gas (LNG) combined-cycle setups.

Natural gas-fired combined-cycle technology — This technology uses natural gas to drive a gas turbine. A heat recovery steam generator is then used to take waste heat from the gas turbine and power a steam turbine. The efficiency of these combined-cycle units is now approaching 60%.

3. History of science

Joule's investigations were driven by a hope of replacing steam engines with electric motors. His early research sought to improve the efficiency of electric motors, and his investigations concerned themselves with the production of heat. Joule's first scientific paper showed that the heat generated by an electrical current is proportional to the square of the current. This was one of the first reports establishing the link between heat and other forms of energy.

In 1850, Joule established the equivalence of heat and mechanical work with his paddle-wheel experiment. This experiment involved the construction of an apparatus in which a weight turned a paddle wheel immersed in water. The water gradually became slightly warmer as friction degraded the mechanical energy into heat. The weights, and the distance they fell, were measured, and the work done was calculated. The rise in temperature of the water, together with the known specific heat capacity of the water, measured the heat produced.

4. Solar cooking applications
Students can visit the *EarthComm* web site to help them answer the questions about solar cookers. Most solar cookers are used today in developing countries where there often is no alternative to the use of solar power. In the United States, solar cookers are used mainly as a novelty, because the fuel necessary to power stoves and grills is readily available and is generally affordable. Whether or not they would be suitable for use in your community will vary. The photograph on page R15 was taken at a restaurant where this large solar cooker is used to cook pizza.

NOTES

ACTIVITY 2— ELECTRICITY AND YOUR COMMUNITY

Background Information

Electricity

All atoms consist of a nucleus around which electrons move in orbits. Electrons are electrically charged particles; the fundamental and smallest unit of electric charge is the charge of an electron. For historical reasons, the electric charge of an electron is defined as negative. The negative charges of the electrons that orbit the nucleus are offset by positive charges of particles, called protons, in the nucleus.

In some kinds of atoms, the electrons are bound in place; in other kinds of atoms— especially the atoms of metallic elements like iron, copper, aluminum, or silver—some of the electrons are not bound to particular atoms and can move from place to place in the material. Materials with immovable electrons are called insulators; materials with some movable electrons are called conductors.

There is a fundamental tendency for electric charge to become distributed evenly through space. If the electric charge is different from one place to another in a conducting material, electrons flow from place to place in such a way as to make the charge the same everywhere. Because in all practical situations the numbers of individual atoms is enormous, the flow of electrons can be viewed as continuous, rather than as the movement of discrete electrons. We sense the flow of electrons as an electric current.

The flow of an electric current is analogous to the flow of water in a pipe. Water flows in a pipe because the water pressure is less at one end of the pipe that at the other end. The rate of flow of water is an outcome of the balance between the pressure difference, which drives the flow, and friction, which retards the flow. In an electric current, the difference in electric change from place to place along the conductor is what drives the current; it is analogous to the difference in water pressure in the water pipe. The flow of electrons in an electric current is retarded by a kind of friction on the atomic scale. In an electric current, this is called the electrical resistance; it is analogous to the friction in the water flowing in a water pipe.

In much the same way that a gravitational potential can be defined for a mechanical system in a gravitational field, an electrical potential can be defined for an electrical system in a field of electric charge. This potential is called the emf; its unit is the called the volt, which is defined as one joule per coulomb (the coulomb being the unit of electric charge). Voltage is thus loosely similar to pressure in a water piping system. The rate of flow of charge in an electrical conductor is called the electrical current. The unit of electrical current is defined as the ampere, or amp for short. One ampere is equal to a current of one coulomb of charge passing a given point in one second. The unit of electrical resistance is defined as the ohm: one ohm is equal to one volt per ampere. This is the consequence of a fundamental law of physics called Ohm's Law: the ratio of voltage to current in any conductor is a constant, which is called the resistance.

When an electric current moves electric charge from one place to another, work is done. Just as in mechanical systems, the rate at which that work is done is called the power in the electrical system. In the meter-kilogram-second system of physical units, power is measured in joules of energy per second, called watts.

For an electric current to be maintained, a difference in electric charge between the two ends of the conductor must somehow be maintained. In electrical generating plants, this is done with generators. Generators work on the principle of induction. It is known that when an electrical conductor is moved in a magnetic field, an electric current is generated in the conductor. Generators are arranged so that a shaft with a very large number of windings of a conducting wire is rotated in a space surrounded by large permanent magnets. As the shaft is rotated, an electric current is generated in the windings. The windings are tapped and the electricity is transmitted through a system of conducting wires to the end use of the electricity.

In essence, the electric current flows from one end of the windings in the generator, through the system of wires and appliances in the system, and back to the other end of the windings in the generator at the generating plant. The electrical potential (that is, the voltage) at the upstream end of the windings is maintained at a level higher than the electrical potential of the Earth (called the ground potential), and the electrical potential at the downstream end of the windings is maintained at exactly the ground potential. The black (hot) wire that leads from the street to your home appliances is at higher than ground potential, and the white (neutral) wire that leads back to the generating plant is at exactly ground potential. That's why, in a properly wired electrical system, you can touch the white wire with impunity, but if you touch the black wire, you are likely to act as a conductor of electric current to a grounded surface, to your detriment.

More Information – on the Web

Visit the *EarthComm* web site www.agiweb.org/earthcomm to access a variety of links to web sites that will help you deepen your understanding of content and prepare you to teach this activity. Many of the sites also contain images that you can download.

Goals and Assessment

Clarify that the goals indicate what students should understand and be able to do as a result of the activity. Make sure students understand that Chapter Assessments are based upon these goals.

Goal	Location in Activity	Assessment Opportunity
Compare energy resources used to generate electricity in the United States to other countries.	**Investigate Part A**	Answers to questions are accurate, and graphical representation of data is correct.
Identify the major energy sources used to produce electricity in the United States and your state.	**Investigate Part A Investigate Part B Digging Deeper; Understanding and Applying What You Have Learned Question 1**	Answers are reasonable, and accurately reflect available data.
Identify trends and patterns in electricity generation.	**Investigate Part A Investigate Part B Digging Deeper**	Answers are reasonable, and accurately reflect available data.
Understand the difference between electric power and electric energy.	**Digging Deeper; Check Your Understanding Question 1**	Answers are reasonable, and closely match those given in Teacher's Edition.
Be able to describe commonly used methods of generating electric power.	**Investigate Part B Digging Deeper; Check Your Understanding Questions 2 and 4**	Answers are reasonable, and accurately reflect available data.

Chapter 1

NOTES

Earth's Natural Resources Energy Resources

Activity 2 Electricity and Your Community

Goals

In this activity you will:

- Compare energy resources used to generate electricity in the United States to other countries.

- Identify the major energy sources used to produce electricity in the United States and your state.

- Identify trends and patterns in electricity generation.

- Understand the difference between electric power and electric energy.

- Be able to describe commonly used methods of generating electric power.

Think about It

Electricity is a key part of life in the United States. Factories, stores, schools, homes, and most recreational facilities depend upon a supply of electricity. The unavailability of electricity almost always makes the news!

- What is electric energy?
- What are some consequences of not having electricity when it is needed?

What do you think? Record your ideas in your *EarthComm* notebook. Be prepared to discuss your responses with your small group and the class.

Activity Overview

To understand which energy resources are used in the United States and throughout the world, students use a data table that shows world net electricity generation by type. They obtain their state's electricity profile to identify trends in electric power generation and consumption. **Digging Deeper** reviews the resources used to generate electricity, including fossil fuels and nuclear energy, hydroelectric power, renewable resources, and geothermal energy. The reading is supplemented by maps that illustrate the distribution of energy sources used to produce electricity in the United States.

Preparation and Materials Needed

Part A

No advance preparation is required for **Part A** of the investigation.

Part B

Students will need Internet access to complete **Part B** of the investigation. If you do not have Internet access in your classroom or cannot schedule time in the school computer lab, obtain your state electricity profile before class and distribute copies of the needed information.

Materials

Part A
No additional materials needed.

Part B
  • Internet access (or copies of your state electricity profile)*

Think about It

Student Conceptions

Students are not likely to understand the difference between electric power and electric energy. If they are having trouble thinking of some consequences of not having electricity, ask them to produce a list of some of the different ways that they use electricity. (The cartoon on page R16 offers numerous cues.) Then ask the students to consider how it would affect them if they could no longer use electricity in these ways.

Answers for the Teacher Only

Electric energy is the amount of work that can be done by an electric current. It is different from electric power, which is the rate at which electricity does work. The consequences associated with not having electricity will vary, depending upon opinions and needs.

*The *EarthComm* web site provides suggestions for obtaining these resources.

Assessment Tool

Think about It Evaluation Sheet
Use this evaluation sheet to help students understand and internalize the basic expectations for the warm-up activity.

Chapter 1

NOTES

Investigate

Part A: Global and United States Electricity Generation

1. Use the data table showing world net electricity generation by type to answer the questions on the following page.

World Net Electricity Generation by Type, 1998 (in billion-kilowatt hours)					
Region Country	Fossil Fuels	Hydro	Nuclear	Geothermal and Other[1]	Total
North America					
Canada	148.7	328.6	67.7	6.1	551.1
Mexico	134.2	24.4	8.8	5.4	172.8
United States	2550.0	318.9	673.7	75.3	3617.9
The Caribbean & South America					
Bolivia	2.0	1.4	0.0	0.1	3.5
Brazil	15.6	288.5	3.1	9.7	316.9
Puerto Rico	17.0	0.3	0.0	0.0	17.3
Venezuela	21.6	52.5	0.0	0.0	74.0
Western Europe					
France	52.9	61.4	368.6	2.8	485.7
Spain	93.2	33.7	56.0	3.5	186.4
Switzerland	2.3	33.1	24.5	1.1	61.1
United Kingdom	235.3	5.2	95.1	6.2	341.9
Eastern Europe & Former U.S.S.R.					
Bulgaria	20.2	3.3	15.5	0.0	39.0
Romania	27.5	18.7	4.9	0.0	51.1
Russia	530.1	157.9	98.3	0.0	786.3
Middle East					
Cyprus	2.8	0.0	0.0	0.0	2.8
Saudi Arabia	116.5	0.0	0.0	0.0	116.5
Africa					
Angola	0.5	1.4	0.0	0.0	1.9
Egypt	47.1	12.1	0.0	0.0	59.2
Ethiopia	0.0	1.6	0.0	0.0	1.6
Morocco	11.6	1.7	0.0	0.0	13.4
Far East & Oceania					
China	880.2	202.9	13.5	0.0	1096.5
Japan	571.3	91.6	315.7	23.8	1002.4
Malaysia	52.5	4.8	0.0	0.0	57.3
Mongolia	2.5	0.0	0.0	0.0	2.5
Nepal	0.1	1.1	0.0	0.0	1.2
Total	**5535.7**	**1645.1**	**1745.4**	**134**	**9060.3**

[1]Geothermal and Other consists of geothermal, solar, wind, wood, and waste generation.
Source: Adapted from the Energy Information Administration/International Energy Annual 1999 report.

Blackline Master Energy Resources 2.1
World Net Electricity Generation by Type

World Net Electricity Generation by Type, 1998 (in billions of kilowatt hours)					
Region Country	Fossil Fuels	Hydro	Nuclear	Geothermal and Other	Total
North America					
Canada	148.7	328.6	67.7	6.1	551.1
Mexico	134.2	24.4	8.8	5.4	172.8
United States	2550.0	318.9	673.7	75.3	3617.9
The Caribbean & South America					
Bolivia	2.0	1.4	0.0	0.1	3.5
Brazil	15.6	288.5	3.1	9.7	316.9
Puerto Rico	17.0	0.3	0.0	0.0	17.3
Venezuela	21.6	52.5	0.0	0.0	74.0
Western Europe					
France	52.9	61.4	368.6	2.8	485.7
Spain	93.2	33.7	56.0	3.5	186.4
Switzerland	2.3	33.1	24.5	1.1	61.1
United Kingdom	235.3	5.2	95.1	6.2	341.9
Eastern Europe & Former U.S.S.R.					
Bulgaria	20.2	3.3	15.5	0.0	39.0
Romania	27.5	18.7	4.9	0.0	51.1
Russia	530.1	157.9	98.3	0.0	786.3
Middle East					
Cyprus	2.8	0.0	0.0	0.0	2.8
Saudi Arabia	116.5	0.0	0.0	0.0	116.5
Africa					
Angola	0.5	1.4	0.0	0.0	1.9
Egypt	47.1	12.1	0.0	0.0	59.2
Ethiopia	0.0	1.6	0.0	0.0	1.6
Morocco	11.6	1.7	0.0	0.0	13.4
Far East & Oceania					
China	880.2	202.9	13.5	0.0	1096.5
Japan	571.3	91.6	315.7	23.8	1002.4
Malaysia	52.5	4.8	0.0	0.0	57.3
Mongolia	2.5	0.0	0.0	0.0	2.5
Nepal	0.1	1.1	0.0	0.0	1.2
Total	**5535.7**	**1645.1**	**1745.4**	**134.0**	**9060.3**

Earth's Natural Resources Energy Resources

a) List the three countries that generate the most electricity.

b) What type of electricity generation is used most by these areas? Which is used least?

c) List the three countries that generate the least electricity.

d) What type of electricity generation is used most by these areas? Which is used least?

e) How do the resources for electricity generation differ between the top and bottom regions? How do you account for the differences?

2. Consider electricity generation by fuel type.

a) Without including the United States, rank global electricity generation by fuel type from highest to lowest.

b) Rank United States electricity generation by fuel type from highest to lowest.

c) How does the electricity generation in the United States match global electricity generation? Why do you think this is so?

3. Devise a way to represent the data for the United States graphically (either a pie chart or a histogram plot).

a) Complete your graph in your *EarthComm* notebook.

Part B: Investigating Electricity Generation in your State

1. Go to the *EarthComm* web site to visit the Department of Energy's Energy Information Administration web page. Click on your state to obtain your "state electricity profile." Use the information at the site to answer the following questions:

a) Is your state a net importer or net exporter of electricity? What does this mean?

b) What energy resources are used to generate electricity in your state?

c) Which energy resource does your state depend upon the most to generate electricity?

d) What are the trends in energy resource use for electric power generation in your state over time?

e) How has the generation of electricity changed in your state over time? How much has it grown in the most recent 10-year period? What is the annual rate of growth?

f) How does the growth in electricity generation in your state compare to the rate of population growth in your state? How might you explain the relationship?

g) How do the types of resources used by your state for electricity generation compare with the averages for the world and the United States, which you looked at in Part A of the investigation?

h) What sector (residential, commercial, industrial, or other) consumes the most energy for electricity in your state? What sector consumes the least?

R 18

Investigate

Part A: Global and United States Electricity Generation

Teaching Tips

Make photocopies of the data table on page R17 of the student text, **Blackline Master Energy Resources 2.1, World Net Electricity Generation by Type** to distribute to your students, so that they do not have to flip back and forth to answer the questions. You can also visit the *EarthComm* web site and download the complete data set, which includes electricity generation for all major countries of the world.

1. a) The United States, China, and Japan generate the most electricity.

    b) All three of these countries rely most heavily on fossil fuels to generate electricity; they rely least on geothermal and other sources for electricity generation.

    c) From the data in the text, Nepal, Ethiopia, and Angola generate the least electricity. Certain other countries not listed in the table generate even less electricity.

    d) Hydroelectric power is used most by these areas. None of these countries make use of nuclear or geothermal electricity generation, and fossil fuels are used very little, if at all.

    e) If they are having difficulty answering this question, students can visit the *EarthComm* web site to learn more about each of these countries. They will find that the principal difference between the top electricity-generating countries and the bottom electricity-generating countries is socio-economic.

2. a) Not including figures for use in the United States, fossil fuel is used most for global electricity generation, followed by hydroelectric power, nuclear, and finally geothermal and other.

    b) In the United States, fossil fuel is used most for electricity generation, followed by hydroelectric power, nuclear, and finally geothermal and other.

    c) The United States generates over one-third of the total world electricity.

3. Graphical representations of electricity generation by fuel type for the United States will vary. Two examples are shown below.

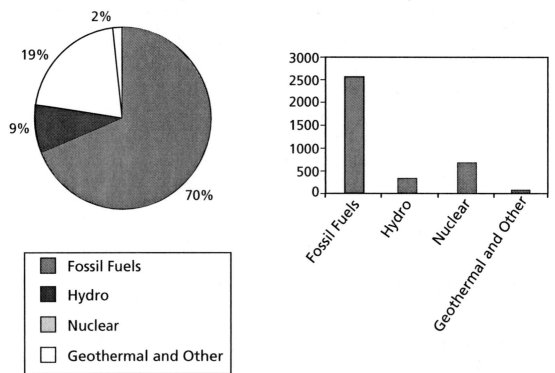

1998 United States Electricity Generation (in billions of kilowatt hours)

Part B: Investigating Electricity Generation in Your State

1. The answers to these questions will vary depending on which state you live in. The information needed to answer the questions can be obtained on the Department of Energy's Energy Information Administration web page. You can access this page from the *EarthComm* web site.

NOTES

i) What resources does your state use most for residential and commercial purposes? for transportation? for industry?

2. Review the data and figures for the electricity profile for your state.

   a) How do they help you to meet your **Chapter Challenge?**

Reflecting on the Activity and the Challenge

In this activity, you looked at the different types of resources that are used to generate electricity. You also investigated the trends in energy resource use and electricity generation in your state. Knowing where and how your state uses energy resources to generate electricity will help you to think about how to meet the energy needs of an expanding population in the **Chapter Challenge.**

Digging Deeper

MEETING ELECTRICITY NEEDS

Energy Resources

As the world's population increases and there is continued comparison to the current western European, Japanese, and North American living standards, there is likely to be demand for more energy. Several different energy resources can be used either for heating or cooling, electricity generation, industrial processes, and fuels for transportation. Energy resources available in the world include coal, **nuclear fission,** hydroelectric, natural gas, petroleum, wood, wind, solar, refuse-based, biomass, and oceanic (tides, waves, vertical temperature differences). In addition, nuclear fusion has been proposed as the long-term source, although research progress toward making it a reality has been very slow.

Geo Words

nuclear fission: the process by which large atoms are split into two parts, with conversion of a small part of the matter into energy.

Figure 1 Nuclear power can come from the fission of uranium, plutonium, or thorium or the fusion of hydrogen into helium.

R 19

2. Students should recognize that completing this part of the investigation will help them address the issue of present energy uses and consumption rates in their community. Completing **Step 1** of **Part B** will also help them consider how the hypothetical population increase will change the energy needs of the community.

Assessment Tools

EarthComm **Notebook-Entry Checklist**
Use this checklist as a quick guide for student self-assessment and/or an opportunity to quickly score student work. Add further criteria specific to your classroom needs or to this particular investigation.

Investigate Notebook-Entry Evaluation Sheet
Point out the criteria listed on this evaluation sheet that are relevant to this particular investigation. Encourage students to internalize the criteria by making them part of your assessment conversations as you circulate around the classroom.

Reflecting on the Activity and the Challenge

Have students read this brief passage and share their thoughts about the main point of **Activity 2** in their own words. Hold a class discussion about how this investigation relates to the **Chapter Challenge**; focus on their responses to **Step 2** in **Part B**. Students should now be aware of the different types of natural resources used to generate electricity in their community.

Digging Deeper

Assign the reading for homework, along with the questions in **Check Your Understanding** if desired.

Earth's Natural Resources Energy Resources

Geo Words

electric power: power associated with the generation and transmission of electricity.

electric energy: energy associated with the generation and transmission of electricity.

fossil fuel: fuel derived from materials (mainly coal, petroleum, and natural gas) that were generated from fossil organic matter and stored deep in the Earth for geologically long times.

geothermal energy: energy derived from hot rocks and/or fluids beneath the Earth's surface.

photovoltaic energy: energy associated with the direct conversion of solar radiation to electricity.

turbine: a rotating machine or device that converts the mechanical energy of fluid flow into mechanical energy of rotation of a shaft.

Generating Electric Energy

Energy resources are used to generate electricity, an energy source with which you are very familiar. **Electric power** is the rate at which electricity does work—measured at a point in time. The unit of measure for electric power is a watt. The maximum amount of electric power that a piece of electrical equipment can accommodate is the capacity or capability of that equipment. You can check the tags or labels on electrical appliances for this information, such as a "1200-W hair dryer" or "40-W stereo receiver". **Electric energy** is the amount of work that can be done by electricity. The unit of measure for electric energy is the watt-hour. A 1200-W hair dryer used for 15 min would require (theoretically) 300 Wh of electric energy.

Fossil fuels supply about 70% of the energy sources for electricity in the United States. Coal, petroleum, and natural gas are currently the dominant fossil fuels used by the electrical power industry. When fossil fuels are burned to generate electricity, a variety of gases and particulates are formed. If these gases and particulates are not captured by some pollution control equipment, they are released into the atmosphere. Other sources of energy can also be converted into electricity, including **geothermal energy**, solar thermal energy, **photovoltaic energy**, and biomass. These alternative sources of energy have many advantages over fossil fuels, as you will explore in Activity 5.

Fossil Fuels and Nuclear Energy

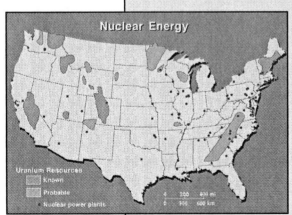

Figure 2 The location of nuclear power plants and uranium resources in the continental United States.

Most of the electricity in the United States is produced in steam **turbines**. A turbine converts the kinetic energy of a moving fluid (liquid or gas) to mechanical energy. In a fossil-fueled steam turbine, the fuel is burned in a boiler to produce steam. The resulting steam then turns the turbine blades that turn the shaft of the generator to produce electricity. In a nuclear-powered steam turbine, the boiler is replaced by a reactor containing a core of nuclear fuel (primarily enriched uranium). Heat produced in the reactor by fission of the uranium is used to make steam. The steam is then passed through the turbine generator to produce electricity, as in the fossil-fueled steam turbine. *Figure 2* shows the nuclear-power plant locations and uranium resources available in the United States.

Teaching Tips

Nuclear power can come from the *fission* of uranium, plutonium, or thorium, or the *fusion* of hydrogen into helium. Power from fission is a reality; power from fusion is still in the research stage, and may or may not ever become a practical method for large-scale generation of electricity. Today, most nuclear power is derived from the fission of uranium. The fission of one atom of uranium produces 10 million times the energy produced by the combustion of an equivalent mass of carbon from coal.

Make an overhead transparency of the map shown in *Figure 2* on page R20 using **Blackline Master Energy Resources 2.2, Nuclear Energy.** Incorporate the overhead into a class discussion on the location of nuclear power plants and uranium resources in the continental United States Where are the most of the power plants located? Is there any relationship between the location of uranium resources and power plants? Why do the students think this is the case? Is there a nuclear power plant near your community?

Hydroelectric Power

Water is the leading **renewable energy source** used by electric utilities to generate electric power. **Hydroelectric power** is the result of a process in which flowing water is used to spin a turbine connected to a generator. Hydroelectric plants operate where suitable waterways are available. Many of the best of these sites have already been developed. (*Figure 3* shows the hydroelectric plants developed in the United States.) Seventy percent of the hydroelectric power in the United States is generated in the Pacific and Rocky Mountain States. The two basic types of hydroelectric systems are those based on falling water and natural river current. In a falling-water system, water accumulates in reservoirs created by dams. This water then falls through conduits and applies pressure against the turbine blades to drive the generator to produce electricity. In the second system, called a run-of-the-river system, the force of the river current (rather than falling water) applies pressure to the turbine blades to produce electricity. Because they do not store water, these systems depend upon seasonal changes and stream flow.

Geo Words

renewable energy source: an energy source that is powered by solar radiation at the present time rather than by fuels stored in the Earth.

hydroelectric power: electrical power derived from the flow of water on the Earth's surface.

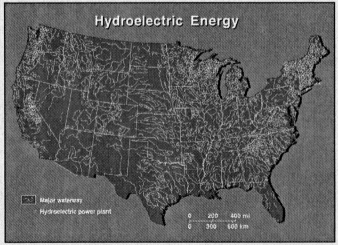

Figure 3 The location of hydroelectric power plants relative to major waterways in the continental United States.

Generating electricity using water has several advantages. The major advantage is that water, a renewable resource, is a source of cheap power. In addition, because there is no fuel combustion, there is little air pollution

R 21

Teaching Tip

Make an overhead transparency of the map in *Figure 3* on page R21, using **Blackline Master Energy Resources 2.3, Hydroelectric Energy.** Incorporate this overhead into a discussion on the location of hydroelectric power plants relative to major waterways in the continental United States. Ask your students to explain the relationship between the two.

Blackline Master Energy Resources 2.2
Nuclear Energy

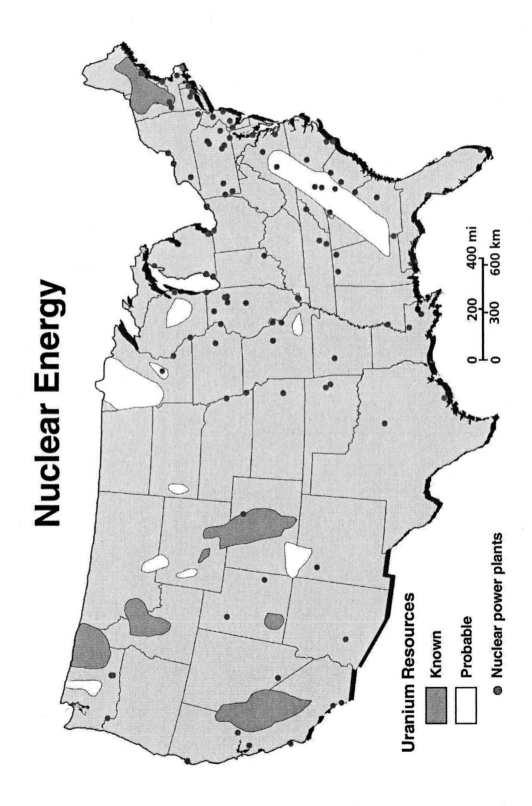

Nuclear Energy

Uranium Resources

▓ Known

☐ Probable

● Nuclear power plants

400 mi
600 km
200
300
0
0

Blackline Master Energy Resources 2.3
Hydroelectric Energy

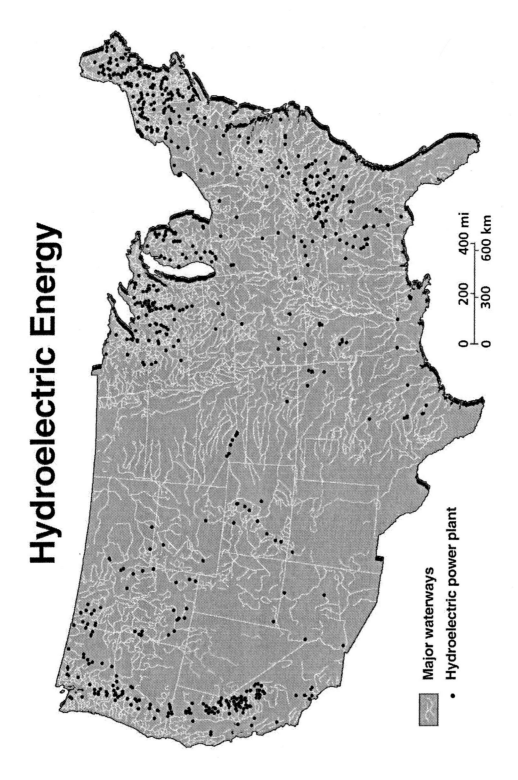

Hydroelectric Energy

400 mi
600 km

200   400 mi
300   600 km

0   0

Major waterways
• Hydroelectric power plant

Earth's Natural Resources Energy Resources

in comparison with fossil fuel plants and limited thermal pollution compared with nuclear plants. They can start quickly because they do not need to wait for water to be heated into steam. Also, the flow of water can be adjusted to make quick changes in power output during peak demands for electricity. Like other energy sources, the use of water for generation has limitations, including environmental impacts caused by damming rivers and streams, which affects the habitats of the local plants and animals. Another disadvantage to some hydroelectric power plants is that they depend upon the flow of water which varies with seasons and during droughts.

Other Resources Used to Generate Electricity

Currently, renewable resources (other than water) and geothermal energy supply less than 1% of the electricity generated by electric utilities. They include solar, wind, and biomass (wood, municipal solid waste, agricultural waste, etc.). Although geothermal energy is a nonrenewable resource, we will not run out of it. Geothermal power comes from heat energy buried beneath the surface of the Earth. Most of this heat is at depths beyond current drilling methods. In some areas of the country, magma—the molten matter under the Earth's crust from which igneous rock is formed by cooling—flows close enough to the surface of the Earth to produce steam. That steam can then be harnessed for use in conventional steam-turbine plants. *Figure 4* shows the locations of geothermal plants within the United States. Most are found in the western United States where magma is close enough to the surface to supply steam.

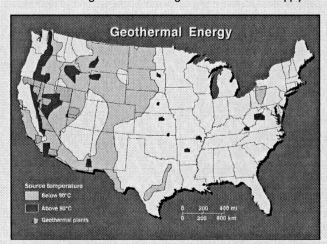

Figure 4 The location of geothermal power plants and source temperatures in continental United States.

Check Your Understanding

1. What is the difference between electric energy and electric power?

2. Compare and contrast steam turbine versus hydroelectric power generation. How are they similar? How are they different?

3. From an Earth systems perspective, what are the advantages and disadvantages of hydroelectric power?

4. In your own words, explain why biomass and wind are called renewable energy sources.

Teaching Tips

The word geothermal literally means Earth plus heat. Heat or thermal energy from the Earth is by far the most abundant resource on Earth. The thermal energy in the upper part of the Earth's crust amounts to nearly 50,000 times the energy of all oil and gas resources in the world. The problem lies in harnessing this energy.

Make an overhead transparency of the map in *Figure 4* on page R22, using **Blackline Master Blackline Master 2.4, Geothermal Energy**. Incorporate the overhead into a discussion of the location of geothermal power plants and source temperatures in the United States. Is your community located in an area where geothermal energy could be used as a viable resource?

Check Your Understanding

1. Electric power is the rate at which electricity does work, measured at a given time. Electric energy is the amount of work that can be done by electricity.

2. Both steam turbine and hydroelectric power generation rely on a spinning turbine that converts the kinetic energy of a moving fluid (liquid or gas) to the rotational mechanical energy of the turbine. With hydroelectric power generation, the moving fluid is water (a liquid), whereas with a steam turbine the moving fluid is steam (a gas).

3. Hydroelectric power is a renewable resource that involves no fuel combustion, which means little air pollution and limited thermal pollution (atmosphere, hydrosphere, and biosphere). However, hydroelectric power often requires the damming of rivers and streams, which affects the habitats of local plants and animals (biosphere) and floods large areas of the land.

4. A renewable resource is defined as an energy resource that is powered directly or indirectly by solar radiation at the present time, rather than by fuels stored in the Earth. Biomass consists of combustible or partly combustible materials like wood, municipal solid waste, and agricultural products, which are all tied to a food chain. The base of the food chain is photosynthetic organisms that derive their energy from the Sun.

Assessment Tool

Check Your Understanding Notebook-Entry Evaluation Sheet
Use this sheet to evaluate the extent to which students understand the key concepts explored in **Activity 2** and explained in **Digging Deeper**, and to evaluate the students' clarity of expression.

Assessment Opportunity

Use (or rephrase) the questions in **Check Your Understanding** for a brief quiz to check comprehension of key ideas and skills. Use the quiz (or a class discussion) to assess your students' understanding of the main ideas in **Digging Deeper** and **Investigate**. A few sample questions are provided below:

Question: Explain the difference between electric power and electric energy. How is each measured?
Answer: Electric power is the rate at which electricity does work, measured at a given time. The unit of measure for electric power is the watt (W). Electric energy is the amount of work that can be done by electricity. One of the units of measure for electric energy is the watt-hour (Wh).

Question: What does it mean if an energy resource is renewable? Give some examples of renewable resources.
Answer: A renewable energy source is one that is powered directly or indirectly by solar radiation at the present time, rather than by fuels stored in the Earth. Water, solar, wind, and biomass are all renewable resources.

Chapter 1

NOTES

Blackline Master Energy Resources 2.4
Geothermal Energy

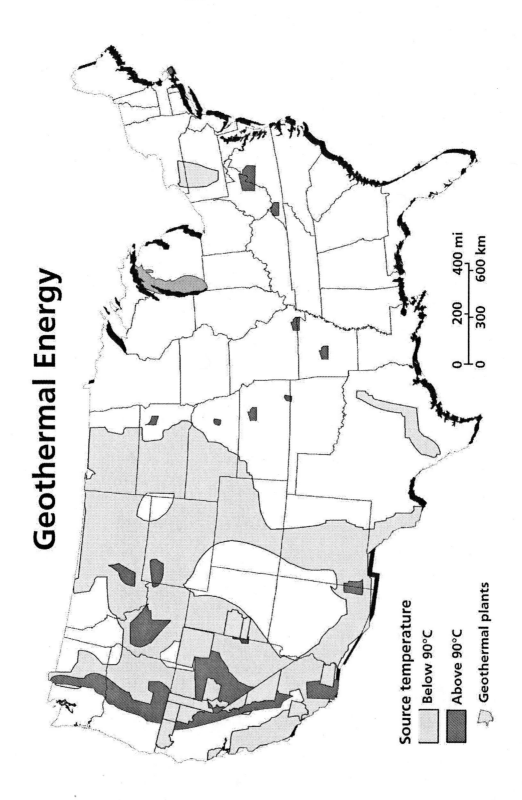

Geothermal Energy

Source temperature

Below 90°C

Above 90°C

Geothermal plants

400 mi
600 km
200
300
0
0

Blackline Master Energy Resources 2.5
Electricity Flow

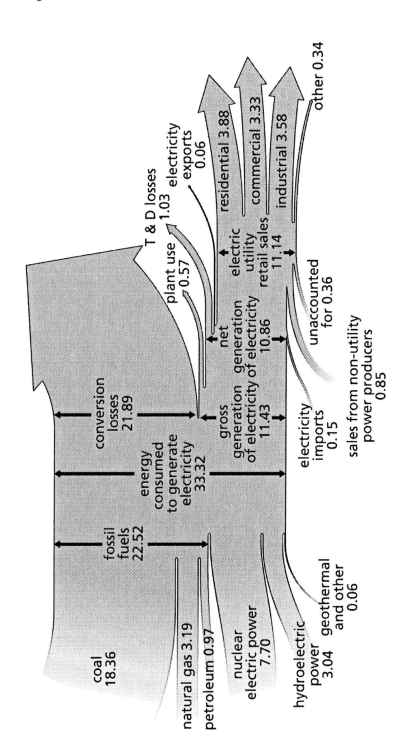

Electricity flow – electric utilities 1999
(quadrillion Btus)

Understanding and Applying What You Have Learned

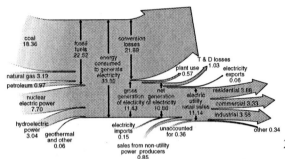

Electricity flow - electric utilities 1999
(quadrillion Btus)

1. Use the diagram above to answer the following questions about the generation of electricity in the United States:

   a) Which of the energy sources listed above are nonrenewable? What percentage of energy consumed to produce electricity in 1999 did they represent?

   b) Which of the energy sources listed above are renewable? What percentage of energy consumed in 1999 did they represent?

   c) In Activity 1 you learned about energy efficiency. Compare the value for energy consumed to produce electricity versus net generation of electricity in 1999.

   What is the overall efficiency of electricity generation systems in the United States?

   d) Conversion losses accounted for 21.89 "quads" of the energy consumed to generate electricity. Explain what this means in relation to your answer to question 3(c) above.

2. Converting one form of energy into another comes with a loss in usable energy (waste heat). The "net efficiency" of a multi-step process is the product of the efficiencies of each step. For example, if each step in a two-step process was 50% efficient, the net efficiency would be $(0.50 \times 0.50 = 0.25)$ 25%.

   a) Use the information below to calculate the net efficiency of a home water heater.

   b) The lowest efficiency of this process is the generation of electricity. Why do you think this is so?

3. Each year, the United States uses more than 71 trillion Btus of solar energy. One million Btu equals 90 lb. of coal or eight gallons of gasoline.

Step	Efficiency (in percent)
Production of coal	96
Transporting coal to power plant	97
Generating electricity	33
Transmission of electricity (from plants to homes)	85
Heating efficiency of the hot water heater	70
Net Efficiency	?

R 23

EarthComm

Teaching Tip

Use **Blackline Master Energy Resources 2.5, Electricity Flow** to make an overhead transparency of the diagram on page R23. Incorporate this overhead into a class discussion centered on the questions raised in **Understanding and Applying What You Have Learned.**

Understanding and Applying What You Have Learned

1. **a)** Fossil fuels (coal, natural gas, and petroleum) and nuclear power are the nonrenewable energy resources shown in the diagram. They account for 90.7% [(30.22 quadrillion Btu/33.32 quadrillion Btu) x 100] of the total energy consumed to produce electricity in 1999.

   b) Hydroelectric power and geothermal and other are the renewable energy resources shown in the diagram. They account for 9.3% [(3.1 quadrillion Btu /33.32 quadrillion Btu) x 100] of the total energy consumed to produce electricity in 1999.

   c) Students can use the equation introduced in **Activity 1** (page R13 of the student text) to calculate efficiency (% efficiency = electricity generated divided by energy consumed). They should plug the numbers from the diagram into the equation and find that the overall efficiency of electricity generation systems in the United States is 34.3% [(11.43 quadrillion Btu gross generation of electricity/33.32 quadrillion Btu of energy consumed to generate electricity) x 100] .

   d) Students should relate their answers to the Second Law of Thermodynamics, which states that that heat cannot be completely converted into a more useful form of energy, and that when converting any form of energy into another form, there is always a decrease in the amount of useful energy.

2. Students should use the percent efficiencies given in the table at the bottom of page R23 to answer the questions.

   a) The net efficiency would be 18%: (0.96 x 0.97 x 0.33 x 0.85 x 0.70 = 0.18)

   b) This is the step in which the Second Law of Thermodynamics has its largest effect. Most of the chemical energy released by the burning of the fossil fuel is converted into waste heat rather than being converted to useful mechanical energy.

3. Note that this question carries over to page R24.

   a) Remind students that 1 metric ton = 1000 kg = 2204.62 lb. (The term ton can be confusing. In addition to metric tons, there are short tons, equal to 2000 pounds, and long tons, equal to 2240 pounds.) Therefore, 71 trillion Btu of solar energy is equivalent to 2,898,459 metric tons of coal [71,000,000,000,000 Btu x (90 lb/1,000,000 Btu) x 1 metric ton/2204.62 lb].

Teaching Tip

Another way of showing this is first to calculate that 40 lb is equal to about 0.0408 metric tons. Then set up an equation:

$$\frac{\text{71 trillion Btu of solar energy}}{x \text{ metric tons of coal}} = \frac{\text{1 million Btu of coal energy}}{\text{0.0408 metric tons of coal}}$$

If your students have any experience with elementary algebra, they can easily solve this equation. The actual calculation is easier if the numbers are expressed in scientific notation:

$$\frac{7.1 \times 10^{13}}{x} = \frac{10^6}{4.08 \times 10^{-2}}$$

$$x = \frac{7.1 \times 10^{13} \times 4.08 \times 10^{-2}}{10^6} = 2.897 \times 10^6 = 2,897,000$$

Students' answers would be slightly different if the value obtained by dividing 90 by 2204.62 is expressed with more digits to the right of the decimal point. This would be a good place to impress upon your students that it makes no sense to carry more digits because the figure of 71 trillion Btu of solar energy per year is a very appropriate figure anyway.

b) 71 trillion Btu of solar energy equals 568,000,000 gallons of gasoline [71,000,000,000,000 Btu x (8 gallons/1,000,000 Btu)].

c) From **Part a)** we know that the current usage equates to 2,898,459 metric tons of coal. To get the answer we must take this number and multiply it by 1.05 ten times. The consumption each year is shown in the table below. This solar energy consumption for the 10-year period is equivalent to 38,279,331 metric tons of coal. This would equate to a savings of an additional 9,294,741 metric tons of coal compared to if the solar energy consumption did not increase over that time period.

Year	Equivalent Tons of Coal
current usage	2,898,459
1	3,043,382
2	3,195,551
3	3,355,329
4	3,523,095
5	3,699,250
6	3,884,212
7	4,078,423
8	4,282,344
9	4,496,461
10	4,721,284
Total	38,279,331

NOTES

Earth's Natural Resources Energy Resources

a) How many tons of coal per year does this equate to?

b) How many gallons of gasoline does this equate to?

c) If the United States increased its use of solar energy 5% annually, how many tons of coal would this equate to during the next 10 years?

Preparing for the Chapter Challenge

Write a short paper in which you review the energy resources used in your community and how their use affects your community. Be sure to include information on how energy resources are used to generate electricity and fuel in your community, which of the current means of producing power your community relies on most, and which seems the least important. This paper will help you to think about what resources your community is using before the population increase.

Inquiring Further

1. **Storage of solar energy**

   Electricity from solar energy offers a clean, reliable, and renewable source of energy for home use. Batteries are used to store electrical energy for use at night or when bright sunlight is not available. Research recent advances in technologies used to store solar energy. How do car batteries perform in comparison to solar batteries in solar systems? Present your report to the class.

2. **Energy from the oceans**

   Three novel ways have been proposed to use the energy of the oceans for electrical generation: tides, waves, and vertical temperature differences. Choose one of these, and research the theory and techniques involved in its use for generating electricity.

3. **Other methods of generating electricity**

   Research one of the following methods of generating electricity. How do they work? Where are they being used? What are the ideal conditions for using these alternatives? What are the advantages of these methods over fossil fuels? What are the disadvantages or limitations? Visit the *EarthComm* web site for sources of information.

   • photovoltaics

   • geothermal

   • solar thermal

Preparing for the Chapter Challenge

This section will allow students to begin thinking about what energy resources their community is currently relying on. Encourage students to incorporate the data they obtained through their state electricity profile into their papers. You may want to review with your students how this paper will allow them to address the following points in the assessment rubric for the **Chapter Challenge** (see **Assessment Rubric for Chapter Report on Energy Resources** on pages 16 and 17):

• current energy uses and consumption rates in your community
• one example that illustrates the difference between renewable and nonrenewable energy resources

Inquiring Further

1. Storage of solar energy

A battery converts chemical energy into electricity. It has two or more cells connected in series, or in parallel. All cells consist of a liquid, paste, or solid electrolyte, and a positive electrode and a negative electrode. The electrolyte conducts ions (charged atoms); one of the electrodes will react, producing electrons, while the other will accept electrons. When the electrodes are connected to a device to be powered, an electrical current flows.

Some solar electricity systems come with special solar batteries. Other systems use ordinary car batteries. A common storage battery, which can be recharged by reversing the chemical reaction that causes electricity to flow, is the lead-acid battery. The lead-acid battery has plates made of lead, mixed with other materials, submerged in a sulfuric acid solution. There are many different sizes and designs of lead-acid batteries, but the most important designation is whether they are deep-cycle batteries or shallow-cycle batteries. Shallow-cycle batteries, like the type used as starting batteries in automobiles, are designed to supply a large current for a short time and to withstand mild overcharge without losing electrolyte. Unfortunately, they cannot tolerate being deeply discharged. If they are repeatedly discharged more than 20%, they will have a very short life. These batteries are not a good choice for a photovoltaic system.

Deep-cycle batteries, on the other hand, are designed to be discharged repeatedly by as much as 80% of their capacity, making them a good choice for power systems. Even though they are designed to withstand deep cycling, these batteries will have a longer life if the cycles are shallower. All lead-acid batteries will fail prematurely if they are not recharged completely after each cycle. Letting a lead-acid battery stay in a discharged condition for many days at a time will cause sulfation of the positive plate and a permanent loss of capacity. Sealed deep-cycle lead-acid batteries are maintenance free. They never need watering or an equalization charge. They cannot freeze or spill, so they can be mounted in any position.

2. Energy from the ocean

Tides – Tides are the daily rise and fall of sea level, caused by the gravitational forces of the Moon and Sun. Tidal power stations are constructed in areas like estuaries and bays with a large tidal range. To generate electricity using tidal energy, a dam or dike is built across the mouth of the water body. As the tide outside the dike rises, water flows through a turbine in the wall to produce electricity. As the tide falls, water flows out through the turbine in the other direction. The turbine is a two-way system that can produce electricity on both incoming and outgoing tides. Some tidal electricity generation stations can produce up to 320 megawatts of electricity.

Waves – Wave power results from the harnessing of energy transmitted to waves by winds moving across the ocean surface. The energy of motion of the near-surface water as the waves pass by is harnessed mechanically to turn a turbine to produce electricity.

Vertical Temperature Differences – This means of generating electricity makes use of the temperature difference between the warm surface waters of the ocean and the cold deep waters. For this technique to be viable, the temperature difference between the surface water and water at a depth of 1000 m needs to be about 20°C.

3. Other methods of generating electricity

A photovoltaic (PV) cell consists of semiconducting material that absorbs sunlight and converts it directly into electricity by using solar energy to knock electrons loose from their atoms. This allows the electrons to flow through the material to produce electricity. Geothermal energy is the heat from the Earth. Geothermal energy often heats water, which can be used directly to heat buildings, grow plants in greenhouses, etc. Solar heating involves the use of solar collectors to trap the Sun's rays to produce heat.

A major advantage of using these energy sources is that there are no emissions of greenhouse gases or acid gases. Also, the energy available from the Earth and Sun each year is many times greater than the worldwide energy demand. A disadvantage is that the availability of these resources varies with location, and, in the case of solar energy, with time of day and season. Students can visit the *EarthComm* web site to find additional information.

NOTES

ACTIVITY 3—ENERGY FROM COAL

Background Information

Coal

Coal is a special kind of sedimentary rock that consists mostly of combustible material in the form of carbon. Coal originates in sedimentary environments called coal swamps. A coal swamp is a low-lying area of lush vegetation to which little mineral sediment is supplied. A typical coal swamp is located in a tectonically stable coastal environment in a perennially warm and humid climatic zone, far away from sources of mineral sediment.

A typical coal-forming environment is an interdistributary bay located between active distributary channels in a major river delta. Plants take root in earlier plant material, and when they die, they form the substrate for later plant growth. Slow subsidence assures deposition of thick successions of organic sediment. As the plant material is buried more and more deeply, it is compacted, thus reducing pore space. The noncarbon constituents, mainly pore water and pore gases, are driven out, resulting in more nearly pure carbon. Decay is minimized because the material is rapidly sealed off from the oxygen-rich near-surface environment. The end result is a low-porosity deposit of nearly pure carbon, which we call coal.

There is a continuous spectrum of coal composition as a function of coalification during burial, ranging from peat, through lignite and bituminous coal, to anthracite coal. The position of the coal along that spectrum is called the rank of the coal.

Geologic Maps

A geologic map shows the distribution of bedrock that is exposed at the Earth's surface or buried beneath a thin layer of surface soil or sediment. Geologic maps are more than just maps of rock types: most geologic maps show the locations and relationships of rock units.

Each rock unit is identified on the map by a symbol of some kind, which is explained in a legend or key, and is often given a distinctive color as well. Part of the legend of a geologic map consists of one or more columns of little rectangles, with appropriate colors and symbols, identifying the various rock units shown on the map. There is often a very brief description of the units directly in this part of the legend. The rectangles for the units are arranged in order of decreasing age upward. Usually the ages of the units, in terms of the standard relative geologic time scale, are shown as well.

All geologic maps convey certain other information as well. They show the symbols that are used to represent such features as folds, faults, and attitudes of planar features like stratification or foliation. They have information about latitude and longitude, and/or location relative to some standard geographic grid system. They always have a scale, expressed both as a labeled scale bar and as what is called a representative fraction— 1:25,000, for example—whose first number is a unit of distance on the map and whose second number is the corresponding distance on the actual land surface.

All geologic maps (except perhaps special-purpose maps that show all the details of an area that might be the size of a small room!) involve some degree of generalization. Such generalization is the responsibility of the geologist who is doing the mapping. Obviously, it is not practical to represent features as small as a few meters wide on a map that covers many square miles: the line depicting the feature on the map would be far finer than the finest possible ink line. The degree of generalization necessarily increases as the area covered by the map increases. You could easily see this for yourself if you had access to a geologic map of some small area together with the corresponding geologic map of the entire state: the state map would show far less detail of the small area than the full map of that same small area.

More Information – on the Web
Visit the *EarthComm* web site www.agiweb.org/earthcomm to access a variety of links to web sites that will help you deepen your understanding of content and prepare you to teach this activity. Many of the sites also contain images that you can download.

Goals and Assessment

Clarify that the goals indicate what students should understand and be able to do as a result of the activity. Make sure students understand that Chapter Assessments are based upon these goals.

Goal	Location in Activity	Assessment Opportunity
Classify the rank of coal using physical properties.	**Investigate** Part A **Digging Deeper**	Data table is complete, and accurately represents coal samples.
Interpret a map of coal distribution in the United States.	**Investigate** Part B	Answers to the questions are clearly based on map and available data.
Understand that fossil fuels represent solar energy stored as chemical energy.	**Investigate** Part A **Digging Deeper; Check Your Understanding** Question 1	Answers to questions are reasonable, and closely match those in Teacher's Edition.
Understand what coal is made of and how coal forms.	**Investigate** Part A **Digging Deeper; Check Your Understanding** Questions 2 and 4	Answers to questions are reasonable, and closely match those in Teacher's Edition.
Be able to explain in your own words why coal is a nonrenewable resource.	**Digging Deeper; Check Your Understanding** Question 1	Answers to questions are reasonable, and closely match those in Teacher's Edition.

NOTES

Activity 3   Energy from Coal

Goals

In this activity you will:

- Classify the rank of coal using physical properties.

- Interpret a map of coal distribution in the United States.

- Understand that fossils fuels represent solar energy stored as chemical energy.

- Understand what coal is made of and how coal forms.

- Be able to explain in your own words why coal is a nonrenewable resource.

Think about It

Coal is the largest energy source for electricity in the United States. Known coal reserves are spread over almost 100 countries. At current production levels, proven coal reserves are estimated to last over 200 years.

- How does coal form?
- Why is coal referred to as "stored solar energy"?

What do you think? Record your ideas in your *EarthComm* notebook. Be prepared to discuss your responses with your small group and the class.

Activity Overview

Students examine samples of different types of coal and produce a data table that reflects the physical properties of the different samples. The coal samples are burned (as a demonstration), and students note the speed of ignition, color of the flame, and odor associated with each sample. Then they look at a map that shows the distribution of coal resources and coal-fired power plants in the United States. Students compare the coal distribution map of the United States to a geologic map of their community or state to determine what types of rocks are associated with coal deposits. They use the Internet to determine how much their state relies on coal as an energy resource. **Digging Deeper** explains how different types of coal form and why coal is essentially stored solar energy.

Preparation and Materials Needed

Part A

To prepare for **Part A** of the investigation, you will need to find samples of different types of coal, including anthracite, bituminous, subbituminous, and lignite.

Part B

Well in advance, you will need to obtain copies of your state or community geologic map. The *EarthComm* web site contains several suggestions to help you find the needed map.

Students will also need Internet access to complete **Part B** of this investigation. If you do not have Internet access in your classroom or cannot schedule time in the school computer lab, you can obtain the needed Energy Consumption, Prices, and Expenditures profiles for your state before class and make copies for distribution.

Materials

Part A
- Samples of coal: anthracite, bituminous, subbituminous, lignite
- Bunsen burner (for teacher demonstration)
- Magnifying glass or hand lens

Part B
- Geologic map of your state or community*
- Internet access (or printouts of your state's Energy, Consumption, Prices, and Expenditures profiles)*

*The *EarthComm* web site has suggestions for obtaining these resources.

Think about It

Student Conceptions

The level of understanding that your students have about how coal forms will vary depending on their prior knowledge of how rocks form. You may want to show them a sample of peat and a sample of anthracite coal. Ask them to consider first what the peat is made of. They will most likely say that peat is plant matter. Then ask them to consider what processes would be necessary to change the peat to anthracite. If students are able to identify plant material as being the primary original material from which coal is made, they should be able to deduce why coal can be referred to as stored solar energy (plants get the energy they need to make food from the Sun).

Answers for the Teacher Only

Most coal starts out as peat, which is a porous deposit of partly decomposed plant material. If the plant material is buried by sediment, it is subjected to compaction and temperature increase, which alter it physically and chemically, to eventually form coal. Increasing time, pressure, and temperature result in the formation of coal with progressively greater percentages of carbon.

Assessment Tool

Think about It Evaluation Sheet
Use this evaluation sheet to help students understand and internalize the basic expectations for the warm-up activity.

Blackline Master Energy Resources 3.1
Coal Resources

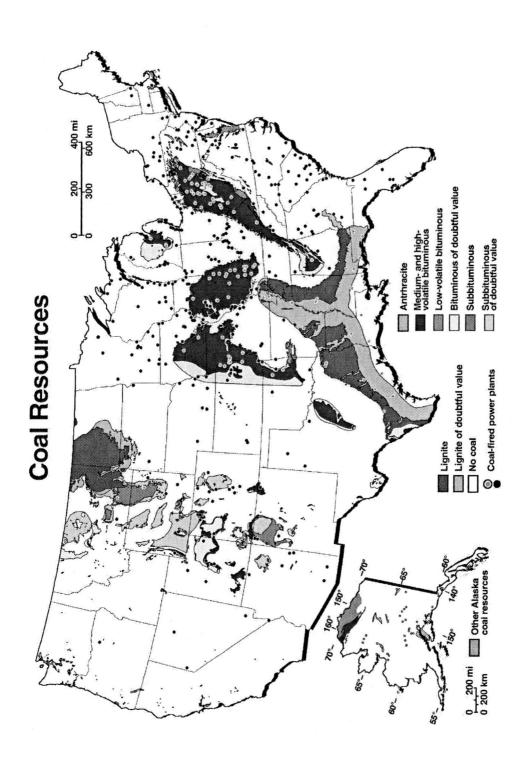

Coal Resources

Legend:
- Anthracite
- Medium- and high-volatile bituminous
- Low-volatile bituminous
- Bituminous of doubtful value
- Subbituminous
- Subbituminous of doubtful value
- Lignite
- Lignite of doubtful value
- No coal
- Coal-fired power plants

Other Alaska coal resources

400 mi / 600 km
200 / 300
0 / 0

0 200 mi
0 200 km

Earth's Natural Resources Energy Resources

Investigate

Part A: Types of Coal

1. Obtain a set of samples of four different types of coal. Examine the samples. Look for evidence of plant origin, hardness, luster (the way light is reflected off the surface of the sample), cleavage (tendency to split into layers), and any other characteristics that you think distinguish the four types of coal. Discuss similarities and differences with members of your group.

   a) In your *EarthComm* notebook, create a data table and record your observations.

2. Use your completed data table, to answer the following questions:

   a) How might the samples be related?

   b) Put the samples in order from least compacted to most compacted.

   c) Which sample do you think contains the most stored energy? Why do you think so?

3. The next step of this activity will be a demonstration. You will observe your teacher igniting a small piece of coal held over a Bunsen burner. The sample will be removed from the flame and you will observe how the sample burns. Four types of coal will be tested.

   a) Note the speed of ignition, color of the flame, speed of burning, and odor. Note any other characteristic that you think distinguishes the samples. Summarize your observations in a data table or chart.

   b) In your *EarthComm* notebook, summarize the major differences you observed.

   c) Review your answer to Step 2(c). Have your ideas changed? Explain.

 Wear goggles. Clean up all loose pieces of coal. Wash your hands after you are done.

R 26

Investigate

Part A: Types of Coal

1. Provide your students with the coal samples. You may wish to review some of the terms, like hardness, luster, or cleavage. (These terms are also defined in the **Digging Deeper** section of **Activity 2** of the **Mineral Resources** chapter, pages R105 to R109.)

   a) Students should produce a data table that will help them keep track of the physical characteristics of coal. Have students set up the data table as shown below.

Sample	Observations and Evidence of Features				
	[Feature]	[Feature]	[Feature]	[Feature]	[Feature]
1					
2					
3					
4					

Observation and Evidence of Features					
Sample	Name	Color	Hardness	Luster	Other
1	lignite	brownish-black	variable, softest of the 4 samples	dull	some visible pieces original texture of plant matter (wood pieces)
2	sub-bituminous	brownish-black to black, can be grayish	variable, harder than sample #1, but softer than #'s 3 or 4	dull	no visible plant matter, perhaps slight indication of layers
3	bituminous	black	variable, harder than #2 but softer than #4	more shiny (vitreous) luster than #'s 1 or 2	can see some indication of layers and cleavage
4	anthracite	black	variable, hardest of all 4 samples (generally between 3 and 4 on the Mohs scale)	semi-metallic luster. Most shiny of all 4 samples	some conchoidal fracture

2. a) Answers will vary depending upon the information that students put into their tables.

   b) The samples in order from least to most compacted are: lignite, subbituminous coal, bituminous coal, and anthracite.

   c) Answers will vary. Students should justify their responses. For example, the higher the rank of coal, the greater the heat energy value—as measured in Btus (British thermal units)—so anthracite contains the most stored energy, whereas lignite contains the least stored energy.

3. Ignite a small piece of each type of coal so that your students can observe how each type burns.

   a) Students should set up a data table to record their observations. They can add to the table they started in **Step 1(a)** above, or start a new one. Lower-rank coals are ignited more readily and burn faster than higher-rank coals. The flame of higher-rank coals is smaller and more bluish. Odor depends on what else is in the coal. The gas released by burning of low-sulfur coal of any rank smells less bad than that of high-sulfur coal.

   b) Answers will vary.

   c) Answers will vary depending on students' original ideas.

NOTES

Part B: Coal Resources

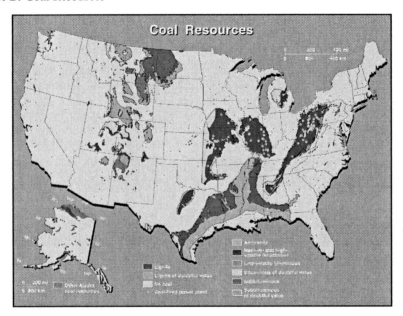

1. Examine the map of coal resources in the United States.

   a) Summarize the major trends, patterns, or relationships in the distribution of coal and the locations of coal-fired power plants.

2. Refer to the map to answer the following questions about coal resources in the United States. Record your responses in your *EarthComm* notebook.

   a) How many states contain coal deposits? Which states are they?

   b) What are the main types of coal present in the following regions: east-central, southeast, and west-central?

   c) Can you infer from the map which state has or produces the most coal? Why or why not?

   d) Measure how far your community is from a source of coal.

   e) Why do you think coal is used in these plants rather than petroleum or natural gas?

   f) What type of coal is closest to your community?

3. Refer to the map to answer the following questions about coal-fired power plants in the United States. Record your responses in your *EarthComm* notebook.

   a) Is there a coal-fired power plant in your state? If so, where is it located?

R 27

Assessment Tools

EarthComm Notebook-Entry Checklist
Use this checklist as a quick guide for student self-assessment and/or an opportunity to quickly score student work. Add further criteria specific to your classroom needs or to this particular investigation.

Investigate Notebook-Entry Evaluation Sheet
Point out the criteria listed on this evaluation sheet that are relevant to this particular investigation. Encourage students to internalize the criteria. by making them part of your assessment conversations as you circulate around the classroom.

Part B: Coal Resources

1. **a)** Student answers will vary, but in general they should notice that although coal-fired power plants are scattered throughout the United States, there are greater numbers of plants around areas that are located near medium- and high-volatile bituminous coal deposits.

2. **a)** Coal deposits are found in the following states: Montana, North Dakota, South Dakota, Wyoming, Utah, Colorado, New Mexico, Arizona, Texas, Oklahoma, Kansas, Nebraska, Iowa, Missouri, Arkansas, Louisiana, Alabama, Mississippi, Tennessee, Indiana, Illinois, Ohio, Pennsylvania, West Virginia, Kentucky, Virginia, Georgia, and Florida. There are some mineable coals in Massachusetts and Rhode Island as well, but they are not shown on the map.

    b) The main type of coal found in the east-central region of the United States is medium- and high-volatile bituminous coal. In the southeastern region, lignite is most common, and in the west-central area, most of the deposits are subbituminous or lignite.

    c) Students could make some reasonable assumptions about which states have more coal. They should keep in mind, however, that the map shows only the surface area of coal deposits, and does not indicate the depth of these deposits.

    d) Answers will vary depending upon where you live.

    e) Answers will vary depending upon where you live. Most likely, coal is used in the plant rather than natural gas or petroleum because the plant is located closer to a coal deposit than to an oil or gas field.

    f) Answers will vary depending upon where you live.

Earth's Natural Resources Energy Resources

b) How far from your community is the closest coal-fired power plant? Where is the plant located?

c) In what part of the country are the most coal-fired power plants found? Why do you think this is so?

d) Identify the states that have coal-fired power plants, but do not have coal deposits.

e) How might distance from coal deposits affect the cost of electricity produced by a coal-fired power plant?

4. Examine a geologic map of your community or state. A geologic map shows the distribution of bedrock at the Earth's surface. The bedrock shown on the map might be exposed at the surface, or it might be covered by a thin layer of soil or very recent sediment. Every geologic map contains a legend that shows the kinds of rocks that are present in the area.

   a) Write down or note the names of the different rock types present in your community, the names of the rock units, their ages, and their locations.

5. Compare the geologic map to the map that shows the distribution of coal deposits in the United States.

a) Are any of the rocks present in your community associated with coal deposits? What kinds of rocks are they? How old are the rocks?

b) Are any of the rocks present in your community likely to yield coal deposits in the future? Which ones? How do you know?

6. Go to the *EarthComm* web site to visit the Department of Energy's Energy Information Administration web page. Click on your state to open a new web page. Use the "Energy Consumption, Prices, and Expenditures" profiles page to answer the following questions:

a) Approximately what percentage of primary energy consumption for residential purposes comes from coal in your state?

b) What percentages of primary energy consumption for commercial, industrial, and transportation purposes come from coal in your state?

c) How would an increase in the population of your community affect this usage?

Reflecting on the Activity and the Challenge

In this activity you described four types of coal and observed how the different types of coal react when burned. You then examined maps to determine the distribution of coal deposits in the United States and also to determine if coal deposits are found near your community. You looked at the location of coal-fired power plants relative to coal deposits and thought about how the distance between the two can affect the cost of electricity production. These activities will help you to determine how coal is used as an energy resource in your community and the impacts of its use.

3. **a)** Answers will vary depending upon where you live.
   b) Answers will vary depending upon where you live.

   c) Most of the coal-fired power plants are in the east-central region of the United States Students should note that this is also where most of the medium- and high-volatile bituminous deposits are located.

   d) New Hampshire, Massachusetts, New York, Connecticut, New Jersey, South Carolina, Georgia, Michigan, Wisconsin, Minnesota, Nevada, California, Oregon, and Washington all have coal-fired power plants, but they do not have coal deposits.

   e) The farther a coal deposit is from a coal-fired power plant, the higher the cost of electricity produced by the plant. This is due to costs associated with transporting the coal from the deposit to the plant.

Teaching Tip

If this is the first time that your students are working with geologic maps, you may want to spend some time reviewing them with your students.

4. **a)** Answers will vary depending upon where you live.

5. **a) - b)** Coal deposits will be associated with sedimentary rocks. Whether or not these rocks are found in your community will vary.

6. **a) - c)** All of the information students need can be found in the Department of Energy's Energy Consumption, Prices, and Expenditures profile for your state, which can be found online. Visit the *EarthComm* web site to access the necessary page, which displays a series of data tables and graphs. Answers are dependent upon what state you live in, and will therefore vary.

Reflecting on the Activity and the Challenge

Students described the physical features of four different types of coal and the distribution of coal deposits throughout the United States and within their own state. They should now have a sense of what types of coal contain the most stored energy and burn most cleanly, and are therefore the most useful as energy resources. They should be aware of where coal deposits are found relative to their community and how dependent their community is upon coal for power generation. Discuss with students the consequences of using natural resources that are found far from the community versus those that are obtained locally. They should recognize that the longer the distance that a resource must be transported, the more expensive it will be. Ask students to consider what factors determine how their community uses coal as an energy resource and what implications the hypothetical population increase proposed in the **Chapter Challenge** might have on the community's use of coal.

Digging Deeper

COAL AS A FOSSIL FUEL

There are only four primary sources of energy available for use by humankind. They are solar radiation, the Earth's interior heat, energy from decay of radioactive material in the Earth, and the tides. You can assume that starlight, moonlight, and the kinetic energy of meteorites hitting the Earth are so small that they can be neglected. The energy in coal is energy from solar radiation that is stored as chemical energy in rock.

The energy in coal originates as solar energy. Plants in the biosphere store this solar energy by a process called **photosynthesis**. During photosynthesis, green plants convert solar energy into chemical energy in the form of **organic (carbon-based) molecules**. Only 0.06% of the solar energy that reaches the Earth is stored through photosynthesis, but the amount of energy that is stored in the Earth's vegetation is enormous. Photosynthesis yields the carbohydrate glucose (a sugar) and water, as shown in the equation below:

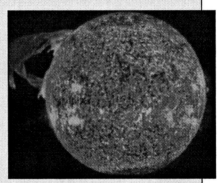

Figure I Solar radiation is one of the primary sources of energy on Earth.

$$6CO_2 \quad + \quad 6H_2O \quad \xrightarrow{\text{Sunlight used}} \quad C_6H_{12}O_6 \quad + \quad 6O_2$$

| Carbon dioxide removed from atmosphere and biosphere | Water removed from hydrosphere | Solar energy stored as chemical energy | Glucose (sugar) for biosphere | Oxygen released to atmosphere |

The energy stored in glucose (a sugar) and other organic molecules can be released when broken in a reaction with oxygen (oxidized). **Oxidation** occurs through **respiration, decomposition,** or combustion (burning).

$$6O_2 \quad + \quad C_6H_{12}O_6 \quad \xrightarrow{\text{Chemical energy released}} \quad 6H_2O \quad + \quad 6CO_2$$

| Oxygen removed from atmosphere and biosphere | Glucose from biosphere | Heat released to organism by respiration or to environment by decomposition or combustion | Water returned to hydrosphere | Carbon dioxide returned to atmosphere and biosphere |

Geo Words

photosynthesis: the process by which plants use solar energy, together with carbon dioxide and nutrients, to synthesize plant tissues.

organic (carbon-based) molecules: molecules with the chemical element carbon as a base.

oxidation: the chemical process by which certain kinds of matter are combined with oxygen.

respiration: physical and chemical processes by which an organism supplies its cells and tissues with oxygen needed for metabolism.

decomposition: the chemical process of separation of matter into simpler chemical compounds.

Digging Deeper

Assign the reading for homework. The questions in **Check Your Understanding** can be provided as a homework assignment.

Assessment Opportunity

Reword or restructure the questions in **Check Your Understanding** for a brief quiz. Use the quiz (or a class discussion of the questions in the textbook) to assess your students' understanding of the main ideas in the reading and the activity.

Teaching Tip

Chemical equations of the kind given here express how two or more chemical compounds react with one another to form products of different composition. Other equations show how a single compound can dissociate into two or more compounds, or how those compounds can combine into a single compound. Chemical reactions can proceed in either direction. The compounds that are being consumed are called the reactants, and the compounds that are being produced are called the products. For a chemical equation to be valid, it must be balanced: the numbers of atoms of each element must the same on both sides of the equation. For example, in the first equation shown, there are six carbon atoms on each side, twelve hydrogen atoms, and eighteen oxygens.

Earth's Natural Resources Energy Resources

Geo Words

sediments: loose particulate materials that are derived from breakdown of rocks or precipitation of solids in water.

lithosphere: the rigid outermost shell of the Earth, consisting of the crust and the uppermost mantle.

nonrenewable resource: an energy source that is powered by materials that exist in the Earth and are not replaced nearly as fast as they are consumed.

You can see from the reactants (on the left sides of the two equations) and the products (on the right sides of the two equations) that photosynthesis and oxidation are the reverse of each other.

The energy that enters the biosphere by photosynthesis is nearly equal to the energy lost from the biosphere by oxidation, as shown in *Figure 2*. Most of the carbon in the biosphere is soon returned to the atmosphere (or to the ocean) as carbon dioxide, but a very small percentage is buried in **sediments**. It is protected from oxidation and becomes part of the **lithosphere** (in the form of peat, coal, and petroleum) for future use as fossil fuels. The rate of storage of energy used by organisms for photosynthesis is very slow. However, these processes have been active throughout much of Earth's history. Therefore, the amount of energy stored as fossil fuels is enormous (32×10^{20} Btus). Fossil fuels are consumed far faster than they form, and therefore they are classified as **nonrenewable resources**.

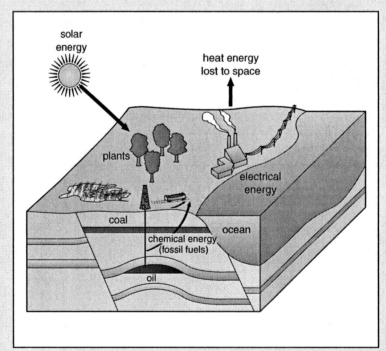

Figure 2 The flow of energy from the Sun, to plants, to storage in fossil fuels, and loss back into space.

Blackline Master Energy Resources 3.2
Flow of Energy from the Sun to Fossil Fuels

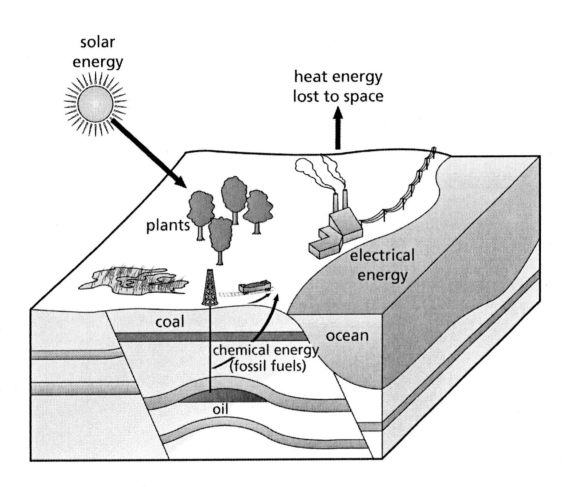

The Formation of Coal

Most coal starts out as **peat**. Peat is an unconsolidated and porous deposit of plant remains from a bog or swamp. Structures of the plant matter, like stems, leaves, and bark can be seen in peat. When dried, peat burns freely, and in some parts of the world it is used for fuel. Today, most peat comes from peat bogs that formed during the retreat of the last ice sheets, between ten thousand and twenty thousand years ago.

Coal, by definition, is a combustible rock with more than 50% by weight of carbonaceous material. Coal is formed by compaction and hardening of plant remains similar to those in peat. The plant remains are altered physically and chemically through a combination of bacterial decay, compaction, and heat. Most coal has formed by lush growth of plants in coastal fresh-water swamps, called coal swamps, in low-lying areas that are separated from any sources of mud and sand. Plants put their roots down into earlier deposits of plant remains, and they in turn die and serve as the medium for the roots of even later plants. In such an environment, the accumulation of plant debris exceeds the rate of bacterial decay of the debris. The bacterial decay rate is reduced because the available oxygen in organic-rich water is completely used up by the decay process. Thick deposits of almost pure plant remains build up in this way. Coal swamps are rare in today's world, but at various times in the geologic past they were widespread.

For the plant material to become coal, it must be buried by later sediment. Burial causes compaction, because of the great weight of overlying sediment. During compaction, much of the original water that was in the pore spaces of the plant material is squeezed out. Gaseous products (methane is one) of alteration are expelled from the deposit. The percentage of the deposit that consists of carbon becomes greater and greater. As the coal becomes enriched carbon, the coal is said to increase in rank. (See *Figure 3*.) The stages in the rank of coal are in the following order: peat, lignite, sub-

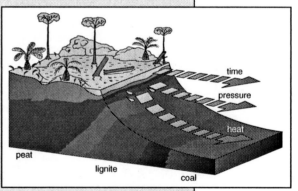

Figure 3 Increasing time, pressure, and heat result in the formation of progressively higher ranking coal.

bituminous coal, bituminous coal, anthracite coal, and finally graphite (a pure carbon mineral). It is estimated that it takes many meters of original peat material to produce a thickness of one meter of bituminous coal.

Geo Words

peat: a porous deposit of partly decomposed plant material at or not far below the Earth's surface.

coal: a combustible rock that had its origin in the deposition and burial of plant material.

Teaching Tip

Use **Blackline Master Energy Resources 3.2, Flow of Energy from the Sun to Fossil Fuels** to make an overhead transparency of the illustration in *Figure 2* on page R30. Incorporate this overhead into a discussion on the carbon cycle and the flow of energy on Earth. Discuss with students why it is reasonable to say that fossil fuels can be thought of as stored solar energy. Use this opportunity to discuss why fossil fuels are considered to be nonrenewable resources.

Teaching Tip

Figure 3 on page R31 illustrates the progressive formation of different types of coal. Coal types range from lignite (the lowest grade coal) to bituminous (the most abundant type of coal) to anthracite (the highest-grade coal). The change in grade reflects the temperatures reached following burial. The higher the temperature, the higher the grade (also called the rank) of coal produced. With increasing grade, the heat energy value (for example, in Btu) of coal increases, because the percentage of carbon relative to oxygen and hydrogen increases, and in many cases (not all) the percentage of sulfur decreases.

Teaching Tip

Use **Blackline Master Energy Resources 3.3, Environments of the Past** to make an overhead transparency of the illustration in *Figure 4* on page R32. Incorporate this overhead into a discussion on how rocks reflect the environments of the past. Encourage students to think about how this can help them understand where coal deposits are likely to be found in the future.

Earth's Natural Resources Energy Resources

Geo Words

sedimentary rock: a rock, usually layered, that results from the consolidation or lithification of sediment.

As shown in *Figure 4*, coal is always interbedded with other **sedimentary rocks**, mainly sandstones and shales. Environments of sediment deposition change with time. An area that at one time was a coal swamp might later have become buried by sand or mud from some nearby river system. Eventually, the coal swamp might have become reestablished. Upon burial, the plant material is converted to coal, and the sand and mud form sandstone and shale beds. In this way, there is an alternation of other sedimentary rock types with the coal beds. Some coal beds are only centimeters thick, and are not economically important. Coal beds that are mined can be up to several meters thick.

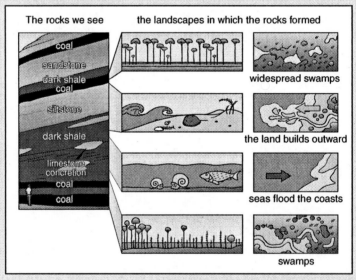

Figure 4 Rocks reflect the environments of the past.

Types of Coal

The type of coal that is found in a given region depends partly on the composition of the original plant material (together with any impurities that were deposited in small quantities and at the same time). But mainly the type of coal depends on the depth and temperature of later burial. Coal that is buried very deeply attains high rank, because of the high pressures and temperatures associated with deep burial. Low-rank coals (lignite and sub-bituminous) have not been buried as deeply.

Blackline Master Energy Resources 3.3
Environments of the Past

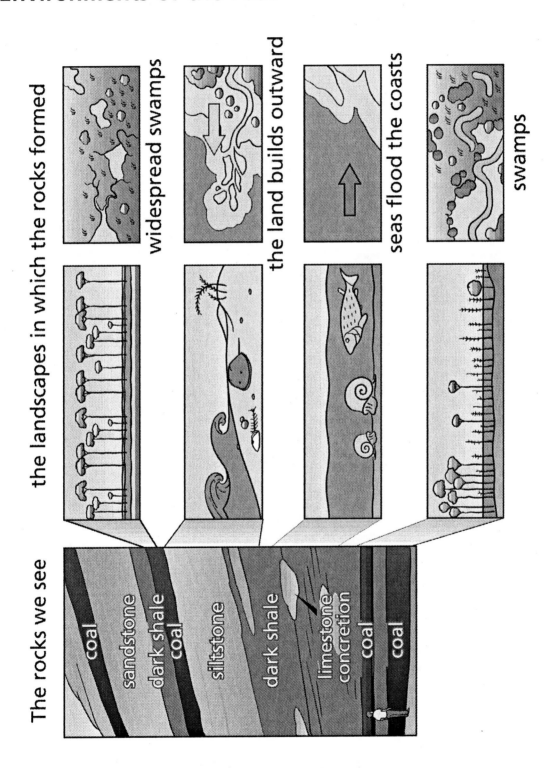

Coal varies greatly in composition. One of the most important features of coal composition is sulfur content. Sulfur is important because it is released into the atmosphere as sulfur dioxide when the coal is burned. The sulfur dioxide then combines with water in the atmosphere to form sulfuric acid, causing acid rain. The sulfur content of coal can range from a small fraction of 1% to as much as 5%, depending mainly on the sulfur content of the original plant material. The carbon content of coal increases with increasing rank of coal, from lignite to anthracite, as illustrated in *Table 1*. The heat content is also an important feature of coal. The greater the heat content, the smaller the mass of coal that needs to be burned to produce the needed heat. The heat content of coal increases with the rank of the coal, because it depends mainly on the carbon content.

Table 1 Percentages of Carbon and Volatile Matter in Coal			
Coal Rank	Carbon Content (%)	Volatile matter (%)	Btus per pound
Lignite	25–35	***	4000–8300
Sub-bituminous	35–45	***	8300–13,000
Bituminous	45–86	14–31	10,500–15,000
Anthracite	86–98	0–14	15,000

The carbon content of coal supplies most of its heat energy per unit weight. The amount of energy in coal is expressed in British thermal units (Btu) per pound. A Btu is the amount of heat needed to raise the temperature of one pound of water one degree Fahrenheit. Peat can be used as a source of fuel, but it has a very low heat content per pound of the fuel burned. Lignite (also called brown coal), the least buried and usually the youngest type of coal, is used mainly for electric power generation. Sub-bituminous coal is a desirable heat source because of its often low sulfur content. Sub-bituminous coal is found in the western United States and Alaska. Bituminous coal is the most abundant coal in the United States, with a large deposit in the Appalachian Province of the East. Bituminous coal is used mainly for generating electricity and making coke for the steel industry. Anthracite, found in a very small supply in the eastern United States, has been used mainly for home heating.

Some of the constituents of coal are not combustible. The part of coal that is not consumed by burning is called ash. Most of the ash consists of sand, silt, and clay that was deposited in the coal swamp along with the vegetation. The purest coal has only a small fraction of 1% ash. The ash content of usable coal can be much higher. Some of the ash remains in the combustion chamber, and some goes up the flue. Ash in coal is undesirable because it reduces the heat content slightly. It also must be removed from the combustion chamber and discarded.

Check Your Understanding

1. Explain how solar energy is stored and released by the processes of photosynthesis and oxidation.

2. Describe, in your own words, the formation of coal, from plant matter to anthracite.

3. Why is the heat content of coal (in Btu/pound) important in determining how the coal is used?

4. What is the origin of the sulfur content of coal, and why is sulfur an undesirable constituent of coal?

Check Your Understanding

1. During photosynthesis, green plants convert solar energy into chemical energy in the form of organic molecules. The chemical energy stored in the organic matter is released when these compounds are broken down in a reaction with oxygen through respiration, decomposition, or combustion.

2. Coal results from the rapid deposition of plant material accumulating in an oxygen-poor environment that inhibits decay. The peat that is formed in this environment is then changed to coal by pressure associated with burial. Increased temperatures and pressures associated with deeper burial result in progressively higher-grade coal.

3. The greater the heat content of coal, the smaller the mass of coal that needs to be burned to produce the needed heat.

4. The sulfur content of coal depends upon the sulfur content of the original plant matter. Sulfur dioxide released into the atmosphere when coal is burned combines with water in the atmosphere to form sulfuric acid, causing acid rain.

Assessment Tool

Check Your Understanding Notebook-Entry Evaluation Sheet
Use this sheet to evaluate the extent to which students understand the key concepts explored in **Activity 1** and explained in **Digging Deeper**, and to evaluate the students' clarity of expression.

Earth's Natural Resources Energy Resources

Understanding and Applying What You Have Learned

1. How much impact do you think the proximity of your community or state to a coal deposit has on the use of coal for energy generation? Support your response with the information you collected in the investigation.

2. Discuss what happens to the carbon content of peat if it is allowed to decay in the presence of oxygen.

3. Use what you have learned about coal in this activity to provide two reasons why it would be a better idea to burn anthracite in a home fireplace than lignite.

4. What are the advantages to using coal to meet energy needs? What are the disadvantages?

Preparing for the Chapter Challenge

You have learned how coal deposits form, and you looked at the distribution of coal deposits in the United States and your community. You also considered how reliant your community is on coal as an energy resource. Write a paper in which you consider how the number of coal deposits in the United States, along with the time required to produce a coal deposit, shape the way that your community uses coal today and how it will use coal in the future, in light of projected population growth.

Inquiring Further

1. **Model the formation of coal**

   Line a plastic shoebox (or two-liter bottle with the top cut off) with plastic wrap. Pour water into the container to a depth of four inches. Spread about two inches of sand on the bottom. Drop small leaves, sticks, and pieces of fern on the sand. Let it stand for two weeks. Record what you observe as changes in color and decomposition occur. Gently sift fine sand or mud on top of the plant layer to a depth of two inches. Wait two weeks and drain any remaining water. Let it sit and dry for another two weeks. Remove the "formation" out of the container. Slice it open to see how you have simulated coal where the plants were, and fossil imprints from the plant leaves.

   Wash your hands after handling any material in this activity. Wear goggles. Complete the activity under adult supervision.

2. **Plants associated with coal**

   Research the ancient plants whose remains formed the United States coal deposits.

Understanding and Applying What You Have Learned

1. The answer to this question will vary slightly depending on where you live. In general, the closer a community is to a coal deposit, the more it will use the deposit. Students can use the data they collected in **Part B** of **Investigate** to support their response. To help them answer the question, you may want to encourage them to compare the coal-use data for their state with data from another state. For example, if you live in a state like Vermont that does not have coal deposits, have your students consider how your state's use of coal as an energy resource differs from coal use in a state like West Virginia.

2. In the presence of oxygen, bacterial decay of peat increases. The decay process consumes organic material from the plant material, thereby decreasing the carbon content of the peat.

3. Answers will vary. The lower content of volatiles in anthracite means that it will burn hotter and cleaner, with a lower concentration of unburned hydrocarbons and particulates, than lignite. A smaller mass of anthracite is required to produce heat. Also, the sulfur content of most anthracite coal is lower than in lower-grade coals.

4. Answers will vary. One possible advantage of using coal to meet energy needs is that coal is relatively abundant in the United States; therefore, it is cost-effective and will likely last for centuries. One possible disadvantage of using coal to meet energy needs might be the environmental concerns regarding:
   • disruption of the land by mining operations
   • release of the greenhouse gas, carbon dioxide, when the coal is burned
   • release of sulfur to the atmosphere when the coal is burned

Preparing for the Chapter Challenge

The short paper students are asked to write will prompt them to consider how their community uses coal as an energy resource today and how the community might use coal in the future. Encourage students to think about what would happen if the coal deposits they rely on were depleted: how would this change the reliance the community has on coal for power generation, and why.

Review the following assessment criteria (see **Assessment Rubric for Chapter Report on Energy Resources** on pages 16 and 17) with your students so that they can be certain to include the appropriate information in their short papers:
   • the current energy uses and consumption rates in your community
   • how energy needs will change if the population of your community increases by 20%
   • whether your community will be able to meet these needs and at what costs

Inquiring Further

1. Model the formation of coal

To make removal of the deposit easier, put a flat, rigid plate in the bottom of the shoebox before making the deposit. This will allow you to lift the entire deposit out of the box. Sectioning works best when the deposit is still slightly damp. Use a large, sharp knife to make a vertical cut. There will be some smearing from the cut. You can clean the cut by brushing the surface gently with a stiff brush. You can also etch the face of the deposit with a hair dryer.

2. Plants associated with coal

Much of the coal in the United States was formed about 300 million years ago during the Pennsylvanian time period. The climate of the United States during the Pennsylvanian was tropical and humid. Forests during this time consisted of seed-bearing woody trees, bushes, and herbaceous plants. The dominant plants reproduced by spores rather than by seeds. Go to the *EarthComm* web site to find out more about Pennsylvanian Period flora and fauna.

NOTES

ACTIVITY 4 — COAL AND YOUR COMMUNITY

Background Information

Coal Mining

Coal has been mined for use by humankind for many centuries. High-rank coal tends to be located deep in the earth, necessitating techniques of deep mining. In a typical deep coal mine, a vertical shaft is excavated down to one or more thick beds of coal, called coal seams. Coal seams range in thickness from a few centimeters to many meters; most actively mined coal seams are from one meter to a few meters thick. A network of tunnels extending from the shaft is excavated in the coal seam by partially automated machinery that chews away the face of the working seam and transports the coal to the shaft, where it is raised to the surface. As much of the seam is mined out as possible while leaving supporting masses. Even now, deep mining is a dangerous operation, subject to collapses, explosions, and toxic gases. Levels of mine safety vary from country to country.

Much of the lignite and bituminous coal is located close enough to the surface to be mined by surface mining techniques. In surface mining, a trench is excavated to the shallow location of the coal seam. The overlying noncoal material is set aside, the coal is removed, and the noncoal material is replaced. The position of the trench gradually shifts, so that the entire sheet of coal is removed. Various practices of revegetation, however, can restore the area to a semblance of its original condition.

More Information – on the Web
Visit the *EarthComm* web site www.agiweb.org/earthcomm to access a variety of links to web sites that will help you deepen your understanding of content and prepare you to teach this activity. Many of the sites also contain images that you can download.

Goals and Assessment

Clarify that the goals indicate what students should understand and be able to do as a result of the activity. Make sure students understand that Chapter Assessments are based upon these goals.

Goal	Location in Activity	Assessment Opportunity
Investigate the production and consumption of coal in the United States.	**Investigate** Part A **Digging Deeper; Check Your Understanding** Question 2	Graphs are completed correctly; axes are properly labeled. Extrapolations are reasonable.
Investigate how coal sources are explored.	**Investigate** Part B **Digging Deeper**	Completed cross section is reasonable; answers to questions accurately reflect cross section.
Understand methods of coal mining.	**Digging Deeper; Check Your Understanding** Question 1	Answer to question closely matches that given in Teacher's Edition.
Determine how coal is used in your state.	**Understanding and Applying What You Have Learned** Questions 1 – 3	Answers to questions closely match those given in Teacher's Edition.
Evaluate possible practices to conserve coal resources.	**Investigate** Part C **Understanding and Applying What You Have Learned** Question 2	Calculations are done correctly, work is shown. Conservation suggestions are reasonable.

Activity 4   Coal and Your Community

Goals

In this activity you will:

- Investigate the production and consumption of coal in the United States.

- Investigate how coal sources are explored.

- Understand methods of coal mining.

- Determine how coal is used in your state.

- Evaluate possible practices to conserve coal resources.

Think about It

Coal production in the United States decreased in 1999 and 2000, but coal consumption continues to increase.

- Why do you think that the production of coal has decreased?
- Why do you think that the consumption of coal has increased?
- Given that coal is a finite, nonrenewable source of energy, what are some ways to extend the supply of coal?

What do you think? Record your ideas in your *EarthComm* notebook. Be prepared to discuss your responses with your small group and the class.

Activity Overview

Students begin the investigation by graphing data that gives recent trends in coal production in the United States They use their graph to extrapolate the trends in the data to estimate coal production in the year 2050. In the second part of the investigation, students use well-log data to complete a cross section. They answer a series of questions about their completed cross sections to help them understand how geologists explore for coal. Students calculate the financial and environmental consequences associated with burning enough coal to light a 100-W bulb for 10 hours. Finally, they investigate means for conserving coal resources. **Digging Deeper** reviews methods of coal mining and exploration.

Preparation and Materials Needed

Part A

No advance preparation is required to complete this part of the investigation. However, it is recommended that you graph and extrapolate the data, to get a sense of problems your students might encounter.

Part B

You will need to make copies of the cross section on page R39 of the student text, using **Blackline Master Energy Resources 4.1, Cross Section of Core Holes**.

Part C

No advance preparation is required to complete this part of the investigation. If time and computer availability are not issues, you can have your students visit the *EarthComm* web site to research ways to conserve energy. However, use of the Internet is not required to answer the questions.

Materials

Part A
  • Graph paper

Part B
  • Copies of cross section on R39 (see **Blackline Master Energy Resources 4.1, Cross Section of Core Holes**)

Part C
  • Calculator
  • Internet access (optional)*

*The *EarthComm* web site provides suggestions for obtaining the needed information.

Think about It

Student Conceptions

From **Activity 3**, students should have a clear sense of how coal is distributed in the United States. Possible explanations for the fact that production of coal has decreased might include that coal reserves are depleted to the point where production is limited (not actually the case), that the easily accessible deposits have been depleted (not actually the case), or that the demand for coal has diminished. Many possibilities exist that could decrease the demand for coal from the United States. Examples might be the competition of coal producers outside of the United States or a preference for other fuel sources for technological, economic, or environmental reasons. Students are likely to say that consumption of coal has increased because of increased demands for energy with decreased availability of resources like oil and gas. If students respond to the third question by saying that the best way to extend the supply of coal is to use less of it, encourage them to elaborate and come up with suggestions on how this could be done.

Answer for the Teacher Only

The decrease in coal production in 1999 and 2000 was attributable to a decrease in total coal stocks, a lack of excess production capacity at some mines, and a reluctance on the part of some producers to expand production to meet increasing demands in the latter part of the year. The increase in coal consumption was due to increased demands for electricity.

Assessment Tool

Think about It Evaluation Sheet
Use this evaluation sheet to help students understand and internalize the basic expectations for the warm-up activity.

Blackline Master Energy Resources 4.1
Cross Section of Core Holes

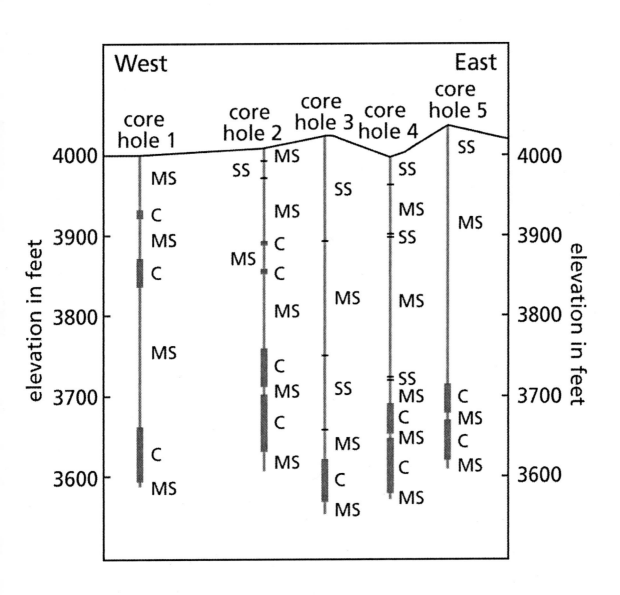

Earth's Natural Resources Energy Resources

Investigate

Part A: Trends in Coal Production and Consumption

Table 1 United States Coal Production, by Region 1991–2000 (in million short tons)				
Year	Appalachian	Interior	Western	Total
1991	457.8	195.4	342.8	996.0
1992	456.6	195.7	345.3	997.5
1993	409.7	167.2	368.5	945.4
1994	445.4	179.9	408.3	1033.5
1995	434.9	168.5	429.6	1033.0
1996	451.9	172.8	439.1	1063.0
1997	467.8	170.9	451.3	1089.9
1998	460.4	168.4	488.8	1117.5
1999	425.6	162.5	512.3	1100.4
2000	420.9	144.7	509.9	1075.5

Source: United States Energy Information Administration, 1996, 1998, 2000.

1. *Table 1* gives recent trends in coal production in the United States during the last 10 years. Using these values, summarize the trends in coal production for the three major coal-producing regions, and the trend in total coal production.

   a) Appalachian

   b) Interior

   c) Western

   d) Total

2. Make one graph showing coal production in the three regions. Leave room at the end of the graph to project coal production for the next 50 years.

   a) Include your graph in your *EarthComm* notebook.

3. Extrapolate the trends in the data to the year 2050. To do this, you will have to produce a best-fit line through the data points and estimate what you think to be the trend in coal production. There is a little guesswork involved in determining how much of the data you think represents the trend. Is it the last four years or the last eight years?

   a) On the basis of your projections only (you are ignoring all other factors that might influence future production), which coal-producing region do you predict will be the first to exhaust its supply of coal? In what year will this happen?

   b) In your group, identify at least three factors that might affect actual coal production. Record your ideas in your notebook.

Investigate

Part A: Trends in Coal Production and Consumption

> **Teaching Tip**
>
> Rather than having students graph the data by hand, have them use a graphing program like Microsoft Excel®. It will save time and produce more accurate graphs that will help increase the accuracy of their extrapolations. On the other hand, construction of a graph is a valuable lesson in itself.

1. a) Coal production in the Appalachian Region of the United States remained relatively constant in the period 1991 to 2000.

   b) Coal production in the Interior Region of the United States saw an overall decrease in the period 1991 to 2000.

   c) Coal production in the Western Region of the United States increased overall during the period 1991 to 2000.

   d) Since 1991, coal production in the United States has increased.

2. a) Note that this graph shows both the data from Table 1 and the extrapolations for **Step 3**.

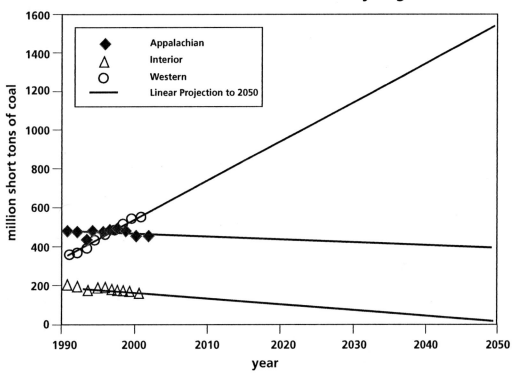

United States Coal Production by Region

y-axis: million short tons of coal

Legend:
◆ Appalachian
△ Interior
○ Western
— Linear Projection to 2050

x-axis: year

Teaching Tip

Extrapolation is the procedure by which a series of data is extended beyond the limits of the existing data. Extrapolation is easiest when the existing data fall closely on a linear trend; then the trend can be extended as a straight line. If the data fall on a curving trend, the choice of how to draw the curvature of the extrapolated curve is not as straightforward. In some cases, that's done just by eye; in other cases, the researcher tries to fit a mathematical curve of some kind to the existing data and then extrapolates that curve mathematically.

Extrapolation is less reliable when there is a lot of scatter in the existing data. Although it might seem obvious, it's also important to note that the reliability of extrapolation decreases with the distance of extrapolation. Extrapolations are never certain, because one can never assume that the factors that determine the data have remained the same outside the range of existing data.

3. a) On the basis of their graphs, students should notice that the rate of production in the Western Region of United States has increased over time, whereas production in the Appalachian and Interior regions has decreased. The assumption that must be made in order to predict which region will run out first is that production is slowing down in the Appalachian and Interior regions because the coal reserves are running out. On the basis of this assumption, students can use the point at which the extrapolation of their graphs reaches zero production to predict when the coal will run out. Students should find that the Interior region will be the first to exhaust its supply of coal. The year that this will happen will vary depending on the student's extrapolation, but it will probably be sometime near the year 2050.

    b) Answers will vary. Because the current coal reserves still exceed production capacity, the most likely factor affecting coal production is demand A wide range of factors can affect the demand for a resource. These include social, technological, economic and environmental issues.

NOTES

4. Examine the data in *Table 2*. Summarize the major trends in coal consumption for each sector from 1991 to 2000, and the total coal consumption.

a) Electric power

b) Coke plants (steel manufacturing)

c) Other industrial plants

d) Residential and commercial users

e) Total

f) What percentage of total coal consumption did electrical power make in 1991? In 2000?

Table 2 United States Coal Consumption by Sector, 1991–2000 (in million short tons[1])					
Consumption by Sector (million short tons)					
Year	Electric Power	Coke Plants	Other Industrial Plants	Residential and Commercial Users	Total
1991	777.2	33.9	75.4	6.1	892.5
1992	786	32.4	74.0	6.2	898.5
1993	820.8	31.3	74.9	6.2	933.9
1994	826.7	31.7	75.2	6.0	939.6
1995	849.8	33.0	73.1	5.8	961.7
1996	896.9	31.7	70.9	6.0	1005.6
1997	922.0	30.2	71.5	6.5	1030.1
1998	937.8	28.2	67.4	4.9	1038.3
1999	946.8	28.1	65.5	4.9	1045.3
2000	970.7	29.3	65.5	4.9	1070.5

Source: United States Energy Information Administration, 1996, 1998, 2000

[1]One short ton = 2000 pounds.

5. Using the data from *Table 2*, make a graph that shows coal consumption for electrical power.

a) Include the graph in your *EarthComm* notebook.

6. Extrapolate the trend in the data to the year 2050, as in Step 3.

a) On the basis of recent trends, how much coal will be needed for electrical power generation in the year 2020? 2050?

7. Coal consumption for electrical power increased at an average rate of 1.65% per year between 1996 and 2000.

a) On the basis of this average, predict coal consumption for electrical power for the years 2010 and 2020 (Hint: Begin by multiplying the value for the year 2000 by 1.0165. This gives you a prediction for the year 2001.)

R 37

4. a) Coal consumption for electric power generation increased from 1991 to 2000.

    b) Coal consumption by coke plants decreased overall from 1991 to 2000.

    c) Coal consumption by other industrial plants decreased from 1991 to 2000.

    d) Coal consumption by residential and commercial users decreased.

    e) Total coal consumption in the United States increased from 1991 to 2000.

    f) In 1991, electric power constituted 87% of the total coal consumption in the United States In 2000, electric power constituted 90.6% of total coal consumption in the United States.

5. a) Note that this graph shows both the data from Table 2 and the extrapolations for **Step 6**.

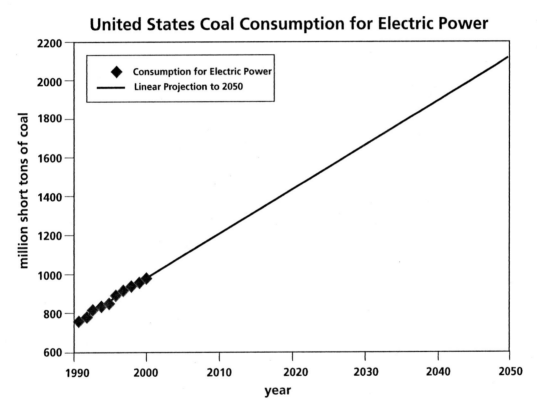

United States Coal Consumption for Electric Power

6. a) Answers will vary depending upon on how students interpret the trends and extrapolate the data.

7. If the increase remains at 1.65%, coal consumption for electric power at 2010 will be 1143.30 million short tons; at 2020, coal consumption for electric power will be 1346.58 million short tons.

Earth's Natural Resources Energy Resources

8. Assume that by conserving electricity the average rate of growth in consumption of coal is cut in half, to 0.825% per year.

a) Predict the amount of coal consumed in 2010 and 2020.

9. Draw a new graph that shows your prediction of coal production and coal consumption for the period 2000 to 2050. Do this by superimposing the curves you fitted in Steps 3 and 6. Also, in a third curve, take into account the prediction you made in Step 8, on the assumption of savings by conservation.

a) Include your graph in your *EarthComm* notebook.

b) Do you predict that consumption will exceed production? If so, how do you think that the shortfall in production will be made up? Or do you predict that production will exceed consumption? If so, do you think that that is a reasonable or likely scenario?

c) In what ways do you think that production and consumption are related? Does production drive consumption? If so, how and to what extent? Or does consumption drive production? If so, how, and to what extent?

Table 3 Logs for Core Holes (elevations are in feet above sea level; ms = mudstone, ss = sandstone).									
Location 1		**Location 2**		**Location 3**		**Location 4**		**Location 5**	
3930–4000	ms	3990–4000	ms	3890–4020	ss	3960–3990	ss	3980–4030	ss
3920–3930	coal	3970–3990	ss	3750–3890	ms	3900–3960	ms	3710–3980	ms
3870–3920	ms	3890–3970	ms	3660–3750	ss	3895–3900	ss	3675–3710	coal
3835–3870	coal	3885–3890	coal	3620–3660	ms	3720–3895	ms	3665–3675	ms
3660–3835	ms	3855–3885	ms	3570–3620	coal	3715–3720	ss	3610–3665	coal
3590–3660	coal	3850–3855	coal	3500–3570	ms	3690–3715	ms	3500–3610	ms
3500–3590	ms	3760–3850	ms		coal	3650–3690	coal		
	ms	3710–3760	coal		coal	3645–3650	ms		
		3700–3710	ms			3590–3645	coal		
		3630–3700	coal			3500–3590	ms		
		3600–3630	ms						

Part B: Coal Exploration

1. In 1998, the Western Region overtook the Appalachian Region as the largest coal-producing region in the United States with 488.8 million short tons produced, up by 8.3% over 1997. The low-sulfur Powder River Basin coal fields in Wyoming dominated growth in coal production in the Western Region. The diagram on the following page shows a cross section of five core holes drilled along an east–west line. *Table 3* provides drilling results for these wells. The results include the elevations and types of rock units in the core holes.

EarthComm

Chapter 1

8. If consumption decreases to 0.825%, coal consumption for electric power at 2010 will be 1053.82 million short tons; at 2020, coal consumption will be 1144.06 million short tons.

9. a) Circulate around the room to see if students need any assistance in creating their graphs. They should have a total of three curves on their completed graphs:

    Curve 1: coal production from 2000 to 2005
    Curve 2: coal consumption from 2000 to 2005
    Curve 3: coal consumption with conservation steps

    b) Answers to these questions will depend upon the earlier predictions made by students.

    c) Consumption and production are related. If coal is not being produced, it cannot be consumed. Consumption is driven by demand, and production is driven by consumption. If production is difficult, expensive, or even impossible, consumption will be low even if there is potential demand.

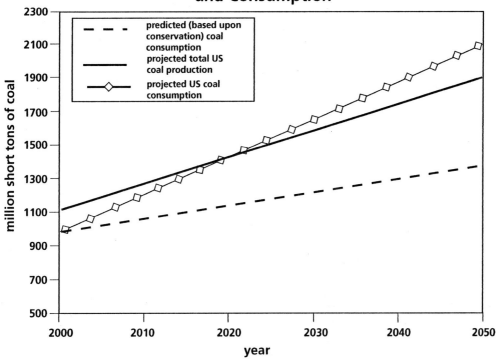

Comparision of Predicted Coal Production and Consumption

Part B: Coal Exploration

1. a)

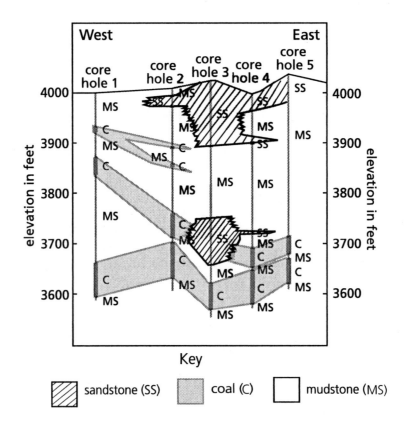

West   East

core hole 1   core hole 2   core hole 3   core hole 4   core hole 5

Key

▨ sandstone (SS)   ▨ coal (C)   ☐ mudstone (MS)

b) Answers will vary. Students should not come up with exactly the same solution.

c) Answers will vary. Students are likely to want to drill between holes 1 and 2, because this is the greatest gap. They should refer to their correlated cross section in justifying their response.

d) The cross section does not give information on the lateral extent of the coal beds. In other words, it shows bed thickness, but it does not show how much surface area the beds cover.

e) The lower coal seam has an average thickness of about 500 ft.

f) Students will need to know that 1 acre-foot = 43,560 ft³. If the coal seam covers 300,000 acres and is 500 ft thick, that means it has a volume of 6,534,000,000,000 ft³, or 150,000,000 acre-feet. Therefore, the seam would yield 265,500,000,000 short tons of coal.

Part C: Conserving Coal Resources

1. a) There are 8760 hours in a 365 day year, so it would take 876 kWh to run a 100-W light bulb continuously for one year.

b) It would cost $52.56 (876 kWh x $0.06) to run a 100-W light bulb continuously for one year.

c) It would take 876 pounds of coal to run a 100-W light bulb continuously for one year.

d) It would take 17,520,000,000 pounds of coal (876 lb x 20,000,000 households) to keep one 100-W light bulb running continuously for one year in 20 million households.

e) Acid rain is a primary environmental effect associated with burning large volumes of coal.

f) $50/month worth of electricity is approximately equal to 833 kWh ($50 month x 1 kWh/$0.06). This would require 10,000 pounds, or 5 short tons (833 pounds x 12 months) of coal.

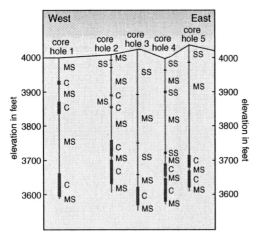

a) Complete the cross section in the diagram. Match up the rock units as you imagine them to be connected in the subsurface. (Hint: The mudstone at the bottom of all five wells is the same unit.)

b) Compare your results with those from other members of your group. How do they compare? How do you explain the differences?

c) Where would you drill your next well in order to determine whether or not your interpretation of the cross section is correct? Explain.

d) Is the information in the cross section sufficient to determine the coal seam with the greatest volume of coal? Why or why not?

e) What is the average thickness of coal in the lower coal seam?

f) Assume that the lower coal seam covers an area of 300,000 acres and that each acre-foot yields 1770 short tons of coal. How many short tons of coal does the lower coal seam contain?

Part C: Conserving Coal Resources

1. A 100-W bulb burning for 10 h uses 1 kWh (kilowatt-hour) of electricity (the same as ten 100-W bulbs burning for one hour).

a) Calculate the kilowatt-hours of electricity used in one year for a 100-W bulb running continuously.

b) Assume that electricity costs $0.06/kWh. (Your teacher may give you a more accurate figure for your community.) Calculate the yearly cost of running the 100-W bulb continuously.

c) Assume that one pound of good-quality coal can produce about 1 kWh of electricity. Calculate the amount of coal required per year to keep the 100-W bulb running continuously.

d) How much coal is required each year to keep a 100-W bulb burning continuously in 20 million households?

e) What do you think are the environmental consequences of burning this much coal?

f) The average electricity bill for a family of four in the United States is about $50 per month (this is for homes where cooking and heating are by natural gas or oil). Estimate the yearly amount of electricity (in kilowatt-hours) that this is equivalent to. How many tons of coal are needed?

2. A variety of researchers are currently seeking methods to reduce our consumption of electricity, much of which is produced from coal. In your group decide on five ways to make

NOTES

homes (or offices) in your community more energy efficient. Go to the *EarthComm* web site for useful links that will help you to explore ways to conserve energy resources. Suggestions are provided below.

- Water heaters — How do you ensure the best energy efficiency of an electric home water heater and reduce energy costs?

- Major home appliances — What are the most efficient ways to use major home appliances and how can you improve their efficiency?

- Home tightening — How can you slow the escape of heat energy from buildings, saving money and making them more comfortable?

- Insulation — How can insulation be used to reduce energy consumption?

- Home cooling — What strategies can you suggest for

keeping a building cool in summer and improving the efficiency of air conditioning units?

a) Record the methods that you decide on in your notebook.

3. For each method you decided on, describe the method of improving energy efficiency, the science behind it, and make sample calculations of cost savings and natural resource savings over 1 year and 10 years (based upon the cost of electricity in your community). Divide up the work in your group.

a) How much electricity would be saved?

b) How much coal would be conserved over the course of a year? Over 10 years?

4. Conservation (reduction in use) is another way to reduce energy usage.

a) In your group, decide on 10 ways to conserve energy.

b) Calculate the savings.

Source: Energy Information Administration, Electric Power Monthly March 2001.

petroleum and other 5%
hydro 7%
coal 52%
gas 16%
nuclear 20%

Figure 1 Percentage share of United States electric power industry net generation.

Digging Deeper

COAL EXPLORATION AND MINING

In 2000, over 52% of electricity in the United States was produced from coal (see *Figure 1*). There was a fall in coal production around the middle of the last century. This was in part due to the decline of the steel industry, which uses coal in steel production. In addition, oil and gas have largely replaced coal for transportation and home heating. Since the early 1960s, however, there has been a steady increase in the production of coal. This is mainly because of the increasing demand for electricity. There has also been a development of major coalfields in the western United States and Canada.

 R 40

2. **a)** Answers will vary. Students can visit the *EarthComm* web site to help with their research. Sample responses are provided below:

   - Water heaters: To ensure the best energy efficiency of an electric home water heater and reduce energy costs, one can use less hot water, turn down the thermostat on the water heater, or get a new, more efficient water heater.

   - Major home appliances: To improve the efficiency of home appliances, purchase those with the ENERGY STAR® label, which have been identified by the EPA and DOE as being the most energy-efficient products in their classes.

   - Insulation: Adding insulation to a home can reduce heating and cooling needs by up to 30%. The most cost-effective way to insulate a home is to add insulation to the attic. Insulation usually comes in four types: batts, rolls, loose fill, and rigid foam boards.

   - Home cooling: To keep a building cool and improve the efficiency of an air conditioning unit, the air conditioning unit should be the proper size for the room that it is cooling. Passive solar cooling techniques, like carefully designed overhangs on buildings, windows with reflective coatings, and the use of reflective coating on exterior walls and the roof, can keep a building cool.

3. **a)** Answers will vary depending upon the methods explored by students.

   b) Answers will vary depending upon the methods explored by students.

4. **a)** Answers will vary. Possible responses include:

   - turning off appliances, televisions, and radios that are not being used, watched, or listened to

   - turning off lights when no one is in the room

   - insulating walls and attics to reduce the amount of energy required to heat or cool a home

   - recycling materials to reduce energy needed to make newspapers, aluminum cans, plastic bottles, and other goods

   b) Answers will vary.

Assessment Tool

EarthComm Notebook-Entry Checklist
Use this checklist as a quick guide for student self-assessment and/or an
opportunity to quickly score student work. Add further criteria specific to your
classroom needs or to this particular investigation.

Investigate Notebook-Entry Evaluation Sheet
Point out the criteria listed on this evaluation sheet that are relevant to this
particular investigation. Encourage students to internalize the criteria by making
them part of your assessment conversations as you circulate around the classroom.

Reflecting on the Activity and the Challenge

This section is absent in Student Edition, but it would be a good idea at this point
to hold a general review about the main concepts explored in **Activity 4**. In **Part A**
(Trends in Coal Production and Consumption), students noted that coal production
rose during the 10-year period of data reported, and that coal production has been
declining in the Appalachian and Interior regions of the United States while increasing
in the Western Region. They extrapolated the trends into the future, which might lead
students to predict that Appalachian coal would cease and Western coal production
would continue to rise. **Question 3(b)** was designed to get students to think about
what factors might affect how much coal we actually produce (discovery of new
deposits, amount of coal available in the ground, etc.). Students also explored
Table 2 on page R37, which provided data on how we use coal. Most of our coal
is used to generate electricity, but some is used to make steel (coke plants) or for other
industries. The United States is rich in coal, so we produce about as much as we use.
As students will learn in **Activity 7**, this is not the case with oil and natural gas.

Part B allowed students to try their hand at coal exploration, and they probably
found a variety of ways to correlate the underground coal seams on their cross
sections. This provides a good opportunity to think about the challenges of estimating
the amount of coal underground, as well as where the coal might be found! You
might discuss how the results from this investigation relate to **Question 3(b)** in **Part A**
(factors that affect coal production).

Digging Deeper

Assign the reading for homework, along with the questions in **Check Your
Understanding** if desired.

Teaching Tip

Use **Blackline Master Energy Resources 4.2, Electricity Generation** to make an overhead transparency of *Figure 1* on page R40. The values in this graph are very slightly different from the values that would be computed on the basis of the table on page R17 in **Activity 2** (World Net Electricity Generation by Type, 1998), probably because the time period to which the data apply is different. Incorporate this overhead into a discussion about coal exploration and mining.

Blackline Master Energy Resources 4.2
Electricity Generation

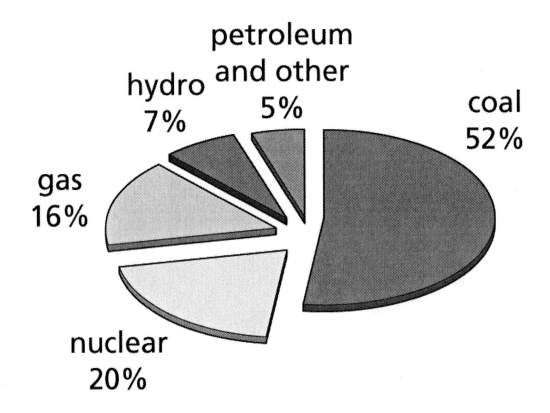

petroleum
and other
5%

hydro
7%

coal
52%

gas
16%

nuclear
20%

Blackline Master Energy Resources 4.3
Mining Methods

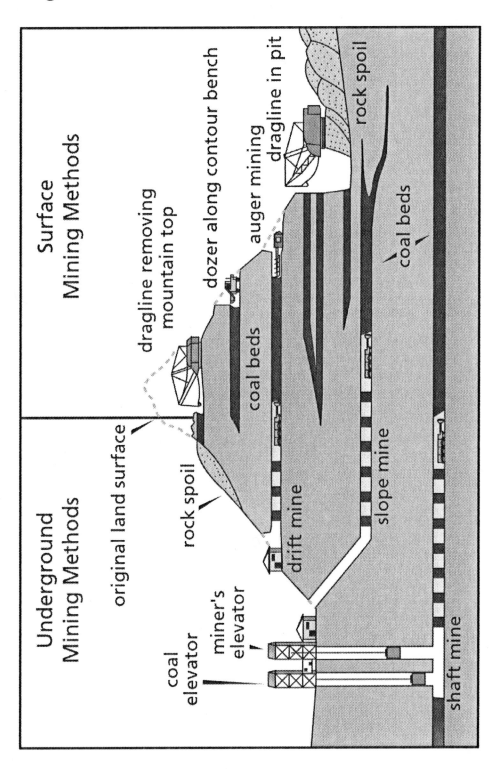

Of all fossil fuels, the largest reserves are contained in coal deposits (92%). Three major factors determine which coals are currently economical for mining. One factor is the cost of transportation to areas where the coal is utilized. Another factor is the environmental concern associated with the mining and use of coal. From a geologic perspective, the quality, thickness, volume, and depth of coal are important in determining whether or not a coal is mined.

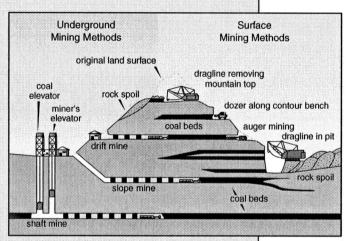

Figure 2 Different methods used to mine coal from the ground.

Figure 2 illustrates underground and surface mining methods (Note: You would not typically see all these mining methods used in one location.) Underground mining methods include drift, slope, and shaft mining.

Surface mining methods include area, contour, mountaintop removal, and auger mining. (An auger is a tool used for boring a hole.) *Figure 3* shows a surface coal mine.

Figure 3 A mine high wall showing layered sandstone and mudstone in a Gulf Coast coal mine. The haul truck provides a sense of scale.

Check Your Understanding

1. What are the major factors that determine whether a coal seam is economically and geologically suitable for mining?

2. What are the major factors that have determined the trends in coal consumption during the 20th century?

Teaching Tip

Ordinarily, underground mining and surface mining are done in different areas, not in the same area as illustrated in *Figure 2* at the top of page R41. The figure is not intended to show an actual mining situation, just to show the two methods in a single figure. You may wish to use **Blackline Master Energy Resources 4.3, Mining Methods** to make an overhead transparency of this diagram.

Check Your Understanding

1. The three most important factors that determine which coals are currently economically and geologically suitable for mining are:

    • the cost of transportation to areas where coal is used

    • the environmental concerns associated with the mining and use of coal

    • the quality, thickness, volume, and depth of coal

2. Coal consumption during the 20th century decreased around the middle of the century following a decline in the steel industry. Coal consumption has since increased because of increased demand for electricity.

Assessment Opportunity

Reword or restructure the questions in **Check Your Understanding** for a brief quiz. Use the quiz (or a class discussion of the questions in the textbook) to assess your students' understanding of the main ideas in the reading and the activity.

Assessment Tool

Check Your Understanding Notebook-Entry Evaluation Sheet
Use this sheet to evaluate the extent to which students understand the key concepts explored in **Activity 4** and explained in **Digging Deeper**, and to evaluate the students' clarity of expression.

Earth's Natural Resources Energy Resources

Understanding and Applying What You Have Learned

1. Think about historical differences in electricity use. Compare your community's electricity use 10 years ago and today.

2. Estimate the total energy use of your community. How much could your community reasonably conserve?

3. Compare electricity usage of a single college student vs. a family of four. Which group uses more electricity per person?

4. Think about regional differences in electricity use. Which part of the country do you think would be the lowest? Which part of the country do you think would be the highest? What are the reasons for your opinions?

Preparing for the Chapter Challenge

Assume that your community is expected to grow 10% in the next 10 years. If electricity usage remains the same, this means that the electrical capacity required to meet this increase must also increase. This may be undesirable and involve the cost of building a new power plant or an increased environmental hazard.

Determine specific steps that could be taken by the community to conserve electricity. How would the reduced demand for electricity offset the growth in the community? What incentives would be given to encourage the community to implement these proposed changes?

Inquiring Further

1. **Investigating a coal mine**

   Investigate the coal mine closest to your community.

   a) What mining method do they use?
   b) What is their annual production?
   c) How has the mine influenced the economy of the community?
   d) Has the mine impacted the environment?
   e) What steps has the mining company taken to reduce the environmental impact of the mine?

   Have your teacher arrange a field trip to this mine and discuss the trip with your class.

2. **Electricity usage in other countries**

   Find out information about electricity usage in other countries. What can you learn from other countries about energy conservation? (For instance, Japan has a high gross national product (GNP) yet very low energy use. How do they manage this?)

3. **Personal electricity usage**

   Learn how to read your electricity meter, and with your family conduct some conservation experiments for a few days (or weeks) and report the results of these experiments to the class.

R 42

EarthComm

Understanding and Applying What You Have Learned

1. Students will need to refer to data they collected in **Activity 2**. Answers will vary.

2. Students should have a sense of how much energy their community uses from previous activities. They can use the conservation estimates they made in this investigation to estimate how much energy their community could conserve. Make sure that students explain how and why this energy conservation could be achieved.

3. Students can take two approaches to answering this question. They could ask college students (e.g., an older sibling) about their monthly electricity consumption and compare this to the use within their family. They should inquire about kilowatt-hours of consumption, not cost, and compare the per person consumption. If students do not have access to quantitative data, have them answer this question as a thought experiment. Have them provide a qualitative analysis of how electricity is used by people in different settings (e.g., living on your own versus living within a family or other group).

4. Answers will vary, depending partly on how students interpret the question. If they interpret it as use per unit land area, thinly populated regions use much less electricity than thickly populated regions. Most students will interpret the question to mean electricity use per capita. Northern areas need winter heating, but most heating is not by electricity. Southern areas need summer air conditioning, and almost all of that is by electricity. Students might also mention regional differences in affluence, with more affluent areas using more electricity than less affluent areas.

Preparing for the Chapter Challenge

Electricity conservation is an important consideration in the **Chapter Challenge,** especially if your community does not have adequate means to extend electricity use far beyond current levels. Reducing the per capita demand for electricity can help a community to meet the needs of a growing population, without investing in additional sources of electricity. Encourage students to be creative in proposing incentives for community members to follow their conservation plan. Students will have to tie the 10% increase over a 10-year period to the proposed 20% increase presented in the **Chapter Challenge.**

Relevant criteria for assessing this section (see **Assessment Rubric for Chapter Report on Energy Resources** on pages 16 and 17) include a discussion on:

- whether your community will be able to meet increased needs

- solutions or alternatives that exist that may help to avoid a potential energy crisis

- one example that illustrates how energy conservation is possible in your community

Inquiring Further

1. Investigating a coal mine
Answers will vary. If possible, arrange for students to visit the coal mine. If the coal mine is not within a distance that makes a field trip feasible, contact the mine for information about their operations. They should be able to provide you with information needed to answer the questions. The Internet can be used as a supplemental resource as well. Check to see if the coal mine in question has its own web site.

2. Electricity use in other countries
Students can visit the *EarthComm* web site to view statistics on international energy consumption. Japanese energy policy aims to:

- reduce energy intensity (use less energy per unit of output)
- use alternative energy resources to achieve a more balanced energy mix with lower dependence on oil and lower CO_2 emissions
- maintain stable, secure energy supplies

Energy prices in Japan are among the highest in the Organization for Economic Cooperation and Development (OECD), which could encourage decreased energy consumption. Additionally, the Japanese government has instituted a plan to reduce residential energy consumption. Aspects of this plan include:

- raising standards on insulation in construction and standards for home appliances
- using waste heat and promoting other underutilized forms of energy through low-interest financing
- decreasing demand for cooling by planting more trees in urban areas

3. Personal electricity use
Answers will vary. Make sure that students have their parents' permission to discuss family electricity use with the rest of the class.

NOTES

ACTIVITY 5— ENVIRONMENTAL IMPACTS AND ENERGY CONSUMPTION

Background Information

The Carbon Cycle

The circulation of the element carbon among various kinds of reservoirs in the Earth system is called the carbon cycle. Carbon exists at, above, and below the Earth's surface in various forms, combined into compounds with other chemical elements. The carbon resides in several distinctive kinds of places or environments, which for budgetary purposes can be viewed as reservoirs. Carbon moves among these reservoirs by various processes along various pathways. These movements, usually called fluxes, are all happening simultaneously.

There are seven major carbon reservoirs in the Earth system. In order of decreasing carbon stocks, these are:
- the solid earth below the surficial soil layer
- the deep ocean below the surface layer
- fossil fuels
- soil
- the atmosphere
- the surface ocean
- plants

The boundaries between these reservoirs are to some extent fuzzy and arbitrary—where does the shallow ocean end and the deep ocean begin? where does soil change downward into rock?—but the differences are real and significant. The quantity of carbon in rock is by far the largest of the reservoirs, something like 7.5×10^{16} tons;

land biota and the surface ocean are the smallest, both being approximately 5×10^{11} tons, or about a hundred thousand times smaller than the deep rock reservoir. The largest flux of carbon, about 10^{11} tons per year, is between the atmosphere and plants during photosynthesis. The flux into and out of the deep rock reservoir, in contrast, is far smaller—not much more than 10^8 tons per year, which is negligible relative to the magnitude of the other fluxes.

Carbon is added to the atmosphere by discharging of volcanic gases, burning of fossil fuels, decomposition of surface organic matter, and outgassing from the ocean surface; carbon is removed from the atmosphere by chemical weathering of rocks, photosynthesis by plants, and absorption into the water of the ocean surface. Exchange of carbon in the form of carbon dioxide between the atmosphere and the oceans involves large fluxes in both directions but only a small net addition of carbon to the oceans.

Acid Rain

Natural rainfall is slightly acidic, with a pH between 5 and 6, because it reacts with carbon dioxide dissolved in the rainwater to form a weak acid called carbonic acid, H_2CO_3, according to the chemical reaction $H_2O + CO_2 = H_2CO_3$. The acidity of rainfall has increased in recent human history, however, because of the great increase in CO_2 concentrations in the atmosphere as a result of burning of fossil fuels (which adds CO_2 to the atmosphere), as well as deforestation (which reduces the removal of CO_2 from the atmosphere). Acidity of rainfall has increased because of another effect as well: combustion of fossil fuels, together with industrial smelting of metal ores, releases sulfur and nitrogen oxide gases into the atmosphere. These form sulfate and nitrate compounds, which combine with water in the atmosphere to form sulfuric acid and nitric acid. Rainfall in areas affected by these anthropogenic acids is commonly in the range of 4 to 5, and locally can be extremely acidic (below 2).

Acid rain has diverse effects on surface ecosystems. Acid rain falling on carbonate bedrock is neutralized as it dissolves some of the carbonate minerals, but acid rain falling on crystalline igneous and metamorphic rocks, like granites and gneisses, releases ordinarily insoluble aluminum, whereupon the aluminum is transported to plant roots. Uptake of aluminum by plants interferes with nutrient absorption, and plants become sickly or even die. The effect is especially severe for coniferous trees. Also, toxic metals released from rock minerals into solution by acid rain are transported by groundwater flow and surface runoff into lakes. The concentrations of toxic metals, together with the increased acidity of the water, are harmful to fish. Many lakes in northeastern United States and eastern Canada, especially those directly downwind of smelting plants and fossil-fuel-burning power plants, have few fish, or even none at all.

More Information – on the Web
Visit the *EarthComm* web site www.agiweb.org/earthcomm to access a variety of links to web sites that will help you deepen your understanding of content and prepare you to teach this activity. Many of the sites also contain images that you can download.

Goals and Assessment

Clarify that the goals indicate what students should understand and be able to do as a result of the activity. Make sure students understand that Chapter Assessments are based upon these goals.

Goal	Location in Activity	Assessment Opportunity
Examine one of the environmental impacts of using coal.	**Investigate; Digging Deeper; Check Your Understanding** Questions 4 – 5 **Understanding and Applying What You Have Learned** Questions 2 and 5	Answers to questions are reasonable, and closely match those given in Teacher's Edition.
Understand how the use of fossil fuels relates to one of Earth's major geochemical cycles—the carbon cycle.	**Digging Deeper; Check Your Understanding** Question 1	Answers to questions are reasonable, and closely match those given in Teacher's Edition.
Understand the meaning of pH.	**Investigate; Digging Deeper; Check Your Understanding** Question 3	Answers to questions are reasonable, and closely match those given in Teacher's Edition.
Understand how weather systems and the nature of bedrock geology and soil affect the impact of acid rain.	**Investigate; Digging Deeper**	Observations are correct, and answers to questions are reasonable.
Determine ways that energy resources affect your community.	**Investigate; Understanding and Applying What You Have Learned** Question 5	
Analyze the positive and negative effects of energy use on communities.	**Investigate; Digging Deeper; Check Your Understanding** Question 4 **Understanding and Applying What You Have Learned** Questions 1 – 2	Answers to questions are reasonable, and closely match those given in Teacher's Edition.

NOTES

Activity 5

Environmental Impacts and Energy Consumption

Goals

In this activity you will:

- Examine one of the environmental impacts of using coal.

- Understand how the use of fossil fuels relates to one of Earth's major geochemical cycles—the carbon cycle.

- Understand the meaning of pH.

- Understand how weather systems and the nature of bedrock geology and soil affect the impact of acid rain.

- Determine ways that energy resources affect your community.

- Analyze the positive and negative effects of energy use on communities.

Think about It

Sulfur dioxide and nitrogen oxides are the primary causes of acid rain. In the United States, about two-thirds of all sulfur dioxide and one-quarter of all nitrogen oxides comes from electric power generation that relies on burning fossil fuels like coal.

- How does acid rain affect the environment?
- What can be done to reduce the amount of sulfur dioxide and nitrogen oxides released into the atmosphere?

What do you think? Record your ideas in your *EarthComm* notebook. Be prepared to discuss your responses with your small group and the class.

Activity Overview

Students examine a map that shows the acidity of rainfall across the United States, and they correlate pH levels of rainwater to the locations of coal-producing regions and coal-fired power plants. Students then watch a demonstration that illustrates the buffering effects of different types of rocks on acid rain. **Digging Deeper** takes an in-depth look at energy resources and their impact on the environment, with a focus on acid rain and coal.

Preparation and Materials Needed

You will need to acquire samples of crushed limestone and granite, as well as crushed rocks that are representative of your community. You will also need to obtain pH indicator solution. This solution can be purchased, or it can be prepared by cutting a red cabbage in half, covering the cabbage in water, bringing the water to a boil, and then allowing the cabbage to steep in the water for about 15 minutes. The water from the cabbage can then be strained and used as an indicator solution:

- very acidic solutions (pH of 2 or 3) will turn the indicator solution pink
- neutral solutions (pH of 7) will turn the indicator solution purple
- very basic solutions (pH of 12) will turn the indicator solution yellow

Other prepared indicator solutions may have a different color scheme. You should familiarize yourself with the color index of the solution you are using.

To prepare the acid rain solution, combine sulfuric acid with water to obtain a pH of about 4.5.

How to Demonstrate the Effects of Acid Rain

1. Cut off the bottoms of two clean 1-L soda bottles. Turn them upside down and support them so that they are stable and there is adequate space to collect the "acid rain" solution. Place crushed granite in one container and the same weight of crushed limestone in the other container. Place a 500-mL beaker beneath each of the inverted soda bottles.

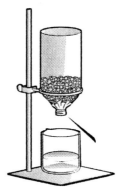
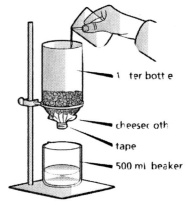

soda bottle
cheesecloth
tape
500 mL beaker

2. Place pH indicator solution in both collecting containers.

3. Slowly pour some of the acid rain solution into each container of crushed rock. You only need to use enough of the "acid rain" solution to make the collected solution clearly visible to the students. Using too much solution can surpass the buffering capacity of the rock, reducing the change in pH. Be sure to save enough "acid rain" solution to test all of the rock types. Have your students observe as the solution infiltrates the crushed rock and as the leachate flows into the collection container.

4. Using a pH meter or water test kit, determine the acidity of the liquid in each collection container.

5. Repeat the investigation using crushed samples of rock from your community.

Teaching Tip

Acids with a pH greater than 2 are not considered hazardous or corrosive by the Environmental Protection Agency. Nonetheless, you should consult your school official or local sewage treatment facility before disposing of the "acid rain" solution down the sink.

Materials
(Used for teacher demonstration; not needed for each group)
- Crushed limestone
- Crushed granite (or clean quartz sand to model sandstone)
- Two 1-L soda bottles
- Two 500-mL beakers
- Dilute sulfuric acid solution (mixed with distilled water to a pH of 4.3 to 4.5)
- pH indicator solution
- Water test kit, pH meter, or pH paper
- Samples of crushed rock from your community
- Ring stands to hold the 1-L soda bottle

Think about It

Student Conceptions

Depending upon where your community is located, your students may or may not be familiar with the effects of acid rain. The term acid rain has negative connotations, however, and students are likely to realize that acid rain is harmful to the environment. From reading the introductory statement, students should know the source of sulfuric acid and nitrogen oxides released into the atmosphere. Suggestions for reducing the release of compounds that will form sulfuric or nitric acid will likely include things like finding sources of fuel other than fossil fuels, and reducing the electric power generation.

Answer for the Teacher Only

Precipitation with a pH value less than 5 is considered acid rain. Acid rain is harmful or fatal to organisms like fish and trees, and has been linked to breathing and lung problems in children and people with asthma. Additionally, acid rain has been found to speed up the natural decay of stone monuments and historical buildings. The 1990 Clean Air Act Amendments mandated substantial reductions in pollutant discharge of compounds that will form sulfuric or nitric acid. To reduce emissions, the EPA issues allowances set below the current level of sulfur dioxide releases to power plants covered by the acid rain program. Plants may only release sulfur dioxide only up to the level of allowances they have been given.

Assessment Tool

Think about It Evaluation Sheet
Use this evaluation sheet to help students understand and internalize the basic expectations for the warm-up activity.

Blackline Master Energy Resources 5.1
Acidity of Rainfall across the United States

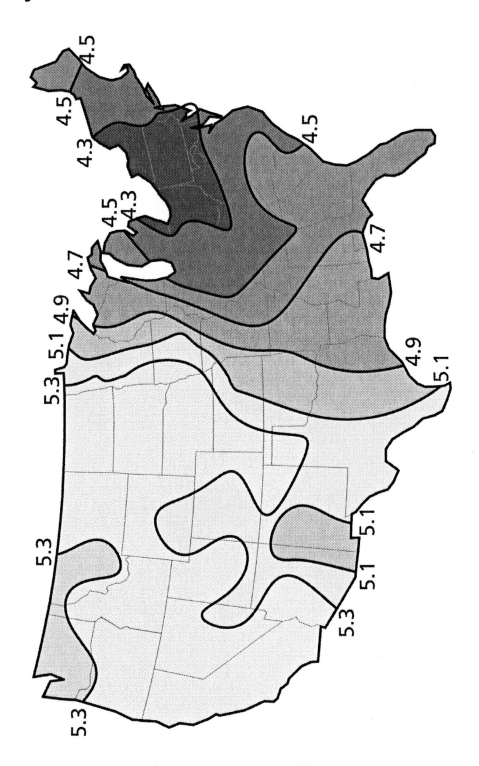

Earth's Natural Resources Energy Resources

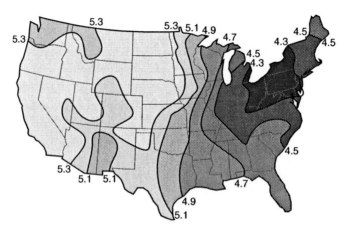

Investigate

1. The map shows the acidity of rainfall across the United States. Normal rainfall is slightly acidic (pH 5.6, where 7.0 is neutral). Carbon dioxide in the atmosphere reacts with water to form a weak acid called carbonic acid. The lower the pH, the more acidic is the rainwater.

   a) What part of the United States has the most acidic rainwater?

   b) How does the pattern of rainwater pH correlate with the locations of coal-producing regions and coal-fired power plants? (See Activity 3.)

   c) What parts of the United States have the least acidic rainwater? Why?

   d) What can you infer about the direction in which wind and weather move across the United States?

2. Your teacher will pour an "acid rain" solution (pH of about 4.5) through samples of crushed limestone and granite. Containers below the crushed rock contain pH indicator solution. A change in the color of the pH indicator reveals a change in the pH.

3. Note your answers to the following questions in your *EarthComm* notebook:

   a) Describe any color changes in the collection containers.

   b) What was the pH of the liquids after passing through the two types of crushed rock?

   c) Which type of rock—limestone or granite—neutralized more of the acid?

   d) Where would you expect a greater environmental impact from acid rain—in a region with granite bedrock or one with limestone bedrock? Explain.

R 44

Investigate

1. **a)** The areas around Ohio, Pennsylvania, and New York have the most acidic rainwater.

    b) Students should refer to the map on page R27. They will find that, in general, more acidic rainwater corresponds to a larger number of coal-producing regions and coal-fired power plants.

    c) The west coast and midwestern region of the United States have the least acidic rainwater. If students look again at the map on page R27, they should notice that there are not as many coal-fired power plants in these areas.

    d) Wind and weather generally move across the United States from west to east.

2. Complete the demonstration by following the instructions provided in the **Preparation** section at the start of **Activity 5**. Make sure that you take your students through the steps and explain to them what you are doing. Students should record their observations throughout the demonstration.

3. **a)** The color of the indicator solution will vary depending upon the solution that you use. For most pH indicators, the solution in the collection container under the crushed granite will remain red, and the solution under the limestone will be purplish.

    b) Exact pH values will vary. The leachate from the granite solution should be close to 4.5, but the pH of the leachate from the limestone should increase.

    c) The limestone has a greater ability to neutralize the acid than the granite.

    d) You would expect that an area with granite bedrock would have a greater environmental impact from acid rain, because granite has no neutralizing effect on acid rainwater. In areas of granite bedrock, therefore, acid precipitation would accumulate in lakes and ponds over time.

    e) This investigation suggests that areas differ greatly in their sensitivity to acid rain, depending upon the composition of the soil and bedrock in the area.

e) What does this investigation suggest about how areas may differ in their sensitivity to acid rainwater?

4. Your teacher will repeat the investigation using crushed samples of rock from your community.

   a) Record all your observations and explain your results.

Reflecting on the Activity and the Challenge

In this activity you examined one of the major environmental impacts of using coal to produce energy. You explored how the extent to which acid rain affects a region depends upon several factors, like the nature of the rock and soil, weather patterns, and the location of coal-fired power plants. This will help you to understand the benefits of understanding how the growing consumption of energy resources impacts Earth systems.

Digging Deeper

ENERGY RESOURCES AND THEIR IMPACT ON THE ENVIRONMENT

Fossil Fuels and the Carbon Cycle

When a fossil fuel is burned, its carbon combines with oxygen to yield carbon dioxide gas (CO_2). Fossil fuels supply 84% of the primary energy consumed in the United States. They are responsible for 98% of United States emissions of carbon dioxide. The amount of carbon dioxide produced depends on the carbon content of the fuel. For each unit of energy produced, natural gas emits about half, and petroleum fuels about three-quarters, of the carbon dioxide produced by coal. Important questions that scientists are trying to answer are what is happening to all of this carbon dioxide, and how does it affect the Earth system?

According to the **First Law of Thermodynamics**, the amount of energy always remains constant. The amount of chemical energy consumed when a fossil fuel is burned equals the amount of heat energy released. Scientists think of matter in a similar way. Matter cannot be created or destroyed. The **Law of Conservation of Matter** helps you to understand what happens when one type of matter is changed into another type. For example, fossil fuels consist mainly of hydrocarbons, which are made up of hydrogen ➡

Geo Words

First Law of Thermodynamics: the law that energy can be converted from one form to another but be neither created nor destroyed.

Law of Conservation of Matter: that law that in chemical reactions, the quantity of matter does not change.

EarthComm

4. Repeat the demonstration using rocks from your community. Be sure to supply your students with samples of the rocks and tell them where the rocks were collected.

　　a) Answers will vary. Encourage students to think about the implications of the experiment in regard to the potential impacts acid rain can have on their community.

Assessment Tools

EarthComm Notebook-Entry Checklist
Use this checklist as a quick guide for student self-assessment and/or an opportunity to quickly score student work. Add further criteria specific to your classroom needs or to this particular investigation.

Investigate Notebook-Entry Evaluation Sheet
Point out the criteria listed on this evaluation sheet that are relevant to this particular investigation. Encourage students to internalize the criteria by making them part of your assessment conversations as you circulate around the classroom.

Reflecting on the Activity and the Challenge

Environmental impacts are an important issue surrounding energy consumption. In a class discussion, review the information about the impacts of coal and acid rain on the environment in a class discussion. Discuss with students the implications of the exercise they have completed with respect to the **Chapter Challenge**. Again, ask students to consider why it is important for them to be able to explain the environmental impacts associated with energy consumption, and why it becomes an even greater concern with an increase in population.

Digging Deeper

Assign the reading for homework, along with the questions in **Check Your Understanding** if desired.

Assessment Opportunity

Use a brief quiz (or a class discussion) to assess your students' understanding of the main ideas in **Digging Deeper** and **Investigate**. A few sample questions are provided below:

Question: Use a sketch to explain how the carbon cycle is related to the consumption of fossil fuels.
Answer: Student sketches should be similar to *Figure 1* on page R46.

Question: How have atmospheric CO_2 concentrations changed in the last 150 years? How do you explain this trend?
Answer: Atmospheric CO_2 concentrations have increased over time because of an increase in the amount of CO_2 being produced by human activity, mainly through the burning of fossil fuels and deforestation.

Question: What part of the United States has the greatest problem with acid rain? Why?
Answer: The northeastern United States has the greatest problem with acid rain because winds blowing from west to east carry acids from the large number of coal-fired power plants in the Midwest to this region. Also, much of the bedrock in the northeastern part of the United States is granite and other crystalline igneous and metamorphic rocks, which have little acid-neutralizing effect.

NOTES

Earth's Natural Resources Energy Resources

and carbon. When a fossil fuel is burned (to heat homes or power cars), the carbon is changed chemically into carbon dioxide. The amount of carbon consumed equals the amount of carbon produced. The carbon never "goes away", but moves from one reservoir within the Earth system to another.

The idea that many processes work together in a global movement of carbon from one reservoir to another is known as the **carbon cycle** (see *Figure 1*). You have already learned about several important reservoirs of carbon. The atmospheric reservoir of carbon is carbon dioxide gas. Biomass contains carbon (mostly carbon in plants and soil). In the geosphere, carbon is found in solids (coal, limestone), liquids (petroleum hydrocarbons), and gas (methane). The ocean is the largest reservoir of carbon (exclusive of carbon-bearing bedrock), found in the form of bicarbonate salts.

Geo Words

carbon cycle: the cycle of carbon in living organisms, involving fixing of carbon dioxide by photosynthesis by plants, consumption by animals, and return to the inorganic state by respiration and by decay of plant and animal tissues after death.

anthropogenic: generated or produced by human activities.

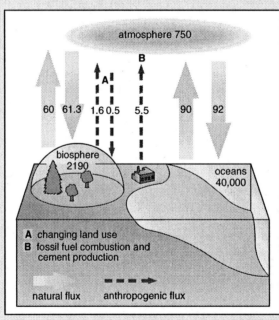

Figure I Global carbon cycle (billion metric tons).

Photosynthesis and respiration dominate the movement (flux) of carbon dioxide between the atmosphere, land, and oceans (the wide blue arrows show this flux in *Figure 1*). The smaller red arrows indicate the flux of carbon related to human (**anthropogenic**) activities. Natural processes like photosynthesis can remove some of the net 6.6 billion metric tons of anthropogenic carbon dioxide emissions produced each year. This means that an estimated 3.3 billion metric tons of this carbon is added to the atmosphere annually in the form of carbon dioxide.

Blackline Master Energy Resources 5.2
Global Carbon Cycle

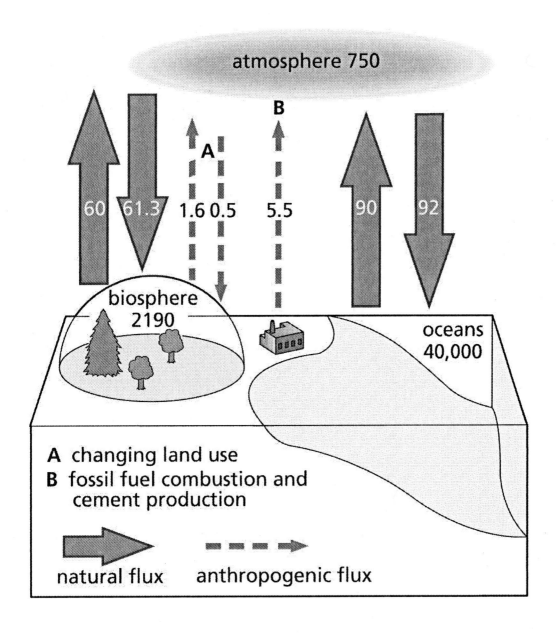

atmosphere 750

B

A

60 61.3 1.6 0.5 5.5 90 92

biosphere
2190

oceans
40,000

A changing land use
B fossil fuel combustion and
   cement production

natural flux   anthropogenic flux

By measuring concentrations of carbon dioxide in the atmosphere over time, scientists have learned that levels of carbon dioxide in the atmosphere have increased about 25% in the last 150 years (see *Figure 2*). Scientists believe that human activity has caused this growth. This coincides with the beginning of the industrial age. The increase is due largely to the burning of fossil fuels and deforestation (*Figure 1*). How does this affect the Earth system? Carbon dioxide is one of several **"greenhouse gases"** (gases that slow the escape of heat energy from the Earth to space). Some scientists believe that the rapid addition of carbon dioxide to the atmosphere is changing the energy budget of the Earth, causing global climate to warm.

Geo Words

greenhouse gases: gases in the Earth's atmosphere that absorb certain wavelengths of the long-wavelength radiation emitted to outer space by the Earth's surface.

acid: a compound or solution with a concentration of hydrogen ions greater than the neutral value (corresponding to a pH value of less than 7).

base: a compound or solution with a concentration of hydrogen ions less than the neutral value (corresponding to a pH value of greater than 7).

neutral: having a concentration of hydrogen ions that corresponds to a value of pH of 7.

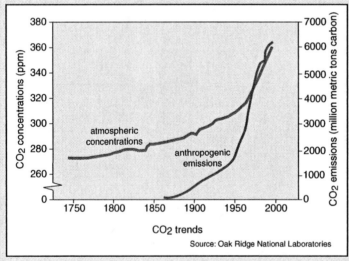

Figure 2 Atmospheric CO_2 concentrations and anthropogenic CO_2 emissions over time.

Coal and Acid Rain

Acidity is a measure of the concentration of hydrogen ions (H^+) in an aqueous (water) solution. A solution with a high concentration of hydrogen ions is acidic. A solution with a low concentration of hydrogen ions is basic. The concentration of hydrogen ions is important, because hydrogen ions are very reactive with other substances. Mixing **acids** and **bases** can cancel out their effects, much as mixing hot and cold water evens out the water temperature. That's because the concentration of hydrogen ions in the mixture lies between that of the two original solutions. A substance that is neither acidic nor basic is said to be **neutral**.

Teaching Tip

Use **Blackline Master Energy Resources 5.2, Global Carbon Cycle** to make an overhead transparency of *Figure 1* on page R46. Incorporate the overhead into a discussion or lecture on the global carbon cycle and how it is tied to energy resources.

Teaching Tip

The graph on page R47 shows a strong correlation between atmospheric CO_2 concentration and anthropogenic CO_2 emissions. A correlation never proves cause and effect, but in this case the anthropogenic emissions are almost certainly the cause of the increase in atmospheric CO_2 concentrations.

Teaching Tip

Use **Blackline Master Energy Resources 5.3, The pH Scale** to make an overhead transparency of the pH scale in *Figure 3* on page R48. Ask your students to consider how the pH of each of the solutions analyzed in the investigation compares to that of the common household objects. Putting these values into a common context will help your students to better understand the implications behind the investigation.

Earth's Natural Resources Energy Resources

The pH scale shown in *Figure 3* measures how acidic or basic a substance is. It ranges from 0 to 14. A pH of 7 is neutral, a pH less than 7 is acidic, and a pH greater than 7 is basic. Each whole pH value below 7 is 10 times more acidic than the next higher value, meaning that the concentration of hydrogen ions is 10 times as great. For example, a pH of 4 is 10 times more acidic than a pH of 5 and 100 times (10 times 10) more acidic than a pH of 6. The same holds true for pH values above 7, each of which is 10 times more alkaline (another way to say basic) than the next lower whole value. For example, a pH of 10 is 10 times more alkaline than a pH of 9.

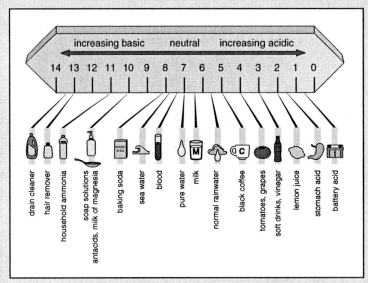

Figure 3 The pH scale.

Pure water is neutral, with a pH of 7.0. Natural rainwater, however, is mildly acidic, with a pH of about 5.7. The reason is that carbon dioxide in the atmosphere dissolves in rainwater, and some of the dissolved carbon dioxide reacts with the water to form a weak acid, called carbonic acid, H_2CO_3. As the carbon dioxide content of the atmosphere has gradually increased in recent decades, because of the burning of fossil fuels, natural rainwater has become slightly more acidic, even aside from the serious problem of acid rain caused by sulfur dioxide.

Blackline Master Energy Resources 5.3
The pH Scale

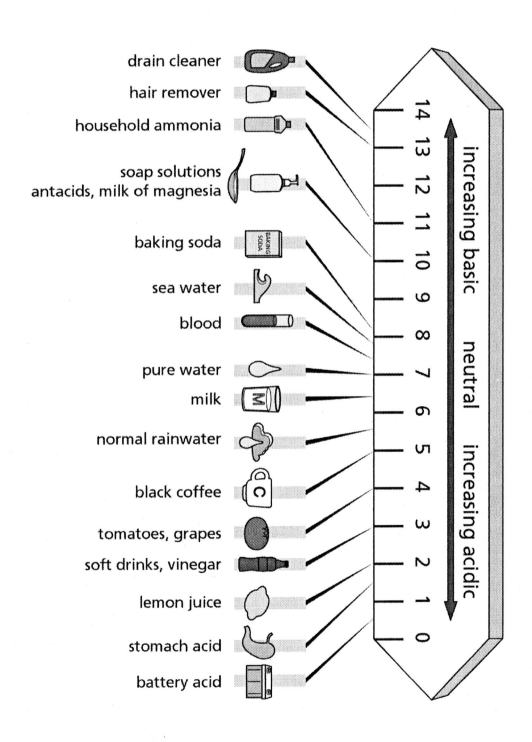

drain cleaner

hair remover

household ammonia

soap solutions
antacids, milk of magnesia

baking soda

sea water

blood

pure water

milk

normal rainwater

black coffee

tomatoes, grapes

soft drinks, vinegar

lemon juice

stomach acid

battery acid

14 13 12 11 10 9 8 7 6 5 4 3 2 1 0

increasing basic   neutral   increasing acidic

You learned in a preceding activity that coal contains as much as 5% sulfur. When the coal is burned, the sulfur is emitted as sulfur dioxide gas, SO_2. The sulfur dioxide then reacts with water in the atmosphere to form sulfuric acid, a strong acid. The last reaction in this process is written as follows:

$$SO_3 + H_2O \rightarrow H_2SO_4$$

Some of the sulfuric acid then dissociates into hydrogen ions and sulfate ions in solution in the water. This reaction can be broken down into two parts, as follows:

$$H_2SO_4 \rightarrow H^+ + HSO_3^- \rightarrow 2H^+ + SO_4^{2-}$$

The H^+ ions reach the Earth's surface dissolved in raindrops, either over land or over the ocean. Burning of fuels also produces nitrogen oxide gases, which react with water to form nitric acid, another strong acid. Some of the nitric acid breaks down to release hydrogen ions, by a reaction similar to that of sulfuric acid. Burning of coal is not the only source of sulfuric acid and nitric acid: petroleum-fueled power plants, smelters, mills, refineries, and motor vehicles produce these acids as well.

Acid rain is especially damaging to lakes. If the pH of lake water becomes too acidic from acid rain, it kills fish, insects, aquatic plants, and plankton. Environmental scientists describe lakes that have suffered heavily from acid rain as "dead," because the entire food web of the lake has been disrupted to the point where little is left alive in the lake.

Acid rain is a serious problem locally in many areas of the United States, but the biggest problem is in the Northeast. The reason is that there are a large number of coal-fueled electric power plants in the Midwest, especially in the heavily populated areas of Ohio, Michigan, and Illinois, and these plants burn mostly the relatively high-sulfur coal that is mined in the Interior coal province (see Activity 3). Winds blow mostly from west to east in the mid-latitudes of North America, and that tends to bring the acids to the Northeast, where large amounts fall with rain.

Figure 4 This marble gravestone from 1921 is being slowly dissolved by acid rain. ➔

R 49

Teaching Tip

Figure 4 on page R49 shows the effect of acid rain on a marble gravestone. Marble consists of the mineral calcite, $CaCO_3$. The calcite dissolves in acidic water according to the following chemical reaction:

$$2H^+ + CaCO_3 = CO_2 + H_2O$$

Acidic water, with a relatively high concentration of hydrogen ions, causes the calcite to dissolve, releasing carbon dioxide and water. The same thing happens with pure water, although much more slowly because of the much lower concentration of hydrogen ions.

Earth's Natural Resources Energy Resources

Geo Words

scrubbing: removal of sulfur dioxide, ash, and other harmful byproducts of the burning of fossil fuels (sulfur dioxide, ash, etc.) as the combustion products pass upward through a stack or flue.

Check Your Understanding

1. Describe the carbon cycle in relation to the consumption of fossil fuels.

2. Explain what is meant by the phrase "The Earth is a closed system with respect to matter." How does this statement relate to the Law of Conservation of Matter?

3. In your own words, explain the concept of pH.

4. Why is acid rain a greater problem in the Northeast than elsewhere in the United States?

5. How is sulfur dioxide removed from the gases emitted from coal-fueled electric power plants?

Another reason for the impact of acid rain in the northeastern United States has to do with the types of rock and soil in that region. In the activity, you learned that limestone has a greater ability to neutralize acid than granite. If the soil upon which acid rain falls is rich in limestone or if the underlying bedrock is either composed of limestone or marble, then the acid rain may be neutralized. This is because limestone and marble are more alkaline (basic) and produce a higher pH when dissolved in water. The higher pH of these materials dissolved in water offsets or buffers the acidity of the rainwater, producing a more neutral pH.

If the soil is not rich in limestone or if the bedrock is not composed of limestone or marble, then no neutralizing effect takes place, and the acid rainwater accumulates in the bodies of water in the area. This applies to much of the northeastern United States where the bedrock is typically composed of granite. Granite has no neutralizing effect on acid rainwater. Therefore, over time more and more acid precipitation accumulates in lakes and ponds.

Coal supplies about half of all electricity generated in the United States. In recent years, as the problem of acid rain has grown, various technologies have been developed to reduce the emission of sulfur dioxide from coal-fueled electric power plants. The general term for these processes is flue-gas desulfurization, also called **scrubbing**.

Several kinds of scrubbers are in use. The most common is the wet scrubber. In wet scrubbing, the flue gas from the power plant is sprayed with a calcium carbonate solution in the form of a slurry. The SO_2 is oxidized to form calcium sulfate. Scrubbers of this kind can remove up to 95% of the SO_2 that is emitted from the power plant. Scrubbers can have a useful byproduct. Certain kinds of scrubbers are now in operation that produce pure calcium sulfate (gypsum) as a byproduct. This gypsum can be used industrially, to make plaster and wallboard, rather than having to be mined and processed.

Advantages and Disadvantages of Energy Resources

Energy resources affect the community in positive and negative ways. Energy resources allow you to maintain a very high quality of life. Energy resources also indirectly provide employment for a large sector of the community. Use of energy resources, however, can have obvious and not-so-obvious negative impacts on the community. One of the most obvious negative impacts is a deterioration of air quality due to the burning of fossil fuels. Not-so-obvious impacts include acid rain, possible global warming, ground-water contamination, and the financial burden on households. It is important to realize that you, both personally and as a member of the community you can control how you use energy resources to potentially reduce the negative impacts.

Check Your Understanding

1. The global carbon cycle involves cycling of carbon that is readily available to the Earth surface system rather than locked away for long periods of time. The burning of fossil fuels adds new carbon to the present carbon cycle. This new carbon then participates in the carbon cycle, causing an increase in atmospheric carbon, oceanic carbon, and biosphere carbon as shown in *Figure 1* on page R46.

2. This phrase means that matter cannot be created or destroyed. The Law of Conservation of Matter states that in chemical reactions, the quantity of matter does not change.

3. pH is defined as the negative of the logarithm to base ten of the hydrogen-ion activity (activity being, essentially, the concentration of hydrogen ions) in a solution. It is a measure of the acidity or basicity of a solution.

4. The northeastern United States has the greatest problem with acid rain because winds blowing from west to east carry acids from the large number of coal-fired power plants in the Midwest to the Northeast, and because much of the bedrock in the northeastern part of the United States is granite or other crystalline igneous and metamorphic rocks, which have little acid-neutralizing effect.

5. Sulfur dioxide is removed from gases emitted from coal-fueled electric power plants through the process of scrubbing, or flue-gas desulfurization.

Assessment Tool

Check Your Understanding Notebook-Entry Evaluation Sheet
Use this sheet to evaluate the extent to which students understand the key concepts explored in **Activity 5** and explained in **Digging Deeper**, and to evaluate the students' clarity of expression.

Understanding and Applying You Have Learned

1. What are some of the positive impacts associated with the use of energy resources?

2. What are some of the negative impacts associated with the use of energy resources?

3. The unit used to measure the amount of SO_2 gas emitted by a coal-fired power plant is pounds of SO_2 per million Btu (lb. SO_2/MBtu).

   a) Assume a 1000-megawatt power plant emits 2.5 lb. SO_2/MBtu. Calculate the amount of SO_2 gas emitted in one year, assuming that the plant operates on average at 75% capacity.

   b) Assume that the coal used has an energy output of 10,000 Btu/pound and use a heat-to-electricity efficiency of 30%. How many tons of coal does the plant use each year?

4. At the present time the United States has about 470 billion tons of coal reserves and the average annual production is about 1.05 billion tons.

   a) Calculate how many years the reserves will last at the present production rate.

   b) What would cause the reserves to run out sooner than this number?

   c) What would cause the reserves to run out later than this number?

5. Assume that you live in a small community in the midwest. The electricity for the community is provided by a 1000-megawatt power plant, and the electricity is produced entirely by the burning of coal. The power plant is located in the community and provides employment to about 75 employees. The coal comes from an underground coal mine about 50 miles away. The coal mine produces about one million tons of coal a year and employs about 150 people. Many of the people that work in the mine also live in your community. The sulfur content of the coal is about 4.5% and the ash content is about 11%. The amendments to the Clean Air Act require that by the year 2010 the power plant must reduce SO_2 emissions by 60%. Also, it is possible that by the year 2010 a "carbon tax" will be levied on electricity produced from coal to address global warming issues. There are a number of ways that this can be done, including retrofitting the power plant to burn natural gas (natural gas emits half the CO_2 gas and almost no SO_2 gas), install a wet scrubber, conserve electricity, etc.

Answer the following questions:

   a) How will the implementation of environmental laws affect your community?

   b) What alternatives does your community have to address changes that could be made to minimize the impact?

   c) How will the community gain from the environmental laws?

R 51

Understanding and Applying What You Have Learned

1. Energy resources allow the maintenance of a high quality of life and indirectly provide employment for a large sector of the community. Students may think of additional positive aspects of energy use.

2. The use of energy resources can result in negative impacts like:
    - deterioration of air quality
    - acid rain
    - possible global warming
    - ground-water contamination
    - financial burden on households

3. **a)** A 1000-MW plant produces 8.760×10^9 kWh per year:

    $$10^6 \text{ kW} \times 8760 \text{ hr/yr} = 8.760 \times 10^9 \text{ kWh/yr}$$

    From the conversion table on page R7, it can be seen that 1 kWh = 3413 Btu. Consequently, a 1000-MW power plant operating at 75% capacity generates:

    $$8.760 \times 10^9 \text{ kWh/yr} \times 3413 \text{ Btu/kWh} \times 0.75 = 2.24 \times 10^{13} \text{ Btu/yr}$$

    Dividing by 10^6 yields 2.24×10^7 MBtu/yr.

    Multiplying by 2.5 lb. SO_2/MBtu yields 5.6×10^7 lb of SO_2 emitted per year.

    b) The plant generates 2.24×10^{13} Btu/yr, as calculated in **Question 3(a)**. If the plant has an output of 10^4 Btu/pound, that means that 2.24×10^9 pounds of coal would be needed if the heat-to-electricity efficiency were 100%:

    $$2.24 \times 10^{13} \text{ Btu/104 Btu/lb} = 2.24 \times 10^9 \text{ lb}$$

    However, the efficiency is only 30%, so the actual number of pounds of coal burned is:

    $$2.24 \times 10^9 \text{ lb/0.3} = 7.47 \times 10^9 \text{ lb}$$

    On the assumption that ton here means short tons (1 short ton = 2000 lb), the number of tons of coal would be:

    $$7.47 \times 10^9 \text{ lb./2000 lb, or } 3.74 \times 10^6 \text{ tons, or 3,740,000 tons}$$

4. **a)** At an average production rate of 1.05 billion tons per year, the United States will take 447.62 years to use all of its coal reserves:

    470 billion tons/1.05 billion tons per year = 447.62 years

    b) Increasing coal production rates would cause the coal reserves to run out sooner.

    c) Decreasing coal production rates would cause the coal reserves to last longer.

5. You may want to review all of the parameters set forth by this question with your students. One idea is to produce a bulleted list on the board.

    a) If environmental laws are implemented:
    - electricity is likely to become more expensive
    - some residents of the community who work in the coal mine will probably lose their jobs if the plant is converted to natural gas
    - air quality in the community is likely to improve
    - acid rain in the community is likely to decrease

    b) Answers will vary. Some possibilities are:
    - conserve electricity
    - apply for state aid
    - develop alternative industries to provide other jobs
    - develop tourism
    - lobby for repeal of the Clean Air Act

NOTES

Earth's Natural Resources Energy Resources

Preparing for the Chapter Challenge

Your challenge is to prepare a plan that will help your community to meet its growing energy needs. Part of your plan must address how the use of energy resources impacts the environment. Based upon what you have learned in this activity, write an essay about ways to reduce the impact of acid rain on communities.

Inquiring Further

1. **The Clean Air Act and environmental regulation**

   Look up information on the amendments to the Clean Air Act having to do with SO_2 emissions and report your findings to the class.

2. **Acid rain and other Earth systems**

   How has acid rain affected your community? How do the processes that lead to acid rain also work to lower the pH of ground water? The latter is a challenge that community officials must face in areas that are mined for coal and some kinds of minerals. Extend your investigation into acid rain by examining acid mine drainage. The *EarthComm* web site will help you to extend your investigations.

3. **Coal mine**

   Investigate the coal mine closest to your community. What are the percent sulfur, weight of SO_2 per million Btu, and percent ash for the coal? What reclamation steps do they take?

4. **Power plant**

   Investigate the power plant closest to your community. If possible, have your teacher arrange a field trip to this plant.

   a) Which fuels do they use?
   b) What are the sources of these fuels?
   c) What is their capacity, in megawatts?
   d) How many people do they employ?
   e) What do they do with the ash (i.e., the solid materials that are caught as they pass up through the flue of the plant)?

Preparing for the Chapter Challenge

When students are preparing their essays, encourage them to think about how the proposed population increase ties in with the effects of acid rain on the community. This will be particularly important if your community relies heavily on coal as an energy source. You may wish to adapt this section by having students consider the implications of a different type of energy consumption that is more relevant to your community, like nuclear power.

For assessment of this section, as contained in the sample rubric (see **Assessment Rubric for Chapter Report on Energy Resources** on pages 16 and 17), students would need to include a discussion of:

- one example that illustrates how the production and consumption of energy affects the environment

Inquiring Further

1. The Clean Air Act and environmental regulation

The 1990 Clean Air Act Amendments mandated substantial reductions in pollutant discharge of sulfur acid-forming compounds. To reduce emissions, the EPA issues allowances set below the current level of sulfur dioxide releases to power plants covered by the acid rain program. Plants may release sulfur dioxide only to the level of the allowances they have been given. Students can visit the *EarthComm* web site to find additional information about the Clean Air Act.

2. Acid rain and other Earth systems

Whether or not your community has been affected by acid rain will vary. Students can learn more about the effects of acid rain by visiting the *EarthComm* web site.

3. Coal mine

Answers will be dependent upon the coal mine closest to your community.

4. Power plant

Answers will be dependent upon the power plant closest to your community.

ACTIVITY 6 — PETROLEUM AND YOUR COMMUNITY

Background Information

Sedimentary Basins

All sediments are deposited upon a preexisting surface of some kind. For a thick succession of widespread sedimentary layers to accumulate, there must be a continuing supply of sediment to the area. One or both of two additional conditions must be met as well: the area must be adjacent to a sediment source at a much greater elevation, and/or the area must undergo continued subsidence to provide space for the sediment to accumulate. Subsidence involves slow lowering of the area relative to surrounding areas. Rates of long-term subsidence might be as great as a few millimeters to a few centimeters per year; although that might seem small, it can add up to thousands of meters of lowering over geologic time spans as short as a few hundred thousand years.

Characteristically, sedimentary basins tend not to have pronounced concavity as they are being filled. If the supply of sediment to the basin is sufficiently great then the basin fills as it forms and the upper surface is fairly level. So long as the sediment supply remains sufficient, the upper surface remains fairly level even as the forces responsible for forming the basin in the first place continue to gradually increases the concavity of its base. In such situations, the basin cannot be recognized as such just by its surface expression.

Most of the sedimentary basins that are important in petroleum exploration have undergone long-term subsidence. Sedimentary basins differ greatly in their origins, depending on the geologic processes leading to long-term subsidence. Many kinds of sedimentary basins are a consequence of the processes of plate tectonics, especially in collisional settings where a volcanic arc is colliding with a continent or where two continents are colliding.

More Information – on the Web

Visit the *EarthComm* web site www.agiweb.org/earthcomm to access a variety of links to web sites that will help you deepen your understanding of content and prepare you to teach this activity. Many of the sites also contain images that you can download.

Goals and Assessment

Clarify that the goals indicate what students should understand and be able to do as a result of the activity. Make sure students understand that Chapter Assessments are based upon these goals.

Goal	Location in Activity	Assessment Opportunity
Recognize the overwhelming dependence of today's society on petroleum.	**Investigate** Part A **Investigate** Part C **Digging Deeper; Check Your Understanding** Question 2	Answers to questions are reasonable, and are relevant to content in text and community data.
Graph changes in domestic oil production and foreign imports to predict future needs and trends.	**Investigate** Part A **Understanding and Applying What You Have Learned** Question 3	Graph is constructed properly, extrapolation of data is reasonable.
Understand why oil and gas production in the United States have changed over time.	**Investigate** Part A **Digging Deeper; Understanding and Applying What You Have Learned** Questions 1–2, and 4	Answers to questions are reasonable.
Understand the origin of petroleum and natural gas.	**Investigate** Part B **Digging Deeper; Check Your Understanding** Question 1	Answers to questions are reasonable.
Investigate the consumption of oil and natural gas in your community.	**Investigate** Part C	Answers to questions are reasonable, and are relevant to community data.

Chapter 1

Activity 6   Petroleum and Your Community

Goals

In this activity you will:

- Recognize the overwhelming dependence of today's society on petroleum.

- Graph changes in domestic oil production and foreign imports to predict future needs and trends.

- Understand why oil and gas production have changed in the United States over time.

- Describe the distribution of oil and gas fields across the United States.

- Understand the origin of petroleum and natural gas.

- Investigate the consumption of oil and natural gas in your community.

Think about It

Oil accounts for nearly all transportation fuel. It is also the raw material for numerous products that you use.

- What percentage of oil used every day in the United States is produced in the United States?
- Where are oil and natural gas found in the United States?

What do you think? Record your ideas in your *EarthComm* notebook. Be prepared to discuss your responses with your small group and the class.

R 53

Activity Overview

First, students plot data on petroleum production, imports, and consumption in the United States They use their graphs to extrapolate the data 20 years into the future. Students then examine a map that shows the distribution of oil and gas production sites, as well as oil and gas fields, to determine whether there is any relationship between the two. They consider their community's location relative to petroleum fields and refineries, and think about how this affects the price of petroleum products. Finally, students examine data to determine trends in petroleum consumption in their community. **Digging Deeper** explores the formation of petroleum and natural gas deposits, traces the history of oil production in the United States, and reviews current petroleum production in the United States.

Preparation and Materials Needed

Internet access is needed to complete **Part C** of the investigation. If computers with Internet access are not available for your students, you will need to download the necessary data before class and make copies to distribute to your students. Visit the *EarthComm* web site to access a link to the correct Energy Information Agency web page.

Materials

Part A
- Graph paper
- Calculator

Part B
No additional materials needed.

Part C
- Internet access (or copies of your state's petroleum and natural gas production, consumption, and distribution data from the Energy Information Agency)*

Think about It

Student Conceptions

Give students a few minutes to think about and respond to the questions. Hold a brief discussion. Students are likely to be aware of issues surrounding petroleum imports and production in the United States from the news media, because it is a popular political topic. Students are likely to say that most of the oil in the United States is found in Texas, and possibly Alaska.

*The *EarthComm* web site provides suggestions for obtaining these resources.

Answer for the Teacher Only

In 1999, the United States produced approximately 42% of the oil that it used; the remaining 58% was imported. The map on page R56 of the student text illustrates the location of oil and gas in the United States. Most of the oil and gas production in the United States takes place in the Gulf states of Texas and Louisiana.

Assessment Tool

Think about It Evaluation Sheet
Use this evaluation sheet to help students understand and internalize the basic expectations for the warm-up activity.

NOTES

Earth's Natural Resources Energy Resources

Investigate

Part A: Trends in Oil Production

Year	Total U.S. Wells (thousands)	Total U.S. Production (thousand barrels/day)	Total Foreign Imports (thousand barrels/day)	Total U.S. Total Consumption (thousand barrels/day)
		Table 1 United States Petroleum Production, Imports, and Total Consumption from 1954 to 1999		
1954	511	7030	1052	8082
1957	569	7980	1574	9559
1960	591	7960	1815	9775
1963	589	8640	2123	10,763
1966	583	9580	2573	12,153
1969	542	10,830	3166	13,996
1972	508	11,180	4741	15,921
1975	500	10,010	6056	16,066
1978	517	10,270	8363	18,633
1981	557	10,180	5996	16,176
1984	621	10,510	5437	15,947
1987	620	9940	6678	16,618
1990	602	8910	8018	16,928
1993	584	8580	8620	17,200
1996	574	8290	9478	17,768
1999	554	7760	10,551	18,311

Source: U.S. Energy Information Agency

1. *Table 1* shows statistics for petroleum every three years from 1954 to 1999. Copy the data table into your notebook. If you have Internet access, you can download the complete set of data (annual data since 1954) at the *EarthComm* web site as a spreadsheet file.

   a) Construct a graph of United States (domestic) petroleum production, foreign petroleum imports, and petroleum consumption. Leave room at the right side of your graph to extrapolate the data for 20 years.

2. What are the major trends in petroleum needs during the last 45 years? Use your graph and the data table to answer the following questions:

   a) Describe how domestic production has changed during the 45-year period. About when did it peak?

   b) Describe how petroleum imports have changed during that period.

   c) Describe how total petroleum consumption has changed over the period.

R 54

EarthComm

Investigate

Part A: Trends in Oil Production

1. a) Note that this graph shows both the data from Table 1 and the extrapolation of the data for **Step 3**.

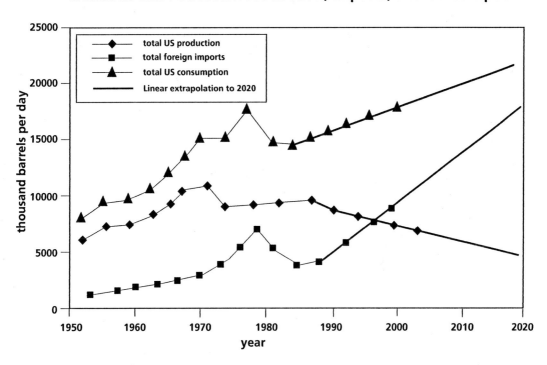

Trends in U.S. Petroleum Production, Imports, and Consumption

2. a) Domestic production increased from 1954 and peaked in 1972. After that time, petroleum production has seen an overall decrease.

b) Petroleum imports increased from 1954 to 1999.

c) Total petroleum consumption increased from 1954 to 1978, then decreased slightly, and increased from 1981 to 1999.

d) In what year did the United States begin to import more petroleum than it produced?

e) What percentage of the total petroleum consumption was met by domestic production in 1954? In 1999? How does this compare to your answer to the first **Think about It** question at the start of this activity?

f) In 1954, the estimated population of the United States was 163 million. In 1999, the estimated population was 273 million. How many barrels of petroleum per person per day were needed in the United States in 1954? In 1999?

g) A barrel of oil is equal to 42 gallons. Convert your answers in Step 2(f) to gallons per person per year needed in the United States in 1954 versus 1999.

h) Has the consumption of petroleum changed at the same rate as the growth in population? What might account for the change?

i) Use the data in the table to calculate the average well yield for oil production in the United States.

3. What do you predict about future needs for and sources of petroleum in the United States?

a) Extrapolate the three curves to the year 2020. Explain your reasoning behind your extrapolations for each of the three curves.

b) On the basis of your extrapolations, how much petroleum will be produced domestically, imported into the United States, and consumed in the year 2010? In 2020?

c) Identify several factors that might make the actual curves for production, import, and consumption different from what you have predicted. Record your ideas in your *EarthComm* notebook.

4. Share your results from Step 3.

a) Set up a data table to record the predictions from each person in the class for the years 2010 and 2020. An example is provided below.

b) Calculate the minimum, maximum, and average for the class.

c) How do you explain the differences in predictions?

Sample Data Table. Results of Class Predictions of United States Oil Production, Imports, and Total Production in 2010 and 2020.

	United States Production		Imports		Total	
	2010	**2020**	**2010**	**2020**	**2010**	**2020**
Student 1						
Student 2						
Student...						
Class Average						

R 55

EarthComm

Chapter 1

2. d) In 1993, the United States began to import more petroleum than it produced.

e) In 1954, nearly 87% of total petroleum consumption in the United States was met by domestic production. In 1999, about 42% of total petroleum consumption in the United States was met by domestic production.

f) In 1954:

(7030 thousand barrels/day + 1052 thousand barrels/day)/
(163 million persons) = 0.050 barrels per person per day.

In 1999:
(7760 thousand barrels per day + 10,551 thousand barrels per day)/
(273 million persons) = 0.067 barrels per person per day.

g) In 1954:
(0.050 barrels per person per day)(42 gallons per barrel) =
2.10 gallons per person per day.

In 1999:
(0.067 barrels per person per day)(42 gallons per barrel) =
2.81 gallons per person per day.

h) Petroleum consumption has increased faster than population. Many factors, most of which could be viewed as involving lifestyle, have contributed to this. People tend to drive more, fly more, use more electrical appliances, and use more air conditioning.

i) In 1954, 511 thousand wells produced 7030 thousand barrels, so the average production per well was:

(7030 thousand barrels)/(511 thousand wells) = 13.8 barrels per well

In 1999, the same figure was 14.0 barrels per well.

3. a) Answers will vary. It would be reasonable for students to conclude that future demand will continue to rise, if only because population will continue to increase. It is also reasonable to conclude that domestic production will continue to decline and foreign imports will continue to rise. Eventually, total foreign and domestic production will have to decrease, as world petroleum reserves are drawn down. Alternative petroleum sources, like tar sands and oil shales, will need to be developed. Alternatively, coal can be converted to petroleum hydrocarbons.

b) Answers will vary depending upon how students do their extrapolations.

c) Answers will vary. Some examples:
- state of the national economy
- decisions on production and pricing on the part of domestic and foreign petroleum producers
- rate of population growth
- government policy decisions
- extent of energy conservation
- development of new technologies, like hybrid or electric motor vehicles
- extent of global warming

4. a) Answers will vary.

b) Answers will vary.

c) Answers will vary.

NOTES

Earth's Natural Resources Energy Resources

Part B: United States Oil and Gas Resources

In Part A you learned that petroleum production has declined in the United States. The questions below will help you to investigate where oil and gas are found in the United States, and where these natural resources are refined into products that communities depend upon.

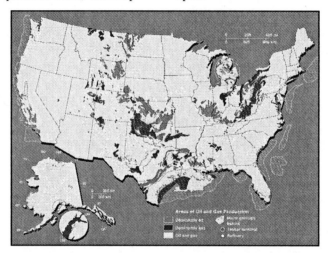

1. Use the map of oil and gas production to explore general trends and patterns in petroleum and gas fields.

   a) Are oil and gas deposits distributed evenly across the country?

   b) Which states have no oil or gas production?

   c) What patterns or trends do you see in oil-producing and gas-producing regions?

   d) Are oil and gas always found together? Give at least two examples.

   e) The map also shows the locations of major sedimentary basins. Sedimentary basins are depressions of the Earth's crust where thick sediments have accumulated. What is the relationship between oil and gas production and sedimentary basins?

2. Refer to the map to investigate where oil and gas are found, refined, and distributed in relation to your community.

   a) How far is your community from the nearest oil field?

   b) How far is your community from the nearest gas field?

   c) How far from your community is the nearest refinery located?

   d) Do you think the factors above might affect the price of petroleum products (gasoline, heating oil, or propane) in your community? Explain why you think this is so.

R 56

Part B: United States Oil and Gas Resources

1. a) No, oil and gas deposits are not evenly distributed. Ask students to describe the distribution of oil and gas fields.

    b) Several states have no oil or gas production, including Washington, Oregon, New Mexico, Idaho, Montana, Wisconsin, Georgia, South Carolina, North Carolina, Virginia, Rhode Island, New Jersey, Connecticut, New Hampshire, Vermont, and Maine.

    c) Students may notice that major areas of oil and gas production tend to occur close to refineries, in populated areas. Also, the production areas are located within major geologic basins.

    d) No, oil and gas are not always found together. Possible examples would include production sites in California, Alaska, and sections of the upper Midwest.

    e) Oil and gas production areas tend to occur within sedimentary basins.

2. Answers to these questions will depend on where your community is located. The farther away your community is located from an oil field, gas field, or refinery, the more expensive petroleum products will be, because of associated transportation costs.

Part C: Oil and Gas Resource Use in Your Community

1. Direct students to the *EarthComm* web site to obtain the needed data.

2. Answers to the questions will be dependent upon your community.

Reflecting on the Activity and the Challenge

In **Activity 6**, students looked at trends of petroleum production and consumption in the United States and in their community. Discuss with students how this ties into the **Chapter Challenge**. Ask them to consider the costs associated with having to transport petroleum products a great distance, and how this factors into the situation created by the continuing population increase.

Assessment Tools

EarthComm **Notebook-Entry Checklist**
Use this checklist as a quick guide for student self-assessment and/or an opportunity to quickly score student work. Add further criteria specific to your classroom needs or to this particular investigation.

Investigate Notebook-Entry Evaluation Sheet
Point out the criteria listed on this evaluation sheet that are relevant to this particular investigation. Encourage students to internalize the criteria by making them part of your assessment conversations as you circulate around the classroom.

Part C: Oil and Gas Resource Use in Your Community

1. Investigate trends in petroleum and natural gas production, consumption, and distribution in your state. Go to the *EarthComm* web site to obtain links to the data sets for your state available at the Energy Information Agency. This will allow you to examine several sources of data that will help you with the **Chapter Challenge**.

2. Develop your own questions as you examine the data and review the **Chapter Challenge**.

   a) How have oil and gas consumption in your state changed during the last 40 years? How can you use the data to predict future needs?

   b) How much petroleum is consumed per capita in your state? How much natural gas is consumed? How can you use this information to predict how much oil and gas your community will need 20 years from now if the population grows by 20%?

   c) How do oil and gas reach your community? If switching to another type of fuel is part of your plan for your community's energy future, how will these resources get to your community if they are not local?

Reflecting on the Activity and the Challenge

In this activity you graphed the recent history of oil consumption in the United States. You used these trends to predict what the future needs might be. You examined a map of oil and gas fields in the United States and learned that these resources are not distributed evenly throughout the country. In Part C, you investigated trends in oil and gas resource use in your community. Think about how your work in this activity will help you to solve the **Chapter Challenge** on the energy future of your community.

R 57

Blackline Master Energy Resources 6.1
Oil and Gas Production

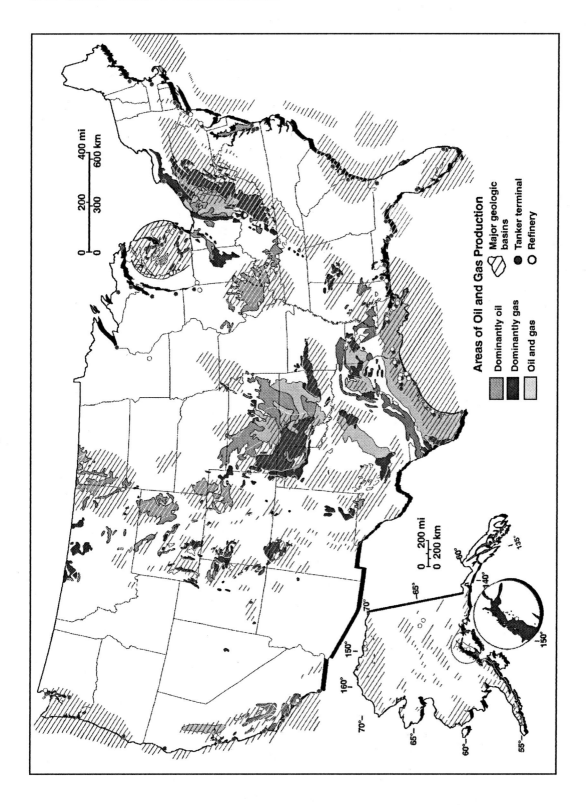

Earth's Natural Resources Energy Resources

Geo Words

petroleum: an oily, flammable liquid, consisting of a variety of organic compounds, that is produced in sediments and sedimentary rocks during burial of organic matter; also called crude.

crude oil: (see petroleum).

natural gas: a gas, consisting mainly of methane, that is produced in sediments and sedimentary rocks during burial of organic matter.

source rocks: sedimentary rocks, containing significant concentrations of organic matter, in which petroleum and natural gas are generated during burial of the deposits.

seal: an impermeable layer or mass of sedimentary rock that forms the convex-upward top or roof of a petroleum reservoir.

reservoir: a large body of porous and permeable sedimentary rock that contains economically valuable petroleum and/or natural gas.

Digging Deeper

PETROLEUM AND NATURAL GAS AS ENERGY RESOURCES

The Nature and Origin of Petroleum and Natural Gas

Petroleum, also called **crude oil**, is a liquid consisting mainly of organic compounds that range from fairly simple to highly complex. **Natural gas** consists mainly of a single organic compound, methane (with the chemical formula CH_4). Oil and natural gas reside in the pore spaces of some sedimentary rocks.

The raw material for the generation of oil and gas is organic matter deposited along with the sediments. The organic matter consists of the remains of tiny plants and animals that live in oceans and lakes and settle to the muddy bottom when they die. Much of the organic matter is oxidized before it is permanently buried. But, enormous quantities are preserved and buried along with the sediments. As later sediments cover the sediments more and more deeply, the temperature and pressure increase. These higher temperatures and pressures cause some of the organic matter to be transformed into oil and gas. Petroleum geologists use the term "maturation" for this process of change in the organic matter. They call the range in burial depth that is appropriate for generation of oil and gas the "oil window" or "gas window" and the regions rich in organic matter subjected to these depths "the kitchen." Depending upon the details of the maturation process, sometimes mostly oil is formed, and sometimes mostly gas.

The mudstone and shale that form the **source rocks** for oil and gas are impermeable. Fluids can flow through them very slowly. The oil and gas rise very slowly and percolate through the source rocks, because they are less dense than water, which is the main filler of the pore spaces. Once they reach fractures and much more permeable rocks like some kinds of sandstone and limestone, however, they can migrate much more rapidly. Much of the oil and gas rise all the

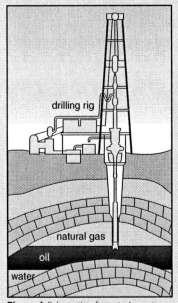

drilling rig

natural gas

oil

water

Figure I Schematic of a petroleum trap.

Digging Deeper

Assign the reading for homework, along with the questions in **Check Your Understanding** if desired.

Assessment Opportunity

Reword or restructure the questions in **Check Your Understanding** for a brief quiz. Use the quiz (or a class discussion of the questions in the textbook) to assess your students' understanding of the main ideas in the reading and the activity.

Teaching Tip

Point out to your students that *Figure 1* on page R58, the schematic of a petroleum trap, is not drawn to scale. The drilling rig might be as much as 20 m high, but the depth of the well might be 1000 m or more.

way to the Earth's surface, to form oil and gas seeps, and escape to the atmosphere or produce tar mats. If the rocks are capped deep in the Earth by an impermeable layer called a **seal**, the oil and gas are detained in their upward travel. A large volume of porous rock containing oil and gas with a seal above is called a **reservoir**. The oil and gas can be brought to the surface by drilling deep wells into the reservoir, as shown in *Figure 1*. Petroleum reservoirs around the world, in certain favorable geological settings, range in age from more than a billion years to geologically very recent, just a few million years old. Oil and gas are nonrenewable resources, however, because the generation of petroleum operates on time scales far longer than human lifetimes.

Geo Words

feedstocks: raw materials (for example, petroleum) that are supplied to a machine or processing plant that produces manufactured material (for example, plastics).

Oil and natural gas are useful fuels for several reasons. They have a very high heat content per unit weight. They cost less than coal to transport and are fairly easy to transport. Oil and gas can also be refined easily to form many kinds of useful materials. Petroleum is used not only for fuel. It is used as the raw material (called **feedstock**) for making plastics and many other synthetic compounds like paints, medicines, insecticides, and fertilizer.

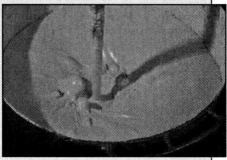

Figure 2 Petroleum is found in many common products, including paint.

The History of Oil Production in the United States

The United States has had several basic shifts in its energy history (see *Figure 3*). The overall pattern of energy consumption in the United States has been that of exponential growth. The burning of wood to heat homes and provide energy for industry was the primary energy source in about 1885. Coal surpassed wood in the late 1800s, only to be surpassed by petroleum and natural gas in the 1950s. Hydroelectric power appeared in 1890

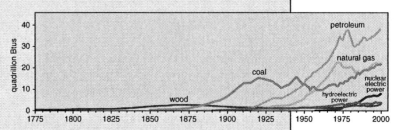

Figure 3 Energy consumption, by source, in the United States since 1775.

Chapter 1

Teaching Tip

Paint (*Figure 2* on page R59) is just one of the many products that contains petroleum. All plastic is made from petroleum and plastic is used almost everywhere: in cars, houses, toys, clothing, and computers. Asphalt used in road construction is a petroleum product, as is the synthetic rubber in the tires. Paraffin wax comes from petroleum, as do fertilizers, pesticides, herbicides, detergents, phonograph records, photographic film, furniture, packaging materials, surfboards, paints, and artificial fibers used in clothing, upholstery, and carpet backing.

Figure 3 on page R59 graphs the history of energy consumption in the United States. The fluctuations in the curves for coal, petroleum, and natural gas can be explained by economics. Consumption of coal, the major energy source in the first half of the 20th century, dipped noticeably during the Great Depression of the 1930s. Consumption of both petroleum and natural gas declined during the oil shocks of the 1970s.

Earth's Natural Resources Energy Resources

Geo Words

sedimentary basin: an area of the Earth's crust where sediments accumulate to great thicknesses.

and nuclear electric power appeared in about 1957. Solar photovoltaic, advanced solar thermal, and geothermal technologies also represent recent developments in energy sources. The most striking of these entrances, however, is that of petroleum and natural gas.

Figure 4 Drilling rigs are used to pump oil to the Earth's surface.

The first successful oil-drilling venture was Colonel Titus Drake's drilling rig in 1859 in northwestern Pennsylvania. Drake reached oil at a depth of 70 ft. Petroleum got major boosts with the discovery of Texas's vast Spindletop Oil Field in 1901. With the advent of mass-produced automobiles, consumption of oil began to grow more rapidly in the 1920s. In the years after World War II, coal lost its position as the premier fuel in the United States. Trucks that ran on petroleum took business once dominated by railroads, which began to switch to diesel locomotives themselves. Labor troubles and safety standards drove up the cost of coal production. Natural gas, which began as a source for lighting (gas lanterns and streetlights), replaced coal in many household ranges and furnaces. By 1947 consumption of petroleum and natural gas exceeded that of coal and then quadrupled in a single generation. No source of energy has ever become so dominant so quickly.

United States Petroleum Production

In 1920, the United States produced more than two-thirds of the world's oil. By 1998, the United States supplied only 12% of the world's oil needs. Oil and gas are found in **sedimentary basins** in the Earth's crust. To find oil, geologists search for these basins. Most of the basins in the United States have already been explored and exploited. In fact, the United States is the most thoroughly explored country in the world. Some people estimate that the United States has less than 35% of its original oil remaining. There is not nearly enough easily recoverable petroleum in the United States to meet the demand, although domestic production will continue for years to come.

Oil and gas resources are not evenly distributed throughout the country. The top ten oil fields account for 33% of all remaining oil in the United States (including a field in California that was discovered in 1899!). Texas has been the largest producing state since the late 1920s, when it surpassed California. In the late 1980s, Alaska rivaled Texas.

Check Your Understanding

1. What is the origin of oil and natural gas?

2. Why do some people refer to the present times as the "petroleum age"?

3. What are the advantages of petroleum and natural gas as fuels?

4. Why are oil and gas considered nonrenewable resources?

Teaching Tip

The photograph in *Figure 4* on page R60 shows a historic drilling rig. To get an appreciation of the scale of the photograph, note the people on the ground and on top of the rig. Visit the *EarthComm* web site to find additional information on the history of oil production in the United States, including information specific to your state.

Check Your Understanding

1. Oil and natural gas are produced when organic matter is deposited along with sediments. As sediment accumulates, layer upon layer, temperature and pressure in lower layers increase. These higher temperatures and pressures cause some of the organic matter to be transformed into oil and gas.

2. Present times are often referred to as the petroleum age because petroleum and natural gas have quickly become the dominant source of energy.

3. Oil and gas are useful fuels because they:
   • have a very high heat content per unit weight
   • cost less to transport
   • are fairly easy to transport
   • can be refined easily to form many useful materials

4. Oil and gas are considered nonrenewable resources because the generation of these fossil fuels operates on time scales far longer than human lifetimes.

Assessment Tool

Check Your Understanding Notebook-Entry Evaluation Sheet
Use this sheet to evaluate the extent to which students understand the key concepts explored in **Activity 6** and explained in **Digging Deeper**, and to evaluate the students' clarity of expression.

Understanding and Applying What You Have Learned

1. What is the world's future supply of petroleum relative to world demand?

2. Is the future decrease in petroleum production likely to be abrupt or gradual? Why?

3. Look at *Table 1* from the investigation.

   a) Calculate the average yield per well in barrels of petroleum per day for each year in the table.

   b) Graph the total number of wells producing and the average well yield over time.

   c) Describe any relationships that you see. How is well yield changing over time?

4. Think back to your projections of future oil production, foreign imports, and consumption.

   a) How might tax incentives for switching to renewable energy sources affect oil and gas consumption rates in the future?

   b) How might a change in the cost of fuels (gasoline, jet fuel, etc.) affect consumption rates in the future?

   c) How might new discoveries of oil affect domestic production?

   d) Identify two other factors that you think might affect one of your three projections.

Inquiring Further

1. **United States oil and gas fields**
   Research one of the top ten United States oil and gas fields. Go the *EarthComm* web site for links that will help you to research the geology and production history of the top ten United States oil fields. Why are these fields so large? What is the geologic setting? When was oil or gas discovered? What is the production history of the field? How much oil and gas remains?

Oil Fields
Alaska (Prudhoe Bay; Kuparuk River), California (Midway-Sunset; Belridge South; Kern River; Elk Hills), Texas (Wasson; Yates; Slaughter), Gulf of Mexico (Mississippi Canyon Block 807)

Gas Fields
New Mexico (Blanco/Ignacio-Blanco; Basin), Alaska (Prudhoe Bay), Texas (Carthage; Hugoton Gas Area [also in Kansas and Oklahoma); Wyoming (Madden), Alabama (Mobile Bay), Colorado (Wattenberg); Utah (Natural Buttes); Virginia (Oakwood).

Understanding and Applying What You Have Learned

1. This is a difficult question for anyone to answer. Demand will almost certainly continue to rise, although perhaps more slowly than in the past. There is a finite quantity of petroleum that can be extracted. Estimates of reserves vary greatly, because many areas have not yet been fully explored. It is generally agreed that production will fall behind demand at some time in the future. Whether that time is 20 years, 50 years, more, or less, is not certain.

2. Answers will vary. Decreases in petroleum will almost certainly be gradual, as readily recoverable reserves are depleted and the cost of production increases.

3. a) See the answer to **Question 2(i)** in **Part A** of **Investigate** for instructions on how to complete the calculation.

Year	Total Wells (thousands)	Average yield per well (barrels)
1954	511	13.8
1957	569	14.0
1960	591	13.5
1963	589	14.7
1966	583	16.4
1969	542	20.0
1972	508	22.0
1975	500	20.0
1978	517	19.9
1981	557	18.3
1984	621	16.9
1987	620	16.0
1990	602	14.8
1993	584	14.7
1996	574	14.4
1999	554	14.0

b)

Comparison of Total Number of Wells and Average Well Yield

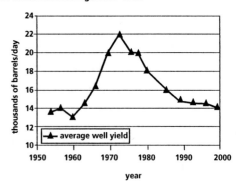

c) The total number of wells producing shows no simple trend. The average well yield generally increases and then decreases.

4. **a)** If the incentives were great enough, they would cause an appreciable decrease in rates of consumption of oil and gas, provided that alternative sources were available.

    b) Rates of consumption would fall with increasing price, but by how much depends upon a number of factors that are difficult to take into account.

    c) Discoveries of new major fields would increase domestic production temporarily, provided that drilling was feasible and allowed.

    d) Answers will vary. Some examples of factors that may affect projections of future oil production, foreign imports, and consumption:
      • a major recession or depression, either in the United States or in other parts of the world
      • political decisions about drilling in environmentally sensitive areas
      • the political situation in foreign oil-producing countries
      • development of economically competitive alternative energy sources
      • development of energy-saving technologies

Preparing for the Chapter Challenge

Activity 6 provides important ideas that relate to the general nature of the **Chapter Challenge**. That is, our country is very dependent upon oil and gas. Our consumption of petroleum continues to grow, yet our production of petroleum has declined over the last 15 years. In order to address the challenge, students are asked to prepare a report that explains how energy resources are formed, discovered, and produced.

Have them identify how the work in this activity will help them to solve the **Chapter Challenge**. Sample questions that you can pose as prompts for their Chapter Report include:

- How would you characterize national trends in oil production and consumption?
- What trends exist in oil and gas production in your state?
- How will we meet the growing demand for oil and gas?
- What have you learned about how oil and gas form?
- Why are oil and gas considered nonrenewable natural resources?

Inquiring Further

1. United States oil and gas fields

Answers will vary depending upon the oil or gas field that students elect to investigate. They can find information about each field on the *EarthComm* web site.

ACTIVITY 7 — OIL AND GAS PRODUCTION
Background Information

Porosity

A porous medium is a material that contains empty spaces, called pore spaces, among its constituent solid materials. The ratio of open space to total volume in some region of the subsurface material is called its porosity. Porosity is a direct measure of the fluid-holding capacity of the material. The porosity of a porous medium is defined as the ratio of the volume of pores to the total volume of the medium, in a representative sample of the medium. The ratio is usually multiplied by 100, to express it as a percentage. The porosity of very porous materials like loose sand or gravel can be as high as 30%. Sedimentary rocks are usually less porous than unconsolidated sands or gravels, because as the sedimentary material is lithified during burial, pore space is reduced by precipitation of new mineral material, called cement (no relation to the Portland cement used in making concrete) in the pore spaces by circulating solutions. Also, the increasing pressure of the overlying material tends to cause the shape of the sediment particles to be changed so as to fit more closely together.

Porosity in sedimentary rocks is of great importance to petroleum geologists, for two reasons: for a body of sedimentary rock to be a good reservoir for petroleum, it must have sufficient porosity to hold the petroleum, and it must have sufficient permeability for the petroleum to flow toward wells.

Permeability

One of the important aspects of porous media like sediments and sedimentary rocks is that fluids (air, water, oil, or gas) can flow through them. It is always important to consider how fast, as well as in which direction, the fluid flows through the medium. The speed of flow is determined by a property of the medium called its permeability. The permeability of a porous medium is a measure of how easily a fluid can be forced through the medium by imposing a difference in fluid pressure from one place in the medium to another place.

Materials with abundant large and connected pore spaces, like sand and gravel, have very high permeabilities, where as materials like clay or solid rock have very low permeabilities. Fine sediments like clay have high porosity but low permeability, because the pore spaces, although abundant, are so small that it is difficult to force the fluid to move.

An easy way to grasp the concept of permeability is to imagine stuffing one of your home water pipes with sand just upstream of a valve or faucet. When you turn the water on, the pressure of your water system tends to force water to flow through the porous sand and out the faucet (provided that you install a screen just upstream of the faucet to keep the sand in the pipe). Very porous media tend to have correspondingly high permeabilities, but two qualifications are needed:

- if the pores in the medium are very small, flow through the pores is slow, because of the increased friction
- in some media, like solid rocks, some—or even most—of the pores are not physically connected, so the fluid cannot flow from pore to pore.

Petroleum Reservoirs

In the petroleum industry, a reservoir is a large body of porous and permeable sedimentary rocks, in the deep subsurface, that contains petroleum and or natural gas. The petroleum hydrocarbons that are generated in organic-rich source rocks tend to migrate slowly upward, because their density is less than the waters that generally fill the pore spaces in sedimentary rocks. If a body of porous and permeable sediment rock is overlain by a layer or mass of rock with very low permeability, the upward movement of the hydrocarbons is diverted or arrested. For porous and permeable rock to be a reservoir, the overlying impermeable rock must have a convex-up geometry, such that some volume of rock underneath can retain the hydrocarbons. In such a situation, the impermeable rock is called a seal. The vertical distance between the crest of the seal and the highest point along its margin is called the closure of the reservoir. Think in terms of an umbrella with its handle tilted slightly off the vertical. If the rain were falling upward rather than downward, the water would accumulate beneath the umbrella until it spilled (upward!) from the part of the margin of the umbrella with the highest elevation. The volume of a petroleum reservoir thus depends on the horizontal area beneath the seal, together with the closure on the reservoir.

Reservoirs vary greatly in volume. The smallest reservoirs that are likely to be economically advantageous might have closures of several meters and a horizontal extent of a few hundred meters, whereas the largest might have closures of many tens of meters and a horizontal extent of several kilometers.

The geometry of reservoirs also varies enormously, depending on the local geologic structure, which in turn is an outcome of geologic processes of sediment deposition, burial, and later deformation. The classic petroleum reservoir is a doubly plunging anticline: a large-scale convex-up fold in the sedimentary layers whose ends are inclined downward. Think in terms of an upside-down canoe. Reservoirs have many other geometries, however, depending upon their origin. The best reservoir rocks are generally sandstones and carbonate rocks (limestone and dolostone). Seals are usually fine-grained sedimentary rocks like mudstones or shales.

More Information – on the Web

Visit the *EarthComm* web site www.agiweb.org/earthcomm to access a variety of links to web sites that will help you deepen your understanding of content and prepare you to teach this activity. Many of the sites also contain images that you can download.

Goals and Assessment

Clarify that the goals indicate what students should understand and be able to do as a result of the activity. Make sure students understand that Chapter Assessments are based upon these goals.

Goal	Location in Activity	Assessment Opportunity
Design investigations into the porosity and permeability of an oil reservoir.	**Investigate** Part A **Investigate** Part C	Experimental design is useful and accurately tests hypotheses; results are useful and can help to answer questions.
Understand factors that control the volume and rate of production in oil and gas fields.	**Investigate** Part A **Investigate** Part B **Investigate** Part C **Digging Deeper; Check Your Understanding** Question 3	Answers to questions are reasonable, and demonstrate knowledge gained through experimentation.
Explore the physical relationships between natural gas, oil, and water in a reservoir.	**Investigate** Part D **Digging Deeper; Understanding and Applying What You Have Learned** Question 3	Observations are accurate, cross section is properly drawn; determination of where oil and gas are found in VanSant Sandstone is accurate.
Use this knowledge to understand the variability in estimates of remaining oil and gas resources.	**Digging Deeper; Check Your Understanding** Question 3	Answer to question is reasonable, and closely matches that given in Teacher's Edition.
Appreciate the importance of technological advances in maximizing energy resources.	**Digging Deeper; Understanding and Applying What You Have Learned** Question 4	Answer to question is reasonable, and it closely matches that given in Teacher's Edition.

NOTES

Earth's Natural Resources Energy Resources

Activity 7   Oil and Gas Production

Goals

In this activity you will:

- Design investigations into the porosity and permeability of an oil reservoir.

- Understand factors that control the volume and rate of production in oil and gas fields.

- Explore the physical relationships between natural gas, oil, and water in a reservoir.

- Understand why significant volumes of oil and gas cannot be recovered and are left in the ground.

- Use this knowledge to understand the variability in estimates of remaining oil and gas resources.

- Appreciate the importance of technological advances in maximizing energy resources.

Think about It

When oil and gas are discovered, petroleum geologists calculate the volume of the oil or gas field. They also need to estimate how much they can recover (remove). This helps them to determine the potential value of the discovery.

- When oil is discovered, what percentage of it can actually be recovered and brought to the surface?
- When will global oil production begin to decline? When will global gas production begin to decline?

What do you think? Record your ideas in your *EarthComm* notebook. Be prepared to discuss your responses with your small group and the class.

Activity Overview

First, students are given a list of materials and are asked to design an experiment to determine whether sand can hold more oil than gravel, or vice versa. This introduces them to the concepts of porosity and permeability. Students then allow oil to drip out of a container of gravel to understand that oil recovery from a reservoir is not 100%. They design an experiment to determine how grain size of sediment affects the ease with which fluid flows through. Finally, students complete an experiment to understand the relationship between oil, natural gas, and water in a reservoir, and they use their results to help them decide where to drill in a hypothetical reservoir. **Digging Deeper** reviews the effects of porosity and permeability on recovery of oil and gas, and also examines estimates of petroleum reserves.

Preparation and Materials Needed

You will need to assemble all of the materials before class. You may want to play around with the materials to see what different types of experiments you can design, so that you can be prepared to guide your students in their experimental design. Note that this is a long investigation with several parts, so plan accordingly. Also, students will have to leave the experiment in **Part B** running overnight. Therefore, you may want to have students do **Parts A** and **B** in one class period, and **Parts C** and **D** in another.

Materials
Part A
- Two 500-mL (16 oz) clear plastic soda bottles with bottoms removed
- Fine cheesecloth
- Electric tape
- Two stands with clamps to hold plastic bottles
- Sand
- Aquarium gravel
- 50-mL graduated cylinder
- 200 mL vegetable oil
- Water
- Calculator

Part B
No additional materials needed.

Part C
- Four 500-mL (16 oz) clear plastic soda bottles with bottoms removed
- Fine cheesecloth
- Electric tape

- Four stands with clamps to hold plastic bottles
- Four 500-mL glass beakers
- Coarse sand
- Fine sand
- Silt/clay
- Aquarium gravel
- 50-mL graduated cylinder
- Water

Part D
- 500-mL clear plastic soda bottle
- Vegetable oil
- Protractor
- Graph paper

Think about It

Student Conceptions

Students are likely to have little knowledge about the average recovery rate of oil from a reservoir. They are likely to think that recovery will be 100%. Answers to the second question will vary. In **Activity 6,** students examined oil and gas production in the United States and noticed the decline in production of oil. This may lead them to think that global oil and gas production are also declining because of diminishing availability of these resources. Students may also think that oil and gas production is driven by demand. Thus, they may state that oil and gas production will decline when demand for these resources declines.

Answers for the Teacher Only

Recovery rates for oil wells often range from 5% to 20%, and up to 80% for natural gas. This means that as much as 95% of the oil discovered might be left behind. New technologies and methods are pushing the recovery rates higher, to the point where recoveries of 40 to 50% for oil are expected. Many factors affect recovery rates, including:
- the viscosity of the oil
- chemical composition (oil with high wax or asphalt content literally plugs up a well)
- the permeability of the rock that holds the oil
- the rate at which oil and gas are pumped from the well
- how and where wells are drilled

Exactly when world oil and gas production will begin to decline will depend upon:
- the rate at which oil and gas are consumed
- our ability to recover oil and gas
- new discoveries

Some scientists estimate that the original recoverable oil endowment of the Earth may have been around 2330 billion barrels, of which we have already recovered one-third. The remaining two-thirds would enable production at our current annual level of 26 billion barrels to continue for about 60 years. More conservative estimates are that world oil production will peak between 2005 and 2010, and that world natural gas production will peak in 2015.

Assessment Tool

Think about It Evaluation Sheet
Use this evaluation sheet to help students understand and internalize the basic expectations for the warm-up activity.

Blackline Master Energy Resources 7.1
Contour Map on Top of VanSant Sandstone

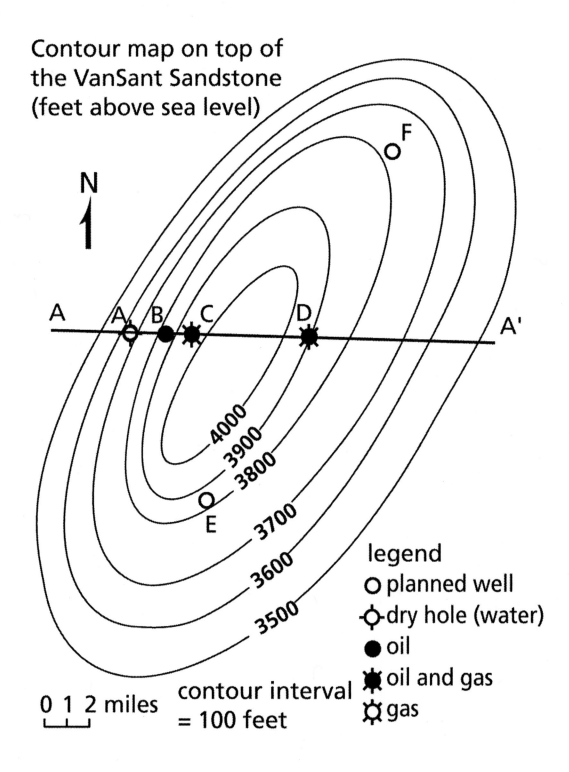

Contour map on top of
the VanSant Sandstone
(feet above sea level)

N

A — A — B — C — D — A'

E

F

4000
3900
3800
3700
3600
3500

legend
O planned well
-O- dry hole (water)
● oil
✹ oil and gas
✿ gas

0 1 2 miles

contour interval
= 100 feet

NOTES

Investigate

Part A: Reservoir Volume

1. In your group, discuss the questions below. Record your ideas in your notebook.

   a) Which material can hold more oil: gravel or sand? Why do you think this is so?

   b) How might this knowledge be useful in oil and gas exploration?

2. Develop a hypothesis related to the question in Step 1. a).

   a) Record your hypothesis in your notebook.

3. Using the following materials, design an investigation to test your hypothesis:
   two 500-mL (16-oz.) clear plastic soda bottles (with bottoms removed)
   fine cheesecloth
   electrical tape
   stands with clamps to hold the plastic bottles
   sand
   aquarium gravel
   50-mL graduated cylinder
   200 mL of vegetable oil
   water

   a) Record your procedure. Note any safety factors.

4. When your teacher has approved your design, conduct your investigation. (Note: At the end of your investigation, do not empty out the sediments and oil. You will need them for Part B.)

   a) Record your results.

5. Use your results to answer the following questions:

   a) What is the porosity (percentage of open space) of your gravel sample?

   Porosity is the ratio of pore space to total volume of solid material and pore space, usually expressed as a percentage by multiplying the ratio by 100. 1 mL = 1 cm^3 (cubic centimeter). For example, suppose you poured 25 mL of sediment into a container and found that 10 mL of water completely filled the spaces between the grains. One milliliter is equal to one cubic centimeter. Expressed as a percentage, the porosity of the sediment would be:

   $$(10 \text{ cm}^3/25 \text{ cm}^3) \times 100 = 40\% \text{ porosity}$$

   b) What is the porosity of your sand sample?

   c) Imagine that a geologist discovers an oil field in a sandy reservoir (in petroleum geology, a reservoir is a porous material that contains oil or gas). The total volume of the reservoir (sand plus oil) is equivalent to 60 million barrels of oil. How many barrels of oil are in the reservoir?

   d) If the reservoir described above were gravel, how many barrels of oil would it contain?

    Wear goggles. Clean up spills. Wash your hands when you are done.

Part B: Recovery Volume

1. Suppose you were to remove the cap at the bottom of the plastic bottle of gravel and oil.

   a) What volume of oil do you think would drip out of the container if you left it out overnight? Explain your answer.

EarthComm

Investigate

Part A: Reservoir Volume

1. **a)** Student responses will vary. Encourage students to explain the reasons behind their choices. The porosity of most sands and gravels is high. Porosity depends partly on the shapes of the sediment particles but even more on the sorting of the sediment. Sorting is a measure of the spread of particles sizes around the average size. A poorly sorted sediment has a wide range of particle sizes, whereas a well-sorted sediment has a narrow range of particle sizes. Natural sands and gravels range from very poorly sorted to very well sorted. A poorly sorted sediment has a smaller porosity than a well-sorted sediment, because the smaller particles tend to fill the spaces among the larger particles.

   b) Students are likely to recognize that it will be more profitable to attempt to recover oil from a reservoir that holds more oil.

2. **a)** Student hypotheses will vary. You may want to review with students that a hypothesis is a statement of the expected outcome of an experiment or observation, along with an explanation of why this will happen. A hypothesis is not a guess; it is based on what the student already knows about something. Students will use the hypothesis to design an experiment.

3. **a)** Student experiments will vary. Encourage students to read through the questions (in **Step 5**) that their experiment should help them answer. A sample design is to tape cheesecloth over the top of two bottles, then screw the cap onto both bottles. Place a known volume of gravel (say 200 mL) into one bottle and an equal amount of sand into the other bottle, and use the graduated cylinder to measure and pour oil into the bottles, one at a time. Getting the oil to flow downward through the sand is a little tricky and requires patience. It helps to unscrew the cap of the bottle at first, which allows air to flow through the grains as oil moves downward. You should try this yourself ahead of time.

4. Be sure to tell your students that you must approve their experiments before they actually carry out the procedures. The most straight forward procedure is to add liquid to the sediment until it is completely saturated, keeping an accurate account of both the volume of sediment and the volume of liquid added. Adding an excess of liquid, such that the level of the liquid exceeds the sediment level, helps to insure complete saturation of the sediment. Note that when the porosity is calculated this excess liquid should be subtracted from the total amount of liquid added . Some students may develop a procedure whereby they saturate the sediment sample and then measure the volume of water that drains out of the sediment relative to the bulk volume of the sediment. Keep in mind that not all of the water will drain out: some is left to wet the sediment particles. Because of the smaller particle size of the sand, the total surface area of sediment particles is

much greater in the sand than in the gravel. Accordingly, a greater volume of water will be retained in the sand after it is drained than will remain in the gravel. This effect distorts the measurement of porosity by such a procedure.

a) Results will vary. Remind students to keep good records of their observations throughout the experiment. Remind them also not to dump out the sediments and oil, because they will be using them in **Part B**.

5. a) Answers will vary depending upon the particular gravel used in the experiment. The results should be almost independent of the volume of gravel used, because porosity is a per unit volume property. You will probably want to help them with the math here, just to make sure that they are headed in the proper direction.

b) Answers will vary depending upon the particular sand used in the experiment.

c) Students will need to assume that the porosity of the sandy reservoir is equal to the porosity of the sand sample they analyzed in their experiments. Therefore, answers will vary.

d) Students will need to assume that the porosity of the gravel reservoir is equal to the porosity of the gravel sample they analyzed in their experiments. Therefore, answers will vary.

Part B: Recovery Volume

1. a) Answers will vary, depending upon student conceptions and also upon how much oil students added to the gravel sample. You may ask students to express their answer as a percentage (total recovered/total added), so that responses can be compared more easily.

NOTES

Earth's Natural Resources Energy Resources

2. Make sure that the cheesecloth is secured to the bottle with electrical tape. Unscrew the cap from the bottom of the container of gravel and oil. Allow the oil to seep through the container and drain into a collection device overnight.

3. Record your answers to the following questions:

   a) What is the volume of oil recovered?

   b) Calculate the percentage of oil recovered.

   c) How do you explain your results?

   d) If your model represented a discovery of 100 million barrels of oil, how many barrels of oil would actually be removed from the reservoir?

4. Brainstorm about ways to improve your results.

   a) How can you recover a greater percentage of oil? If your situation permits, test your method of removing more oil.

Part C: Factors that Affect Oil and Gas Production Rates

1. In your group, discuss your ideas about the following question: How does the grain size of sediment affect the ease with which fluids flow through them?

2. Develop a hypothesis about the question above. You will be testing the rate of flow of water through gravel, coarse sand, fine sand, and silt/clay. Your hypothesis should include a prediction (what you expect) and a reason for your prediction.

   a) Record your hypothesis.

3. Write down your plan for conducting the investigation. An idea about how to set up the equipment is shown in the diagram below.

   a) What is your independent variable?

   b) What is your dependent variable?

   c) How will you control other variables that might affect the validity of your results?

   d) How will you record and keep track of your results?

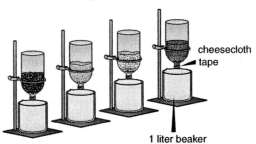

cheesecloth
tape

1 liter beaker

4. When you are ready, conduct your investigation.

   a) Describe your results. Was your hypothesis correct?

   b) Describe the relationship between grain size and permeability.

   c) How do your results relate to the production of oil?

5. If time permits, develop and test a hypothesis about the relationship between the viscosity of a fluid and its rate of flow through material. Viscosity is a term used to describe a fluid's internal friction or resistance to flow. In this case, you might wish to compare the speed at which oil, water, and a third fluid flow through a particular type of sediment.

2. If students did not tape cheesecloth to the mouth of the bottle as part of their experimental design, be sure that they do it now. Have them place a container under the bottle to collect the oil. Make sure that they label their bottles, and have them set the bottles aside overnight.

3. Answers to these questions will vary. Students will find that not nearly all of the oil is actually recovered. For a sediment with a given porosity, the percentage recovery depends on the sorting: see **Step 4** in **Part A** above for an explanation of why the percentage recovery in **Step 3(c)** below is not 100%.

4. a) Answers will vary. One way would be to dissolve the oil by flushing with warm soapy water. A simpler way would be to heat the material, thereby decreasing the viscosity of oil. The lower the viscosity, the more easily the residual oil will drain out of the sediment by gravity. Another way would be to fasten a cord or rope to the bottom end of the container and twirl it around, in order to increase the effect of gravity on the material. Actually doing that would involve considerable difficulty and danger, however.

Part C: Factors that Affect Oil and Gas Production Rates

1. Answers will vary. The ease with which fluid flows through different materials is measured by the permeability of a material. In general, in order for a material to have a high permeability, it must have a high porosity, and the pores should be large and interconnected. For the materials examined in this investigation, the gravel will have the highest permeability and the clay/silt material will have the lowest permeability, even if the porosity is about the same. This is because the pore spaces in the clay/silt are very small compared to the pore spaces in the gravel.

2. a) Student hypotheses will vary. Remind students that they will need to design an experiment to test their hypothesis.

3. Student investigations will vary. Using the diagram on page R64 as a guide, students could pour an equal volume of water into the bottles (one at a time) and record how long it takes for the first water to emerge from the bottom of the bottles. For the material with the highest flow rate (i.e., highest permeability), water will take the shortest time to flow through the bottle. Be sure to check and approve students' designs before permitting them to proceed.

a, b) Answers will vary depending upon experimental design. You may want to remind your students that an *independent variable* is a variable that is imposed by the experimenter, and a *dependent variable* is a variable whose value is determined by the *independent variable* and cannot be varied independently by the experimenter. For the setup shown in the student text, the independent variable is the nature of the sediment used (that might include average particle size and also sorting), and the *dependent variable* is the percentage of water recovered. (In an ideal experiment of this kind, the effect of sorting and average particle size would be investigated separately. This could be done by perhaps using two samples with the same average particle size but different values of sorting, and another two samples with a different average particle size but with the same two values of sorting.)

c) Other variables students might want to consider will vary with their experimental designs. The volume of sediment used should be controlled, as well as the size and shape of the containers used and the method of pouring water into the containers. The temperature of each container should be the same during the experiment.

d) Students may want to set up a data table that records the time elapsed and volume of water collected from each container of sediment.

4. a) Answers will vary depending on the students' hypotheses.

b) For given sorting, permeability increases with grain size.

c) The maximum recovery would be from coarse-grained sedimentary rocks with high porosity and high permeability.

Teaching Tip

Given the potential time and equipment requirements of this exercise, you may wish to discuss the proposed plans for this investigation as a class, and after coming to a consensus, conduct the experiment as a demonstration.

5. The hypothesis would be that flow rate is inversely related to viscosity: the higher the viscosity, the lower the flow rate. This might be tested in an experiment similar to the one shown in the diagram on page R64 in the student text, using long thin glass tubes (in place of the plastic bottles) packed with the same sediment but filled with the different liquids. If the liquids are colored with food coloring, the upper surface of the liquid-saturated zone could be viewed as the liquid surface descends in the tube during drainage. The higher the viscosity of the liquid, the slower the drainage, and the slower the upper surface of the liquid descends in the tube.

NOTES

Part D: Properties of Fluids in Reservoirs

1. Pour water into a 500 mL clear plastic soda bottle until the bottle is about half full. Add vegetable oil until the bottle is three-fourths full. Screw the cap onto the bottle. The water, vegetable oil, and air in this model represent water, crude oil, and natural gas. The bottle represents an oil and gas reservoir.

2. Slowly turn the bottle upside down (if you shake the bottle or turn it quickly, you will create bubbles in the oil, which will interfere with your observations).

   a) In your notebook, sketch a diagram to represent the relationship between the oil, water, and air.

   b) Why do you think that the material stacks in the order that it does?

   c) If a well is drilled into an oil and gas reservoir, what material would it encounter first? second? third?

3. Tilt the bottle at a 45° angle and hold it there. Repeat this with a 10° angle. Look at the bottle from the side, and then from the top.

   a) Draw a diagram of the side view of the bottle when it is tilted at a 45° angle versus a 10° angle.

   b) How does the surface area covered by the oil change with the angle of the oil reservoir (bottle)?

   c) Imagine drilling a vertical well through the bottle. This represents a well being drilled into reservoirs that are sloping at different angles in the subsurface. Would it penetrate a greater thickness of oil when the reservoir is at a 10° angle or when the reservoir is at a 45° angle? Why?

4. Look at the data table below. It shows the results of drilling of four wells. The elevation of the top of each well was 5000 ft. above sea level.

Feature	Well A	Well B	Well C	Well D
Elevation of top of VanSant Sandstone	3650 ft.	3850 ft.	3950 ft.	3900 ft.
Elevation of base of VanSant Sandstone	3450 ft.	3700 ft.	3825 ft.	3800 ft.
Result	Water (dry hole)	Oil at 3850 ft.; water at 3800 ft.	Gas at 3950 ft.; oil at 3875 ft.	Gas at 3900 ft.; oil at 3875 ft.

Part D: Properties of Fluids in Reservoirs

1. You can have students do this themselves, or you may want to do it as a demonstration.

2. a) Students should find that the water would be on the bottom, followed by the oil, and then the gas (air).

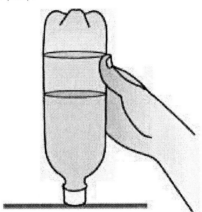

   b) The material stacks itself vertically in the order that it does because of density differences: water is the densest and therefore sinks to the bottom, and air is the least dense and therefore rises to the top. The oil is of intermediate density, so it finds the intermediate position.

   c) The well would encounter gas, oil, and then water.

3. a)

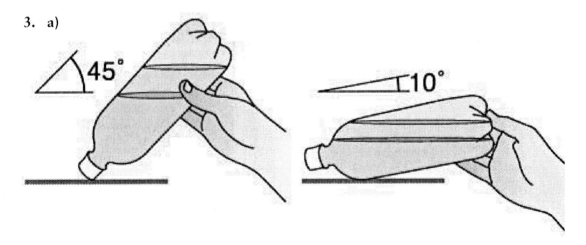

   b) The greater the angle that the bottle is tilted, the larger the surface area covered by the oil.

   c) The well would encounter a greater thickness of oil when the reservoir is at a 45° angle. The reason is that the available volume of oil is spread over a greater horizontal area. Students should be able to see this easily by drawing to-scale sketches of the bottle when it is at different angles.

Earth's Natural Resources Energy Resources

The locations of the wells are shown on the map below. The map shows the elevation of top of the VanSant Sandstone. Symbols on the map show the results of drilling.

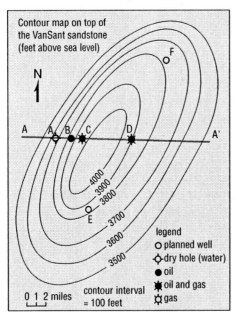

Contour map on top of the VanSant sandstone (feet above sea level)

N

4000
3900
3800
3700
3600
3500

contour interval = 100 feet

0 1 2 miles

legend
○ planned well
◇ dry hole (water)
● oil
✳ oil and gas
✿ gas

a) Draw a cross section across the oil field along the east–west line labeled A–A'. Plot the top of the VanSant Sandstone and the base of the VanSant Sandstone.

b) Use the results of drilling to show the level of gas, oil, and water in the cross section.

c) Use the results and your cross section to map the areal extent of the oil and gas. Color the gas red and the oil green on the map.

d) An example from another oil and gas field is provided below.

A geologist proposes to drill two more wells at locations E and F. Use your map and cross section to decide whether or not you would support the plan.

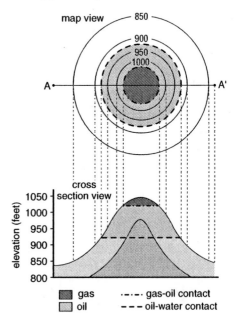

map view

850
900
950
1000

A • A'

cross section view

elevation (feet)
1050
1000
950
900
850
800

■ gas
□ oil
–·–·– gas-oil contact
– – – oil-water contact

Teaching Tip

You may want to make copies of the data table on page R65 so that students do not have to flip back and forth between pages as they construct their cross sections.

4. To draw a cross section, lay out a horizontal line near the top of a sheet of paper and locate the positions of the wells along that line. The positions can be taken from the line labeled A–A' in the left-hand diagram on page R66. Then draw vertical lines downward from those positions, to represent the wells. Find the depths to the top and the bottom of the VanSant Sandstone by subtracting the respective elevations from the 5000 foot elevation of the land surface. Plot the positions using the same scale of distance as you did for the horizontal line. Interpolate the positions of the top and base of the sandstone with smooth, continuous lines.

Cross Section of the VanSant Sandstone

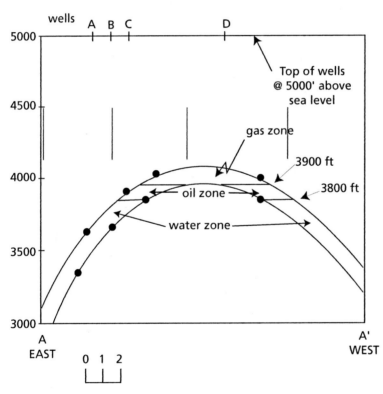

d) Students should look at the map of the VanSant Sandstone (left column of page R66) to answer this question, not the sample map and cross section. Using their cross sections, students should determine that drilling at point E is a good idea, but point F will yield only a dry hole.

Assessment Tools

EarthComm **Notebook-Entry Checklist**
Use this checklist as a quick guide for student self-assessment and/or an opportunity to quickly score student work. Add further criteria specific to your classroom needs or to this particular investigation.

Investigate Notebook-Entry Evaluation Sheet
Point out the criteria listed on this evaluation sheet that are relevant to this particular investigation. Encourage students to internalize the criteria by making them part of your assessment conversations as you circulate around the classroom.

NOTES

Reflecting on the Activity and the Challenge

In this activity, you explored the percentage of open space in various kinds of sediment. You saw that materials hold different amount of oil and that it is not easy to get all of the oil out of the material. You also explored the rate at which fluids flow through different materials. In the last part of the activity, you investigated the relationship between gas, oil, and water, and used the relationships to make predictions about where oil and gas would be in an oil and gas field. As you read further, you will see how the factors you have explored relate to the ability to predict how much oil and gas can be produced to meet future energy needs.

Digging Deeper

PETROLEUM RECOVERY

Porosity and Permeability

Most sedimentary rocks have open spaces in addition to the solid materials. These open spaces are called pores. Sedimentary rocks are the most porous rocks. Think about what a sandstone looks like, on the inside, when it is first deposited by flowing water. The sand particles are in contact with one another in the form of a framework. This framework is much like the way oranges are stacked in the supermarket, except that the sand particles usually have a much less regular shape than oranges. As you learned in the investigation, the **porosity** is defined as the volume of pore spaces divided by the bulk volume of the material, multiplied by 100 to be expressed as a percentage.

The porosity of loose granular material like sand can be as much as 30%–40%. If the sand particles have a wide range of sizes, however, the porosity is less, because the smaller particles occupy the spaces between the larger particles. When the sand slowly becomes buried in sedimentary basins to depths of thousands of meters, where temperatures and pressures are much higher, new mineral material, called **cement**, is deposited around the sand grains. The cement serves to make the sediment into a rock. The cement in sedimentary rocks is not the same as the cement that's used to make concrete, although it serves the same purpose. The addition of the cement reduces the porosity of the rock, sometimes to just a few percent.

Geo Words

porosity: the ratio of pore space to total volume of a rock or sediment, multiplied by 100 to be expressed as a percentage.

cement: new mineral material precipitated around the particles of sediment when it is buried below the Earth's surface.

Reflecting on the Activity and the Challenge

Discuss with your students how each part of the investigation illustrates the nature of producing oil and gas. Remind your students that they must address this issue in their final reports. One of the challenges to ensuring that we can meet our future energy needs is to make the most of the energy resources that we have discovered (students will explore in more detail in the **Digging Deeper** section of the activity). By understanding the nature of the flow of oil and gas in reservoirs, we can maximize the amount of petroleum that we recover from each discovery.

Digging Deeper

Assign the reading for homework, along with the questions in **Check Your Understanding** if desired.

Assessment Opportunity

Reword or restructure the questions in **Check Your Understanding** for a brief quiz. Use the quiz (or a class discussion on the questions in the textbook) to assess your students' understanding of the main ideas in the reading and the activity.

Earth's Natural Resources Energy Resources

Geo Words

permeability: the ease with which a fluid can be forced to flow through a porous material by a difference in fluid pressure from place to place in the material.

The most productive petroleum-reservoir rocks are those in which the pore spaces are mostly connected rather than isolated. When the pore spaces are all connected to one another, fluid can flow slowly through the rock. Fluid flows through a porous rock when the fluid pressure differs from one place to another. Here's a simple "thought experiment" to show the concept. Attach a long horizontal pipe to the base of a large barrel. Stuff the pipe with sand, and put a screen over the downstream end of the pipe. When you fill the barrel with water, the water flows through the sand in the pipe, because the pressure at the bottom of the barrel is high but the pressure at the open end of the pipe is low. The water finds its way down the pipe through the connected pore spaces in the sand. The rate of flow through the porous material in response to a pressure difference is called the **permeability** of the material. For a rock to have a high permeability, it generally has to have high porosity, but the pores must be fairly large and also well connected with one another. Good petroleum reservoir rocks must have high porosity, to hold the oil or gas, and also high permeability, so that the oil or gas can flow through the rock to the bottom of a well. Some reservoir rocks are highly fractured and the oil and gas can move along open fractures as well as through pore spaces.

Fluids also flow in porous rocks because of differences in density. Most sedimentary rocks at depth in the Earth have their pore spaces filled with water. When droplets of oil or bubbles of gas are formed in the rock, during maturation of the organic matter in the rock, they tend to rise upward through the pore water. They do this because both oil and gas are less dense than water. Only when they are trapped and accumulate beneath a seal do they form a petroleum reservoir.

For a mass of rock to form a seal, it has to be impermeable. In its simplest form, it also has to have a convex-upward shape, like an umbrella. Of course, it is usually much more complicated in its geometry than an umbrella. Just imagine raindrops falling upward into the inside of your umbrella, rather than downward to be shed off the umbrella! The vertical distance from the crest or top of the seal to its lower edge, where the petroleum can spill out, is called the closure of the reservoir. Giant petroleum reservoirs can have closures of many tens of meters, as well as horizontal dimensions of thousands of meters. Reservoirs that large can hold hundreds of millions of barrels of oil.

Recovery of Petroleum

Oil and gas are removed from reservoirs by drilling deep wells into the reservoir. At greater depths, the wells are only a few inches in diameter, but they can extend down for as much as several thousand meters. The technology of drilling has become very advanced. Wells can be drilled vertically

NOTES

downward for some distance and then diverted and angled off to the side at precise orientations. Sometimes the oil is under enough pressure to flow out of the well to the ground surface without being pumped, but usually it has to be pumped out of the well. The pumping lowers the pressure in the well, and the oil flows slowly into the well from the surrounding reservoir.

Much of the oil in a reservoir remains trapped in place after years of production. The main reason is that oil adheres to the walls of the pores in the form of coatings. The coatings are especially thick around the points of

Figure 1 Pumps are used to extract oil.

contact of sediment particles, where the pore spaces narrow down to small "throats." Several techniques have been developed to recover some of the remaining petroleum. Use of such techniques is called **secondary recovery**. One of the most common methods of secondary recovery is to inject large quantities of very hot steam down into a well, which mobilizes some of the remaining petroleum by causing it to flow more easily.

Geo Words

secondary recovery: the use of techniques to recover oil still trapped among sediment particles after years of production.

Estimates of Petroleum Reserves

Petroleum geologists attempt to estimate the future recoverable reserves of petroleum in the world. Estimates of this kind are very important, because they are needed to predict when world petroleum production will start to level off and then decline. This is a very difficult task, because it has to include regions of the world which have not yet been explored in detail but which seem to be similar, geologically, to areas already explored. Estimates of recoverable reserves have increased steadily with time, as petroleum has been discovered in regions that were not believed to contain petroleum. That can't go on forever, of course, because there is only a finite volume of petroleum in the Earth. The most recent estimate from the United States Geological Survey gives a good idea of recoverable reserves. Rather than being a single-value estimate, it is based on probability. The 95% probable value is about 2250 billion barrels of oil. That means that the probability that recoverable reserves are greater than 2250 billion barrels is 95%. The 50% estimate is about 3000 billion barrels, and the 5% estimate is about 3900 billion barrels (a one-in-twenty chance). You can see from this great range in values how indefinite such estimates are.

R 69

Teaching Tip

The pump system shown in *Figure 1* on page R69, called a pump jack, consists of an electric motor that drives a gear box that, in turn, moves a lever. The lever pulls a polishing rod up and down, which in turn causes a sucker rod to move up and down. The sucker rod is attached to a pump. The movement of the pump up and down creates a suction that draws oil up the well. In some cases, the oil is too viscous to flow upward; a second hole is drilled into the reservoir and steam is injected. The heat from the steam thins the oil, and the pressure associated with the steam helps to force the oil up the well. This process is called enhanced oil recovery.

Earth's Natural Resources Energy Resources

Figure 2 A drilling platform offshore.

The world's supply of petroleum is finite. Most of the land areas of the world have been intensively explored already, and the major new areas for petroleum exploration and production (called "plays," in the language of the oil business) lie in deep offshore areas of the oceans. It's generally agreed among petroleum geologists that such areas are the last major exploration frontier.

Gradually, over the next few decades, petroleum production will level off and start to decline, although only slowly. Because demand is likely to keep on rising, there will be a growing shortfall in petroleum. That shortfall will have to be made up by use of various alternative energy sources. Solid organic matter contained in certain sandstones, called tar sands, and in certain shales, called oil shales, are already becoming economically favorable to produce, and they will certainly be a major factor in future energy production. Canada is especially rich in tar sands, and there are abundant oil shales in the western United States. Developing fuel-efficient vehicles and giving greater attention to conservation of fuel will also help communities to address the shortfall of petroleum.

Check Your Understanding

1. In your own words, describe the difference between porosity and permeability.

2. What causes the permeability of a sediment deposit to decrease as it is buried deeply and turned into a sedimentary rock?

3. Why is it not possible to extract all of the oil from a reservoir?

4. What are the benefits of predicting oil and gas reserves?

Figure 3 Heavy oil found in tar sands at Vernal, Utah.

R 70

Teaching Tips

Figure 2 on page R70 shows one of the offshore drilling platforms that are used to recover oil located beneath the sea floor. Fluid from the reservoir is pumped through wells to the platform, where it is processed. The processed fluid then flows through a pipeline back down the platform and to the shore.

Figure 3 on page R70 pictures the heavy oil found in tar sands. Tar sands are sands (or sandstones) that contain solidified petroleum hydrocarbons. The rock is mined, and the broken rock material is put through a process by which the solid petroleum is recovered from the rock through dissolution by solvents.

Check Your Understanding

1. Porosity is defined as the ratio of pore space to total volume of a rock or sediment, multiplied by 100 to be expressed a percentage. Permeability is defined as the ease with which a fluid can be forced to flow through porous material by a difference in fluid pressure from place to place in the material.

2. When sand is buried to depths of thousands of meters, new mineral material, called cement, is deposited around the sand grains, reducing the porosity of the rock.

3. Some of the oil remains clinging to the surfaces of the sediment particles.

4. Predictions of oil and gas reserves are needed to anticipate when world petroleum production will start to level off and then decline. Knowing the probable extent of oil and gas reserves influences the search for alternative energy sources, as well as efforts to conserve fossil fuel resources.

Assessment Tool

Check Your Understanding Notebook-Entry Evaluation Sheet
Use this sheet to evaluate the extent to which students understand the key concepts explored in **Activity 7** and explained in **Digging Deeper**, and to evaluate the students' clarity of expression.

Understanding and Applying What You Have Learned

1. a) Why does the permeability of a fine-grained sedimentary rock tend to be less than that of a coarse-grained rock?

   b) Why does the permeability of a rock with particles of approximately equal size tend to be greater than a rock with particles of a wide range of sizes?

2. Why does a porous and permeable sedimentary rock need to have an upper seal to have the potential to be a petroleum reservoir?

3. What do you think replaces the oil in the pore spaces of a reservoir rock when the oil is pumped out? Why?

4. It is very expensive to drill wells and recover oil and gas. Use what you have learned about porosity and permeability to explain why geologists evaluate these features of reservoirs.

Preparing for the Chapter Challenge

Communities depend upon a fairly constant supply of the fuels and other products that are developed from oil and gas. Use what you have learned about oil and gas exploration and production to write a brief essay. The essay should outline the basics of oil and gas exploration and production. It should also help community members to understand the scientific basis for differences in estimates about the future supply of oil and gas.

Inquiring Further

1. **Secondary recovery methods**

   Research modern methods of secondary recovery:

   • carbon dioxide flooding,
   • nitrogen injection,
   • horizontal drilling,
   • hydraulic fracturing.

2. **Protection of sensitive environments**

   Investigate how oil and gas exploration and production companies are using advanced technologies to obtain resources while minimizing environmental impacts. Examples include:

   • The use of slimhole rigs (75% smaller than conventional rigs) to reduce the effect on the surface environment.
   • Using directional drilling technology to offset a drilling rig from a sensitive environment.
   • Using downhole separation technology to reduce the volume of water produced from a well.

3. **Careers in the petroleum industry**

   Investigate careers in the oil and gas industry. No single industry employs as many people as the petroleum industry.

4. **Conserving transportation fuels**

   Investigate methods of reducing consumption of transportation fuels.

Understanding and Applying What You Have Learned

1. a) Even if the porosity of a fine-grained sedimentary rock is the same as that of a coarse-grained rock, the permeability of the fine-grained rock tends to be less because there is much greater surface area of sediment particles per unit bulk volume of the rock. This makes the friction of the flowing oil with the sediment surface correspondingly greater.

    To make this concept more concrete, ask students to assume that they fill a container with 100 spheres of a given size, say a diameter of 1 cm, and then compute how many spheres of a smaller size, say a diameter of 1 mm, it would take to fill a second container with the same volume. Then, they compute the total surface area of the 100 1-cm spheres and the total surface area of the given number of 1-mm spheres, and compare the surface areas in the two cases. They would use simple formulas for the volume of a sphere and for the surface area of a sphere, which you can supply if they do not remember them or have not had prior experience with them. They would find that for the same total volume of spheres, the smaller spheres would have a much greater total surface area.

    b) The permeability of a rock with particles of approximately equal size tends to be greater than that of a rock with particles of a wide range of sizes. In the case of a poorly sorted sediment, in which there is a wide range of particle sizes, the smaller particles tend to fill the spaces among the larger particles, thereby reducing the porosity. Other factors being equal, a rock with lower porosity will also have a lower permeability.

2. If a porous and permeable sedimentary rock did not have an upper seal, less dense droplets of oil or bubbles of gas would rise to the Earth's surface and be lost, and the potential of this rock to be a petroleum reservoir would not be realized.

3. Water replaces the oil, because water rises up from below to fill the pore spaces vacated by the upward-flowing oil.

4. Students should tie their responses to the fact that reservoirs with higher porosities and permeabilities are able to hold more oil. Oil flows from them into the well more readily; therefore, these reservoirs are more profitable to drill.

Preparing for the Chapter Challenge

Remind students that their essays should be tied to their **Chapter Challenge**. Refer students to the rubric that you will be using to grade their final products. From the sample rubric at the front of this Teacher's Edition (see **Assessment Rubric for Chapter Report on Energy Resources** on pages 16 and 17), relevant material for assessment would be a discussion on:

  • how energy resources are discovered and produced

Inquiring Further

1. Secondary recovery methods
Secondary recovery involves the displacement of oil, driving it to wells for extraction.

- **Carbon dioxide flooding** involves pumping of CO_2 in liquid form into an injection well. The CO_2 is soluble in the oil and the resultant oil – CO_2 mixture has a lower density and viscosity facilitating more efficient extraction.
- **Nitrogen injection** involves the injection of nitrogen gas into a reservoir rock. The gas displaces the oil, facilitating its extraction.
- **Horizontal drilling** involves the installation of pipes horizontally. Instead of drilling numerous holes in an area, two or three wells are drilled and pipes are extended horizontally below the Earth's surface.

The two techniques (multiple wells versus horizontal drilling) have the same recovery levels, but horizontal drilling involves less impact on the environment and is more cost-effective.

- **Hydraulic fracturing** involves pumping high-pressure fluid (usually water) down a well, opening fractures in the reservoir rock. This allows oil or natural gas to move freely to a well for extraction.

Students can learn more about secondary recovery methods by visiting the *EarthComm* web site.

2. Protection of sensitive environments
Technologies are continually being developed that will allow the extraction of oil and natural gas while minimizing environmental impacts.

- **Slimhole drilling** reduces the environmental impact of drilling activities, resulting in a 75% reduction in the area of surface that is disturbed and in the amount of waste that is produced. Slimhole drilling is also less expensive to operate than a standard rig.
- In **directional drilling**, an oil well is drilled at an angle rather than straight down. This allows the oil rig to be offset from sensitive environments.
- **Oil production** usually also results in the recovery of water. The oil and water must be separated. Downhole separation involves separating the oil and water while they are still in the borehole. This can result in increased yield from an oil field, thereby reducing the number of drilling platforms required for maximum recovery and reducing the disruption to the environment.

3. Careers in the petroleum industry
Careers in the petroleum industry are highly varied and include positions in:
- exploration (geologists, engineers, environmental scientists, etc.)
- drilling and production (engineers, equipment operators, geologists, etc.)
- refining and petrochemicals (engineers, chemists, technicians, etc.)
- transportation and processing (plant operators, engineers, etc.)
- marketing and distribution (analysts, traders, etc.)
- administration (accountants, economists, analysts, etc.).

4. Conserving transportation fuels

To reduce consumption of transportation fuels, you can buy a new, more fuel-efficient vehicle. Smaller vehicles, those with two-wheel drive and manual transmissions, use less fuel. You can also take steps that do not involve buying a new vehicle. These include:

- taking public transportation
- walking or biking when possible
- keeping your vehicle well serviced
- using air conditioning in your vehicle sparingly
- driving within the posted speed limit
- keeping the tires properly inflated
- not topping off or overfilling when refueling
- replacing the gas cap tightly
- traveling as light as possible

Students can find additional tips through links on the *EarthComm* web site.

ACTIVITY 8— RENEWABLE ENERGY SOURCES – SOLAR AND WIND

Background Information

Solar Energy

The Sun radiates energy in the form of electromagnetic waves in all directions. The Earth intercepts a tiny part of this radiant energy, but the absolute amount received by the Earth is enormous. The solar energy received by the United States is about five hundred times the total energy consumption. Two major problems hinder widespread use of solar energy:

• it does not arrive in concentrated form
• at any given point, it varies with time, both from day to night and from clear weather to cloudy weather

With current technology, it would take two to three dozen one-meter-square solar panels to meet the energy consumption of a single person in the mid-latitude United States. For solar energy to be a significant contributor to the world's energy needs, there is no alternative to covering large areas with solar panels. This can be done either by covering a single large area with solar panels or by installing a large number of solar panels in homes and businesses. In areas with large amounts of available sunshine, this is already being done to generate electricity equivalent to a medium-sized conventional fossil-fueled generating facility. The use of individual solar panels for space heating, water heating, and electricity for small-scale use in isolated regions is common and growing rapidly.

As technological improvements in efficiency of direct conversion of sunlight to electricity improve, solar generation of electricity will grow in the future.

Wind Energy

Winds are the indirect result of latitudinal temperature differences brought about by net radiational heating at low latitudes and net radiational cooling at high latitudes. To maintain a long-term steady heat balance, heat must be transported bodily from low latitudes to high latitudes to counteract the latitudinal imbalance in radiational heating and cooling. Part of this transport is by winds in the atmosphere, and part is by currents in the oceans.

The pattern and typical speed of the Earth's wind systems is highly counterintuitive, as a consequence of the effects of the Earth's rotation. This effect, referred to as the Coriolis effect, makes winds blow largely parallel to the isobars of equal atmospheric pressure rather than perpendicular to the isobars in the direction of decreasing atmospheric pressure. Also, typical wind velocities turn out to be much greater than if there were no effect of the Earth's rotation.

At any given place and time, wind speeds usually increase upward in the atmosphere owing to the friction of the underlying ground surface. Wind speeds always decrease to zero at the ground surface but are nonzero at all finite distances above the surface. The wind speed increases very sharply in the first millimeters above the solid surface, and then much more slowly upward at higher altitudes. Winds are almost invariably turbulent, with instantaneous wind speeds and directions varying seemingly randomly with time on scales of fractions of a second to many minutes. The gustiness of the wind we sense, however, is in large part a consequence of eddying around local obstacles like trees, hills, and buildings. Over extensive level areas like treeless plains, vast empty parking lots, or the ocean surface, the gustiness of the wind is much less than over rough and irregular surfaces.

More Information – on the Web
Visit the *EarthComm* web site www.agiweb.org/earthcomm to access a variety of links to web sites that will help you deepen your understanding of content and prepare you to teach this activity. Many of the sites also contain images that you can download.

Chapter 1

Goals and Assessment

Clarify that the goals indicate what students should understand and be able to do as a result of the activity. Make sure students understand that Chapter Assessments are based upon these goals.

Goal	Location in Activity	Assessment Opportunity
Construct a solar water heater and determine its maximum energy output.	**Investigate** Part A	Solar cooker is constructed properly, data is correctly recorded, calculations are correct, work is shown.
Construct a simple anemometer to measure wind speeds in your community.	**Investigate** Part B	Anemometer is constructed properly, data is correctly recorded, calculations are correct, work is shown.
Evaluate the use of the Sun and the wind to reduce the use of nonrenewable energy resources.	**Investigate** Part A **Investigate** Part B **Digging Deeper; Understanding and Applying What You Have Learned** Questions 1 and 3	Responses to questions are reasonable, and closely match those given in Teacher's Edition.
Understand how systems based upon renewable energy resources reduce consumption of nonrenewable resources.	**Investigate** Part B **Digging Deeper; Understanding and Applying What You Have Learned** Questions 1 and 3	Responses to questions are reasonable, and closely match those given in Teacher's Edition.

NOTES

Earth's Natural Resources Energy Resources

Activity 8

Renewable Energy Sources—Solar and Wind

Goals

In this activity you will:

- Construct a solar water heater and determine its maximum energy output.

- Construct a simple anemometer to measure wind speeds in your community.

- Evaluate the use of the Sun and the wind to reduce the use of nonrenewable energy resources.

- Understand how systems based upon renewable energy resources reduce consumption of nonrenewable resources.

Think about It

The solar energy received by the Earth in one day would take care of the world's energy requirements for more than two decades, at the present level of energy consumption.

- How can your community take advantage of solar energy to meet its energy needs?
- Is your community windy enough to make wind power feasible for electricity generation?

What do you think? Record your ideas in your *EarthComm* notebook. Be prepared to discuss your responses with your small group and the class.

EarthComm

Activity Overview

Students begin the investigation by constructing a simple solar water heater and determining its maximum energy output. Then they look at maps of the United States that show average solar radiation for January and July to determine how the output of their solar water heater compares to the average daily solar radiation received by their community. In the second part of the investigation, students construct a simple anemometer, which is then used to measure wind speeds in their community. **Digging Deeper** looks at how solar and wind energy can be used as an energy resource.

Preparation and Materials Needed

You will need to collect all of the necessary materials before class. Note that **Part A** of the investigation must be completed on a sunny day. In **Part B**, students will need to take the anemometers they construct home for a weekend, so keep this in mind as you prepare your schedule. Also, you will want to provide enough materials for each student to build his or her own anemometer, so that he or she can each take an instrument home and make measurements of wind speed. It is recommended that you construct and try both the solar water heater and anemometer before class so that you can be better prepared to anticipate problems that may arise for your students.

Materials

Part A

- 15 ft. of flexible plastic tubing
- Cardboard box (or lid) trimmed to height of 1 in. (photocopy paper box top, or large pizza box). The inside surface should be painted black or lined with black construction paper.
- Clear plastic wrap or sheet (or sheet of thin clear Plexiglas® cut to box size)
- Ruler
- Tape
- Two glass or plastic containers (2-L minimum)
- Styrofoam® cooler
- Two thermometers
- Siphoning bulb
- Adjustable tubing clamp
- Stopwatch or watch with second hand

Part B

- Five paper cups
- Three wooden dowels, each about 10 in. long
- Modeling clay
- Piece of wood, approximately 10 in. x 10 in. – to serve as a base
- Calculator
- Graph paper

Think about It

Student Conceptions

Students are likely to say that solar panels can be used to generate energy in the community. Beyond recognizing the need for adequate solar radiation, it is unlikely that they will have a strong understanding of the factors that determine how successful the use of solar energy would be in a community. Students may or may not be familiar with how wind power works.

Assessment Tool

Think about It Evaluation Sheet
Use this evaluation sheet to help students understand and internalize the basic expectations for the warm-up activity.

Answers for the Teacher Only

A community might take advantage of solar energy in various ways. In some locations, its local power company might be able to tap into a power grid that relies in part on solar energy. Or, the community may develop its own solar power plant. Individual homes might be equipped with solar space heating or solar water heating; in many areas of the United States, that is feasible using only minimal technology. Use of solar panels to supply electricity in remote areas far from power lines is growing rapidly. Wind generation of electricity, however, is likely to be economically advantageous only where strong winds blow for a sufficient part of the time.

NOTES

Investigate

Part A: Solar Water Heating

1. Set up a solar water heater system as shown in the diagram. Put the source container above the level of the heater so that water will flow through the tubing by gravity. Do the experiment on a sunny day and face the surface of the solar water heater directly toward the Sun. This will enable you to calculate maximum output.

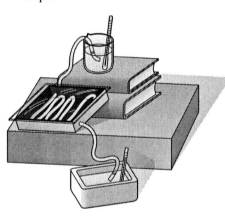

2. In its simplest form, a solar water heater uses the energy of the Sun to increase the temperature of water as the water flows through a coil of tubing.

   A Btu (British thermal unit) is the amount of heat energy required to raise the temperature of one pound of water one degree Fahrenheit. You need to determine the number of Btus of heat generated. If you divide this value by the surface area of the heater and by the time it takes to heat the water, you will have the Btu/ft^2/min

output of your solar water heater. It will help to know that 1000 mL of water weighs about 1 kg, or 2.205 pounds.

a) Decide how you will do the investigation and record the data.

b) Prepare a data table to record relevant data.

c) Think about the conversions that you will need to make in order to end up with Btu/ft^2/min.

3. Run about one liter of cold water through the heater. Refill the source container with two liters of cold tap water and run the experiment.

4. Answer the following questions:

a) What is the output of your solar water heater in Btu/ft^2/min?

b) Convert the output of your solar water heater from Btu/ft^2/min to Btu/ft^2/hour.

c) Assume that you could track the Sun (automatically rotate the device so that it faces the Sun all day long). Assuming an average of 12 hours of sunlight per day, what would its output be per day?

d) Convert this value to kWh/m^2/day by dividing by 317.2.

5. Examine the maps on the following page of average daily solar radiation for the United States. The first map shows the values for January, and the second one shows the values for July. The maps represent 30-year averages and assume 12 hours of sunlight per day and 30 days in a month.

Investigate
Part A: Solar Water Heating

1. Students can use the illustration on page R73 to help them set up their solar heaters. The upper and lower reservoirs need to be large enough so that water can be supplied for a long enough time for the solar heating system to come to equilibrium. The easiest way to start the siphon is to use a siphoning bulb. The other way to start the siphon is to fill the tube with water, seal both ends with your fingers, have someone else arrange the tubing in the heater, immerse the upstream end in the supply reservoir, and release both ends of the tubing at the same time. After the system has come to equilibrium, they should empty the lower reservoir and record the time. To stop the experiment, they abruptly divert the output hose to another container and record the time.

2. a) Students should start with the lower reservoir empty and the upper reservoir full. They record the temperature of the water in the upper reservoir just before starting the siphon. They monitor the temperature of the output water in the lower reservoir until it stops changing (at which time the system is operating at equilibrium), and they record the equilibrium output temperature. They measure the elapsed time during which the flowing water is heated under equilibrium conditions, and they measure the volume of water that passed through the heater.

   b) The students will need to compute the area of the heater that is exposed to the Sun.

   c) To find the heat added to the water, the students will need to know that one cubic foot of water weighs about 62.4 lb. They will probably be measuring the volume of water in either fluid ounces or milliliters. One cubic foot contains about 957 fluid ounces and about 28,316 mL. Then, they can compute the weight of water caught in the lower reservoir.

3. This is likely to be confusing. The important thing is to run a known volume of water through the heater when it is working at equilibrium.

4. Answers to the questions will vary depending upon the equipment that you use.

   a) Use the number of pounds of water heated and the number of degrees Fahrenheit the water temperature increased to compute the number of Btu of heat added. Divide that by the area of the heater in square feet to find the number of Btu added per square foot. Then divide that by the number of minutes of elapsed time to find the number of Btu added per square foot per minute.

   b) Divide the result from **Step 4(a)** by 60.

   c) Multiply the result from **Step 4(b)** by 12.

Earth's Natural Resources Energy Resources

a) How does the average daily solar radiation per month for January compare with that of July? How might you explain the difference?

b) Find your community on the map. Record the values for average daily solar radiation for January and July.

c) How does the daily output of your solar water heater compare to the average daily solar radiation for the two months shown?

d) Percent energy efficiency is the energy input divided by the energy output, multiplied by 100. Calculate the estimated efficiency of your solar water heater. If you have Internet access, you can obtain the average values for the month in which you do this activity. Otherwise, calculate the efficiency for whatever month is closest—January or July.

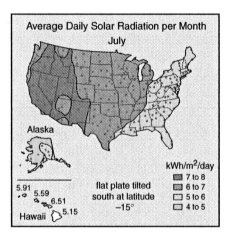

Average Daily Solar Radiation per Month
July

Alaska

Hawaii

kWh/m²/day
- 7 to 8
- 6 to 7
- 5 to 6
- 4 to 5

flat plate tilted
south at latitude
−15°

5.91
5.59
6.51
5.15

6. In your group, brainstorm about ways to increase the efficiency of your solar water heater. Variables that you might explore include the role of insulation, diameter of the tubing, type of material in the tubing, and flow rate of the water through the system. If time and materials are available, conduct another test.

Part B: Harnessing Wind Energy

1. Set up an anemometer as shown in the diagram below.

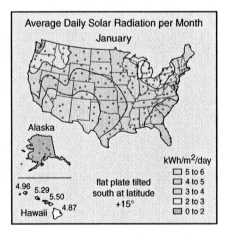

Average Daily Solar Radiation per Month
January

Alaska

Hawaii

kWh/m²/day
- 5 to 6
- 4 to 5
- 3 to 4
- 2 to 3
- 0 to 2

flat plate tilted
south at latitude
+15°

4.96
5.29
5.50
4.87

5. a) In general, average daily solar radiation per month is greater in July than in January.

    b) Answers will vary.

    c) Answers will vary. Several factors are at work here:
       • the efficiency of the solar heater
       • the difference, in the given area, between average solar radiation and the solar radiation on a clear day
       • the clarity of the atmosphere on the day the measurements were taken

    d) Visit the *EarthComm* web site to find the necessary solar radiation data. If you do not have Internet access, you can obtain the data before class and give your students copies (or they can use the information for January or July). The energy efficiency (as a percentage) can then be calculated by dividing the energy output by the energy input and multiplying by 100. For this calculation, the energy output is the calculated output of the solar water heater and the energy input is the average daily solar radiation.

6. The most effective ways to increase efficiency would be to:
       • insulate the bottom and walls of the heater with good insulation material
       • insulate the transparent cover of the heater by doubling it with an air space between the layers
       • paint the tubing a low-gloss black

    It would probably be even better to paint the bottom surface of the tubing black. Students might come up with the idea of coloring the water black with dye or ink, with the idea that black materials absorb solar radiation most efficiently. The greater the flow rate, the greater the efficiency of the heater, because, other factors being equal, the temperature difference between the flowing water and its surroundings is greater, thereby increasing conductive heat transfer from the warm surroundings to the colder water.

Chapter I

2. To find the wind speed, count the number of revolutions per minute. Next, calculate the circumference of the circle (in feet) made by the rotating paper cups. Multiply the revolutions per minute by the circumference of the circle (in feet per revolution), and you will have the velocity of the wind in feet per minute. (This will be an underestimate, because the cups do not move quite as fast as the wind.)

3. Take your anemometer home during a weekend. This will give you a chance to make measurements at different times of the day and over two days.

4. Make a map of your house and the yard around the house. Plan to place the anemometer at different locations and make velocity measurements at different times of the day.

5. Using a watch, count the number of times the colored cups spins around in one minute. Use your calibration to convert this to wind velocity.

   a) Keep a record of the wind speeds you are measuring for the next few days.

6. Measure the wind speed at different times of the day.

   a) Is the wind speed the same in the morning; the afternoon; the evening?

   b) Move your anemometer to another location. Is it windier in other places?

   c) Do trees or buildings block the wind?

7. Back in class, determine the best location for a wind turbine at your home.

   a) On the basis of the data that you have taken, make a plot of wind velocity throughout the day. You will assume that the wind will blow the same way every day. (Is this a good assumption?)

8. Following the example given below, calculate the electricity that can be produced at your home for a wind turbine with a 2-m blade (the cost of this wind turbine is about $3000).

   a) What percentage of your household electricity needs does this represent?

   b) How much wind power could be produced with a 3-m blade? (Cost about $5000.)

Example calculation of wind power:
Wind power per square meter of turbine area is equal to 0.65 times the cube of the wind velocity cubed. For instance, if a wind turbine has a blade 2 m long, then the area that the turbine sweeps is $\pi(1\text{m})^2 = 3.14 \text{ m}^2$.

If a constant wind of 10 m/s (22 mph) is blowing, the power is $(10^3)(0.65)(3.14) = 2000$ W. The efficiency of wind turbines is about 40%, so the actual power that can be produced is about 800 W. If the wind blew a constant 10 m/s every day of the year, this wind turbine could produce 7000 kWh of electricity.

Part B: Harnessing Wind Energy

1. Find a way to attach the cups rigidly to the spokes, and the spokes rigidly to the central spindle cup. Be sure that the rotating assembly is well balanced. The most difficult part is to figure out how to arrange for the rotating assembly to rotate freely on the vertical shaft. To make the anemometer more efficient, cut two rigid, hollow, rubber balls in half and use the four hemispheres in place of the cups.

2. Students can also calibrate their anemometers by having someone drive them in a car on a windless day. They should hold the instrument out the window at arm's length and have the driver proceed at a series of constant speeds, starting at 5 mph. Then students should count the number of turns the cups make in 60 seconds. Repeat counting at 10 mph, 15 mph, and so on, to 35 mph. Students can then produce a data table with wind speed (how fast the car was moving) in one column and the number of cup revolutions per minute in the other. They can plot a graph with number of revolutions per minute on the horizontal axis and wind speed on the vertical axis, and then draw a smooth curve connecting the data points. They can then use the data table or, better yet, the data graph to translate revolutions per minute into wind speed in miles per hour. This method of calibration is more accurate than counting the number of revolutions made by the cups, because the wind exerts a smaller, but not negligible, force on the bottoms of the cups as they move against the wind.

3. Remind students that they are required to make measurements at different locations and different times of day over a two-day period. You may want to set a requirement for the number of measurements students should take, e.g., measurements taken from three different locations each, three times a day. Encourage your students to read all of the questions for this part of the investigation, and they will find that they are asked to compare wind speeds measured in the morning, afternoon, and evening.

4. Remind students that they should make a map of the outsides of their houses. They should indicate on their maps the locations where they are going to make their measurements of wind speed. You may want to provide students with a sample map to give them an idea of what they are being asked to do.

5. You may want to take your students outside and have them practice making measurements of wind speed using their anemometers before they take the instruments home.

    a) Remind students to record all of their measurements.

6. a) Answers will vary. On average, wind speed is least in the morning and evening and greatest in the afternoon, but there is much variability in that regard.

    b) Answers will vary. Wind speed is likely to vary somewhat from place to place, depending upon the geometry of the surroundings.

c) Yes. Wind speed is especially reduced directly behind a large building, and especially when the walls of the building are nearly parallel and perpendicular to the wind direction.

7. a) No, it's not a good assumption. Students should graph their results.

8. Answers to these questions will vary depending upon the wind speed determined in **Step 7**. You may want to go over the sample calculation in the text with your students.

NOTES

Earth's Natural Resources Energy Resources

9. Get together as a class and discuss this activity.

   a) How do your calculations compare with those of other students?

   b) What are the characteristics of the locations of homes that have the highest wind velocities?

   c) What are the characteristics of those with the lowest wind velocities?

   d) Overall, how well suited is your community for wind power?

   e) Which locations in your community are optimally suited for wind turbines?

   f) How realistic is the assumption that two days' worth of measurements can be extrapolated to yearly averages?

   g) For how long should measurements of wind velocity be taken to get an accurate measure of wind velocities in your community?

Reflecting on the Activity and the Challenge

Water heating can consume as much as 40% of total energy consumption in a residence. In this investigation, you determined the energy output of a flat-plate collector, a device used to heat water using solar energy. You also built a simple anemometer in the classroom and used it to measure wind speed around your home at different times of the day. Finally, you calculated whether it would be feasible and cost-effective for you to generate part of your family's electricity needs using wind power. You may wish to consider the use of different forms of solar energy as an alternative energy source for your community.

9. Answers to these questions will vary by class. Have students write their results from **Step 8** on the blackboard for comparison. They should recognize that it is not a very realistic assumption that measurements taken over a two-day period can be extrapolated to yearly averages. How long measurements should be taken to get an accurate measure of wind velocity will vary by community. In some communities, wind is regular and sustained, whereas in others there is considerable variation from day to day.

Assessment Tools

EarthComm Notebook-Entry Checklist
Use this checklist as a quick guide for student self-assessment and/or an opportunity to quickly score student work. Add further criteria specific to your classroom needs or to this particular investigation.

Investigate Notebook-Entry Evaluation Sheet
Point out the criteria listed on this evaluation sheet that are relevant to this particular investigation. Encourage students to internalize the criteria by making them part of your assessment conversations as you circulate around the classroom.

Reflecting on the Activity and the Challenge

Have students read this brief passage and share their thoughts about the main points of **Activity 8** in their own words. Students should now be aware of the different types of natural resources used in their community and the availability of these resources. Discuss with students the alternatives for energy production explored in **Activity 8,** and whether or not these alternatives would be useful in their community. Hold a class discussion about how this activity relates to what is being asked of the students in the **Chapter Challenge**.

Digging Deeper

SOLAR ENERGY

Forms of Solar Energy

Forms of solar energy include direct and indirect solar radiation, wind, photovoltaic, biomass, and others. Tidal energy is produced mainly at the expense of the Earth–Moon system, but the Sun contributes to the tides as well. The three forms of solar energy with the most potential are solar-thermal (direct solar radiation to generate electricity), wind power, and photovoltaic cells.

Solar energy is produced in the extremely hot core of the Sun. In a process called nuclear fusion, hydrogen atoms are fused together to form helium atoms. A very small quantity of mass is converted to energy in this process. The ongoing process of nuclear fusion in the Sun's interior is what keeps the Sun hot. The surface of the Sun, although much cooler than the interior, is still very hot. Like all matter in the universe, the surface radiates electromagnetic energy in all directions. This electromagnetic radiation is mostly light, in the visible range of the spectrum, but much of the energy is also in the ultraviolet range (shorter wavelengths) and in the infrared range (longer wavelengths). The radiation travels at the speed of light and reaches the Earth in about eight minutes. Only an extremely small fraction of this energy reaches the Earth, because the Earth is small and very far away from the Sun. It is far more than enough, however, to provide all of the Earth's energy needs, if it could be harnessed in a practical way. The Sun has been producing energy in this way for almost five billion years, and it will continue to do so for a few billion years more.

Capturing the Sun's energy is not easy, because solar energy is spread out over such a large area. In any small area it is not very intense. The rate at which a given area of land receives solar energy is called **insolation**. (That's not the same as insulation, which is material used to keep things hot or cold.) Insolation depends on several factors: latitude, season of the year, time of day, cloudiness of the sky, clearness of the air, and slope of the land surface. If the sun is directly overhead and the sky is clear, the rate of solar radiation on a horizontal surface at sea level is about 1000 W/m^2 (watts per square meter). This is the highest value insolation can have on the Earth's surface except by concentrating sunlight with devices like mirrors or lenses. If the Sun is not directly overhead, the solar radiation received on the surface is less because there is more atmosphere between the Sun and the surface to absorb some of the radiation. Note also that insolation decreases when a surface is not oriented perpendicular to the Sun's rays. This is because the surface presents a smaller cross-sectional area to the Sun. ➤

Geo Words

insolation: the rate at which a given area of land receives solar energy.

Digging Deeper

Assign the reading for homework, along with the questions in **Check Your Understanding** if desired.

Assessment Opportunity

Reword or restructure the questions in **Check Your Understanding** for a brief quiz. Use the quiz (or a class discussion about the questions in the textbook) to assess your students' understanding of the main ideas in the reading and the activity.

Earth's Natural Resources Energy Resources

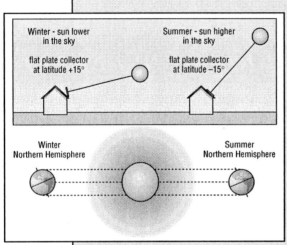

Winter - sun lower
in the sky

flat plate collector
at latitude +15°

Summer - sun higher
in the sky

flat plate collector
at latitude –15°

Winter
Northern Hemisphere

Summer
Northern Hemisphere

Figure 1 The tilt of the Earth on its axis during winter and during summer.

The Earth rotates on its axis once a day, and it revolves in an almost circular but slightly elliptical orbit around the Sun once a year. The Earth's rotation axis is tilted about 23° from the plane of the Earth's orbit around the Sun, as shown in *Figure 1*. It points toward the Sun at one end of its orbit (in summer) and away from the Sun at the other end (in winter). In summer, when the Earth's axis points toward the Sun, the Sun traces a path high in the sky, and the days are long. In winter, the Sun's path across the sky is much closer to the horizon, and the days are short. This has important effects on seasonal insolation and on building designs.

Solar Heating

Home heating is one of the main uses of solar energy. There are two basic kinds of solar heating systems: active and passive. In an active system, special equipment, in the form of a solar collector, is used to collect and distribute the solar energy. In a passive system, the home is designed to let in large amounts of sunlight. The heat produced from the light is trapped inside. Passive systems do not rely on special mechanical equipment, but they are not as effective as active systems.

Water heating is another major use of solar energy. Solar water heating systems for buildings have two main parts: a solar collector and a storage tank. The solar energy is collected with a thin, flat, rectangular box with a transparent cover. As shown in *Figure 2*, the bottom of the collector box is a plate that is coated black on the upper surface and insulated

glazing frame

flow tubes

outlet connection

glazing

inlet connection

enclosure

absorber plate

insulation

Figure 2 Flat plate collector.

R 78

Teaching Tips

Figure 1 on page R78 addresses how the efficiency of flat-plate solar collectors relates to the tilt of the Earth on its axis. Flat-plate solar collectors are most efficient when the Sun shines perpendicular to the surface of the collector. That is partly because then the collector presents the largest surface area to the Sun's rays, and partly because the percent reflection from the transparent surface of the collector turns out then to be at a minimum. Maximum yearly solar radiation can be achieved using a tilt angle approximately equal to the latitude of the site. To optimize performance in the winter, when the Sun's rays strike the Earth's surface at a smaller angle (less directly), the collector can be tilted 15° greater than the latitude. To optimize performance in the summer, when the Sun's rays strike the Earth at a higher angle (more directly), the collector can be tilted 15° less than the latitude. *Figure 1* illustrates how this works.

Figure 2 on page R78 shows a cutaway view of a typical flat-plate solar collector. It shows the insulated sides and bottom, the flow tubes to carry the circulated liquid, and the transparent cover. The glazing works best when it consists of two transparent sheets with a very thin air gap between, as in modern insulating windows.

on the lower surface. The solar energy that strikes the black surface is converted to heat. Cool water is circulated through pipes from the hot collector box to a storage tank. The water is warmed as it passes through the collector box. These collectors are not very expensive, and in sunny regions they can provide most or all of the need for hot water in homes or swimming pools.

Many large commercial buildings such as the one shown in *Figure 3*, can use solar collectors to provide more than just hot water. A solar ventilation system can be used in cold climates to preheat air as it enters a building. The heat from a solar collector can even be used to provide energy for cooling a building, although the efficiency is less than for heating. A solar collector is not always needed when using sunlight to heat a building. Buildings can be designed for passive solar heating. These buildings usually have large, south-facing windows with overhangs. In winter, the Sun shines directly through the large windows, heating the interior of the building. In summer, the high Sun is blocked by the overhang from shining directly into the building. Materials that absorb and store the Sun's heat can be built into the sunlit floors and walls. The floors and walls then heat up during the day and release heat slowly at night. Many designs for passive solar heating also provide day lighting. Day lighting is simply the use of natural sunlight to brighten up the interior of a building.

By tapping available renewable energy, solar water heating reduces consumption of conventional energy that would otherwise be used. Each unit of energy delivered to heat water with a solar heating system yields an even greater reduction in use of fossil fuels. Water heating by natural gas, propane, or fuel oil is only about 60% efficient, and although electric water heating is about 90% efficient, the production of electricity from fossil fuels is generally only 30% to 40% efficient. Reducing use of fossil fuel for water heating not only saves stocks of the fossil fuels but also eliminates the air pollution and climate change gas emission associated with burning those fuels.

Figure 3 Solar collectors on the roofs of these buildings can be used to provide energy for heating water and regulating indoor climates.

Teaching Tip

Figure 3 on page R79 is a photograph of buildings fitted with rooftop solar collectors. Solar panels are simplest to install and use when they are laid flat on the existing roof, as in this figure, although they are more efficient when they are mounted to face directly south at an angle appropriate for the local latitude.

Earth's Natural Resources Energy Resources

Photovoltaics

Generation of electricity is another major use of solar energy. The most familiar way is to use photovoltaic (PV) cells, which are used to power toys, calculators, and roadside telephone call boxes. PV systems convert light energy directly into electricity. Commonly known as solar cells, these systems are already an important part of your lives. The simplest systems power many of the small calculators and wristwatches. More complicated systems provide electricity for pumping water, powering communications equipment, and even lighting homes and running appliances. In a surprising number of cases, PV power is the cheapest form of electricity for performing these tasks. The efficiency of PV systems is not high, but it is increasing as research develops new materials for conversion of sunlight to electricity. The use of PV systems is growing rapidly.

Wind Power

People have been using wind power for hundreds of years to pump water from wells. Only in the past 25 years, however, have communities started to use wind power to generate electricity. The photograph in *Figure 5* is taken from an electricity-generating wind farm near Palm Springs, California. The triple-blade propeller is one of the most popular designs used in wind turbines today. Wind power can be used on a large scale to produce electricity for communities (wind farms), or it can be used on a smaller scale to meet part or all of the electricity needs of a household.

Figure 4 Wind turbines on a wind farm in California.

Teaching Tip

Figure 4 on page R80 is a photograph of wind turbines in California. For the optimum balance between energy output and installation cost of the wind farm, the wind turbines have to be spaced far enough apart that the wake of the nearest turbine upstream has accelerated back to the approximate ambient wind speed. That distance is surprisingly large.

Teaching Tip

Figure 5 on page R81 maps the location of wind-power plants in the United States It should be obvious to students that most wind farms have been sited in areas of relatively high-velocity wind class. This figure also suggests the great potential for wind power: many areas of the United States with high wind-power class have no wind-power installations yet. You may want to use this to lead a discussion with your class using the Blackline Master provided below.

Figure 6 shows typical wind velocities in various parts of the country. California leads the country in the generation of electricity from wind turbines (in 2000, California had a wind-power capacity of 1700 MW, with 20,000 wind turbines). Many other areas in the country have a high potential for wind power as well. These areas include the Rocky Mountains, the flat Midwest states, Alaska, and many other areas. Commercial wind turbines can have blades with a diameter as large as 60 m.

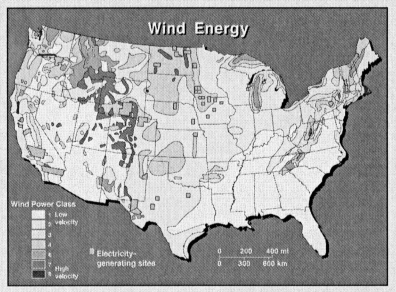

Figure 6 Location of electricity-generating wind-power plants relative to wind-power classes in the United States.

The cost of producing electricity from wind has dropped significantly since the early 1990s. This is due mainly to design innovations. In 2000, the cost of wind power was less than six cents per kilowatt-hour, which makes it competitive with electricity produced by coal-fired plants. The price is expected to decrease below five cents per kilowatt-hour by 2005.

Check Your Understanding

1. How does the Sun produce solar energy?

2. What factors determine the rate at which solar energy is received by a given area of the Earth's land surface?

3. What are the differences between passive solar heating systems and active solar heating systems?

4. Why must solar energy systems occupy a much larger area, per unit of energy produced, than conventional systems that burn fossil fuels?

EarthComm

Blackline Master Energy Resources 8.1
Wind Energy

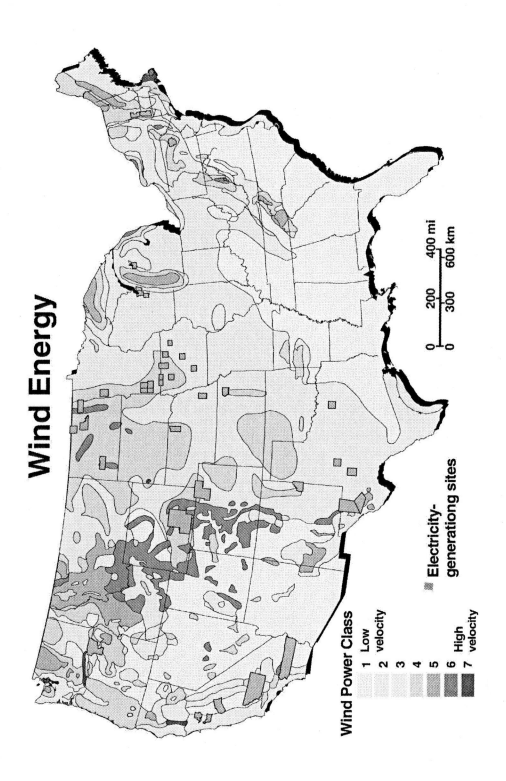

Check Your Understanding

1. Solar energy is produced in the extremely hot core of the Sun, by the process of nuclear fusion.

2. The rate at which solar energy is received by a given area of the Earth's land surface is influenced by:
   - latitude
   - season of the year
   - time of day
   - cloudiness of the sky
   - clearness of the air
   - slope of the land surface

3. Passive solar heating systems do not rely on special mechanical equipment, but produce heat from light that is trapped by the design of a structure. Active solar heating systems use special equipment, in the form of a solar collector, to collect and distribute solar energy. Active heating systems are more effective than passive systems, but they are more expensive to install and maintain.

4. Solar energy systems must occupy a much larger area, per unit of energy produced, than conventional systems that burn fossil fuels. This is because fossil fuels are derived from plant and animal tissues that contain chemical energy that has been concentrated from sunlight by photosynthesis.

Assessment Tool

Check Your Understanding Notebook-Entry Evaluation Sheet
Use this sheet to evaluate the extent to which students understand the key concepts explored in **Activity 8** and explained in **Digging Deeper**, and to evaluate the students' clarity of expression.

NOTES

Earth's Natural Resources Energy Resources

Understanding and Applying What You Have Learned

1. Suppose that you live in a region that requires 1 million Btus (or 293 kWh) per day to heat a home. An average solar water heating system costs $50 per square foot to install with an output of 720 Btu/ft^2/day.

   a) What size of system would be required to meet these heating requirements?

   b) What would the system cost?

   c) Find out how much it costs to heat a home for five months in the winter where you live (November through March). Calculate how many years it would take to recover the cost of the solar-water heating system.

2. Why is metal used for tubing (rather than the plastic tubing that you used in the activity) in flat plate solar collectors? Use what you have learned about heat conduction to explain.

3. Why is wind power more efficient than generating electricity from all the methods that involve heating water to make steam?

4. Why is wind power less efficient than hydropower?

5. A wind turbine is placed where it is exposed to a steady wind of eight mph for the entire day except for two hours when the wind speed is 20 mph. Calculate the wind energy generated in the two windy hours compared with the 22 less windy hours.

6. Looking at the United States map of wind velocities, where does your community fall in terms of wind potential?

7. Assume that you live in a community in eastern Colorado.

   a) Use the wind map above to determine the number of wind turbines that would be needed to produce the electricity equivalent to a large electric power plant (1000 MW). Assume that each of the wind turbines has a blade diameter of 50 feet.

   b) If these wind turbines are replacing a coal-fired power plant that burns coal with a ratio of pounds of SO_2 per million Btu's of 2.5, calculate the reduction in the amount of SO_2 emitted per year.

   c) What are some problems that might be encountered with the wind farm?

Preparing for the Chapter Challenge

Your challenge is to find a way to meet the growing energy needs of your community. You have also been asked to think about how to reduce the environmental impacts of energy resource use. Using what you have learned, describe how some of the energy needs of your community could be accommodated by wind or solar power.

Understanding and Applying What You Have Learned

1. a) Divide the daily Btu requirement (1 million Btu/day) by the daily Btu output per square foot of the system (720 Btu/day per square foot) to obtain the number of square feet needed.

    b) Multiply the result from 1(a) by the $50 per square foot cost.

    c) Answers will vary.

2. Metal is used for tubing in flat-plate solar collectors because metal is an excellent conductor of heat, so it is more efficient at transferring heat from the warmed surface of the tubing to the relatively cool water flowing inside the tubing.

3. To answer this question, students will probably need to review **Digging Deeper** in **Activity 1**. Any method of generating electricity by use of steam to turn a turbine is inherently inefficient because of the large amount of heat wasted in the process.

4. Wind turbines extract a relatively small proportion of the mechanical energy of the wind, because the entire rotor is exposed to the full force of the wind, and the wind force on the parts of the rotor that are moving *into* the wind partly counteract the wind force on the parts of the rotor that are moving *with* the wind. In a hydropower installation, the geometry of the turbine can be designed to minimize the effect.

5. Students should use the sample calculation of wind power on page R75 twice: once for a wind of 8 mph for 22 hours, and once for a wind of 20 mph for 2 hours.

6. Answers will vary depending upon where you live.

7. This question refers to the wind map on page R81 of the student text. In order to answer the question as written, you must convert the wind power class ratings to average wind speed. This data is summarized in the table below. The relationship between wind speed and wind power class depends in part on the height above the ground. The data shown in *Figure 5* on page R81 is based on a height of 50 m (164 ft).

wind power class	average wind speed (meters/second)
1	0-5.6
2	5.6-6.4
3	6.4-7.0
4	7.0-7.5
5	7.5-8.0
6	8.0-8.8
7	8.8-11.9
8	>11.9

As can be seen from the table, each power class represents a range of possible wind speeds. Given the wind speeds, the calculation of power follows the example given on page R75. (Note that the wind power per square meter equals 0.65 the cube of the wind velocity, as noted in **Part B, Step 8** of **Investigate** above).

   a) In order for students to answer **Questions 7 (a)** and **7 (b)**, they will need to know that 1 m = 3.28 ft.

Portions of eastern Colorado fall into Wind Power Class 7. Assuming that the proposed wind turbines are placed in this region, then according to the table given here, this corresponds to an average wind speed of 8.8-11.9 m/s (19.7-26.6 mph). For the purposes of this calculation, we will use the median value of 10.4 m/s.

First, the area of the turbine must be calculated in square meters. The diameter of the turbine is 50 ft. This equates to 15.24 m (50 ft/3.2 ft/m = 15.24 m). Dividing by two yields a radius of 7.62 m. Using the formula for the area of a circle (πR^2), this yields a turbine area of 182.51 m^2 (3.14 x 7.62^2). Using the median wind velocity of 10.4 m/s yields a theoretical power production of 133,444 W (10.4^3 x 0.65 x 182.51).

Considering an efficiency of 40%, this results in an actual power yield of 53,378 W per turbine. 1000 megawatts equals 1 billion watts, so dividing one billion by 53,378 yields 18,734 turbines! If 11.9 m/s wind speed was used in the calculation, the result would be 5002 turbines.

   b) From the conversion table given on page R7, 1 kWh = 3413 Btu. Dividing both sides by hours, one arrives at 1 kW = 3413 Btu/hr. The turbine farm is proposed to generate 1000 MW or 1,000,000 kW. This equates to 3.413 x 10^9 Btu/hr. There are 8760 hrs/yr, so the wind farm would generate 2.99 x 10^{13} Btu/year. Dividing by 1,000,000 to convert to MBtu and multiplying by 2.5 lb/MBtu yields 74,744,700 lb/yr of SO$_2$ saved.

c) Answers will vary. Some possible problems that could occur with the wind farm might include periods of slack wind, finding a location with a large enough percentage of usable land area, maintenance of the many turbines, etc.

Preparing for the Chapter Challenge

By now, students should have a clear sense of how the population increase proposed by the **Chapter Challenge** will affect energy consumption in their community. They should also have an idea of whether or not their community will be able to accommodate the increased demand for energy. Students should explore the alternative energy sources investigated in **Activity 8** to determine whether they could be used to help the community avoid a potential power crisis. Remind students that this paper will be an important component of their **Chapter Challenge**.

Inquiring Further

1. **Solar-thermal electricity generation**

   Research how a solar thermal power plant, like the LUZ plant in the Mojave Desert in California, produces electricity. Diagram how such a system works. What are the kilowatt-hour costs of producing electricity using this method? What does the future hold for power plants of this kind?

2. **Photovoltaic electricity**

   Investigate photovoltaic electricity generation and discuss the results of your investigation with the class.

3. **History of wind energy or solar energy**

   Investigate how people in earlier times and in different cultures have harnessed the Sun or wind energy. What developments have taken place in the past hundred years? How is wind energy being used today? Include diagrams and pictures in your report.

4. **Wind farms**

   Prepare a report on how electricity is generated on wind farms. Describe types of wind generators, types and sizes of wind farms, the economics of electricity production on wind farms, and the locations of currently operating wind farms in the United States. Include diagrams.

Inquiring Further

1. Solar thermal electricity generation

Solar thermal power plants use large reflectors to concentrate solar radiation onto a solar receiver on the ground. The receiver heats water to produce steam, which turns a turbine to generate electricity. The biggest problem associated with solar thermal power plants is that when the Sun is not shining, as at night or on cloudy days, the plants cannot generate electricity. Therefore, most plants make use of hybrid technology, meaning that they make use of the Sun when available and rely on another source, like natural gas, at other times. Students can visit the *EarthComm* web site to find additional information.

2. Photovoltaic electricity

Photovoltaic devices generate electricity directly from sunlight. Photovoltaics are an attractive alternative to electricity generation because they can provide a method of electric power generation that is clean, nonpolluting, and renewable. However, it is expensive: photovoltaic electricity costs between 30 and 40 cents per kilowatt hour, which makes it three to four times more expensive than conventional power generation at a centralized electric utility. Direct your students to the *EarthComm* web site to learn more about photovoltaic electricity.

3. History of wind energy or solar energy

The earliest known use of wind power is the sailboat. The first windmills were developed around A.D. 500 to pump water. Early windmills were used mainly to grind grain. Over time, windmill development moved from the use of simple devices, to heavy material-intensive devices, to the use of light material-efficient devices. The first use of a large windmill to generate electricity was a system built in Cleveland, Ohio, in 1888 by Charles Brush. The development of bulk-power, utility-sized, wind energy conversion systems was begun in Russia in 1931 with the 100-kW Balaclava wind generator. The development of modern vertical-axis rotors was begun in France by G.J.M. Darrieus in the 1920s. Students can find more information and view images of these systems by visiting the *EarthComm* web site.

The use of solar energy dates back more than 100 years, to the middle of the industrial revolution. Solar power plants were constructed to produce steam from the heat of the Sun, which was used to drive the machinery of the time. Henri Becquerel discovered the photovoltaic effect. Becquerel's research was extended by Werner Siemens, among others. Early photovoltaic applications focused on sensing and measuring light; with increased technology, however, the production of photovoltaic energy has become more feasible. Today, solar energy is used in two forms: solar thermal (the heat of the Sun is used to heat water to produce steam to turn a turbine and generate electricity) and photovoltaic (using energy of the Sun directly).

4. Wind farms

Electricity-generating wind turbines come in many shapes and sizes and produce anywhere from a few tens of watts to multiple megawatts of electricity. Most turbines consist of a set of blades connected to a generator. As the blades of the windmill turn, they turn the generator to produce power. The most common type of wind turbine is the horizontal axis type, which has blades like an airplane propeller. In vertical axis turbines, the blades move around a vertical axis (rather than being mounted to the front of the axis). Most wind farms are located in the western regions of the United States, as shown on the map on page R81.

Chapter 1

NOTES

Earth Science at Work

ATMOSPHERE: *Air-Monitoring Technician*
Sophisticated equipment is used by government agencies to monitor air quality. Reports are issued that inform the public as to the type and amount of pollutants present on any given day.

BIOSPHERE: *Environmental Scientist*
Companies are concerned about the impact they may have on the environment. Environmental impact studies are completed before new projects are undertaken.

CRYOSPHERE: *Snowmobile Dealers*
There are approximately 1.6 million registered snowmobiles in the United States. They are often used for recreation, but in some areas, they are the preferred forms of transportation during winter months.

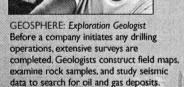

GEOSPHERE: *Exploration Geologist*
Before a company initiates any drilling operations, extensive surveys are completed. Geologists construct field maps, examine rock samples, and study seismic data to search for oil and gas deposits.

HYDROSPHERE: *Turbine Manufacturer*
In the generation of hydroelectricity, falling water strikes a series of blades attached around a shaft, causing a turbine to rotate.

How is each person's work related to the Earth system, and to energy resources?

R 84

NOTES

Energy Resources and Your Community: End-of-Chapter Assessment

1. An object is thrown straight up in the air. At the moment the object reaches its highest point, its kinetic energy
    a) is equal to its potential energy
    b) is highest and its potential energy is lowest
    c) is lowest and its potential energy is highest.
    d) is transformed into heat energy.

2. Power is the
    a) energy output of electric utility company.
    b) rate at which work is done.
    c) force per unit area applied to an object.
    d) strength of an object's kinetic energy.

3. Two covered pots of water sit in a warm oven overnight. The large pot has two liters of water and the small pot has one liter of water. Which of the following statements is true?
    a) The large pot has twice the heat of the small pot and is at twice the temperature.
    b) The large pot has twice the heat of the small pot and is at the same temperature.
    c) The two pots have equal heat, but the temperature of the small pot is twice as high.
    d) The two pots have equal heat and temperature.

4. If the energy input of a heating system is 300,000 Btus and the energy output of the system is 150,000 Btus, what is the efficiency of the system?
    a) 150,000 Btus.
    b) 2 Btus.
    c) 50%.
    d) cannot be answered without knowing how long the system is operating.

5. In the United States, what is our major energy resource used to produce electricity?
    a) fossil fuels.
    b) water.
    c) nuclear.
    d) geothermal.

6. In the United States, what is the leading renewable energy source used to produce electricity?
    a) fossil fuels
    b) water
    c) nuclear
    d) solar photovoltaic
    e) geothermal

7. The process by which solar energy is stored in the biosphere is known as
    a) decomposition.
    b) oxidation.
    c) respiration.
    d) photosynthesis.

8. The greater part of the carbon in the biosphere
    a) becomes stored as a carbon compound in coal.
    b) becomes stored as a hydrocarbon compound in oil and natural gas.
    c) gets buried as organic carbon compounds in sediments.
    d) is soon returned to the atmosphere as carbon dioxide.

9. The amount of energy that enters the biosphere is _____ the energy lost from the biosphere.
    a) 10% of
    b) less than 30% of
    c) nearly equal to
    d) nearly twice that of

10. Fossil fuels are considered to be _____ resources because they are consumed _____ than they form.
    a) renewable / far faster than
    b) renewable / far slower than
    c) nonrenewable / far faster than
    d) nonrenewable / far slower than

11. Which of the following statements about coal resources is true?
    a) Each state has coal resources.
    b) The United States has an unlimited supply of coal resources.
    c) Coal production is increasing in each of the three major United States coal basins.
    d) The amount of energy per unit mass depends upon the type of coal.

12. Fossil fuels are responsible for 98% of United States emissions of
    a) acid rain.
    b) carbon dioxide.
    c) ozone.
    d) oxygen and nitrogen.

13. According to the First Law of Thermodynamics, the amount of chemical energy consumed when a fossil fuel is burned
    a) is less than the amount of heat energy released.
    b) is greater than the amount of heat energy released.
    c) cannot be equal to the amount of heat energy released.
    d) is equal to the amount of heat energy released.

14. The global movement of carbon from one reservoir to another is known as
    a) the Law of Conservation of Matter.
    b) the carbon cycle.
    c) global warming.
    d) the greenhouse effect.

15. Which of the following neutralizes acid rain?
    a) soil or bedrock with a low pH.
    b) soil or bedrock that is slightly more acidic than the acid rainwater.
    c) soil or bedrock that is rich in granite.
    d) soil rich in lime or bedrock rich in limestone.

16. The advantage to switching from fossil fuels to photovoltaics is that
    a) photovoltaic systems are 100% efficient.
    b) photovoltaic systems are inexpensive.
    c) it would reduce production of carbon dioxide.
    d) it would stop global warming.

17. The amount of oil currently produced in the United States _____ the amount of oil imported from foreign countries.
    a) is less than
    b) is equal to
    c) is greater than
    d) has always been greater than

18. Which of the following statements about oil and gas resources in the United States is true?
    a) Every state has at least some oil and gas deposits.
    b) Oil and gas deposits accumulate in sedimentary basins.
    c) There is no trend or pattern to the distribution of oil and gas deposits.
    d) Oil and gas are always found together.

19. Between 1850 and 2000, the primary energy sources used in the United States has shifted from
    a) wood to coal to petroleum.
    b) petroleum to nuclear power to coal.
    c) petroleum to coal to nuclear power.
    d coal to petroleum to solar.

20. The amount of oil produced from a reservoir
    a) is equal to the amount of oil in the reservoir.
    b) is controlled by the porosity and permeability of the reservoir.
    c) depends upon pressure in the reservoir.
    d) is a, b, and c.
    e) is b and c.

21. Designing a home to allow maximum sunlight in winter and trapping the heat inside is an example of which kind of solar system?
    a) solar water heating
    b) passive solar heating
    c) photovoltaic solar
    d) active solar

22. Transforming light energy directly into electricity is known as
    a) photovoltaic conversion.
    b) passive solar conversion.
    c) solar thermal collecting.
    d) flat-plate collecting.

23. Which of the following control(s) the amount of solar energy received at a given location?
    a) insolation.
    b) elevation.
    c) longitude.
    d) a, b, and c.
    e) a and c.

Answer Key

1. c; 2. b; 3. b; 4. c; 5. a; 6. b; 7. d; 8. d; 9. c; 10. c; 11. d; 12. b; 13. d;
14. b; 15. d; 16. c; 17. a; 18. b; 19. a; 20. e; 21. b; 22. a; 23. a

Teacher Review

Use this section to reflect on and review the investigation. Keep in mind that your notes here are likely to be especially helpful when you teach this investigation again. Questions listed here are examples only.

Student Achievement

What evidence do you have that all students have met the science content objectives?

_____

Are there any students who need more help in reaching these objectives? If so, how can you provide this?_____

_____

What evidence do you have that all students have demonstrated their understanding of the inquiry processes?_____

_____

Which of these inquiry objectives do your students need to improve upon in future investigations? _____

_____

What evidence do the journal entries contain about what your students learned from this investigation? _____

_____

Planning

How well did this investigation fit into your class time?_____

_____

What changes can you make to improve your planning next time? _____

_____

Guiding and Facilitating Learning

How well did you focus and support inquiry while interacting with students?

_____

What changes can you make to improve classroom management for the next investigation or the next time you teach this investigation? _____

_____

How successful were you in encouraging all students to participate fully in science learning?_____

_____

How did you encourage and model the skills values, and attitudes of scientific inquiry? _____

_____

How did you nurture collaboration among students?_____

_____

Materials and Resources

What challenges did you encounter obtaining or using materials and/or resources needed for the activity? _____

_____

What changes can you make to better obtain and better manage materials and resources next time? _____

_____

Student Evaluation

Describe how you evaluated student progress. What worked well? What needs to be improved? _____

_____

How will you adapt your evaluation methods for next time?_____

_____

Describe how you guided students in self-assessment. _____

_____

Self Evaluation

How would you rate your teaching of this investigation? _____

_____

What advice would you give to a colleague who is planning to teach this investigation? _____

_____

NOTES

2

Mineral Resources ...and Your Community

EARTH'S NATURAL RESOURCES
CHAPTER 2

MINERAL RESOURCES...
AND YOUR COMMUNITY

Chapter Overview

The **Chapter Challenge** for **Mineral Resources and Your Community** is for students to help a local company use mineral resources, preferably from the community, to create a beverage container. Students are asked to prepare a report that explains what mineral resources are, how they are used and extracted, and what impact their use has on the environment—all relative to the proposed beverage container. They begin the chapter by surveying the materials that are currently being used for beverage containers in their community. They learn to identify minerals. They learn how mineral resources are found, extracted, and processed, and they explore the environmental impacts of these activities. By the end of the chapter, students have a better understanding of what makes minerals a valuable resource and how the use of mineral resources affects the other Earth systems.

Chapter Goals for Students

- Understand how mineral resources are tied to other Earth systems.
- Participate in scientific inquiry and construct logical conclusions based on evidence.
- Recognize that minerals are an indispensable natural resource whose use and impact on the environment needs to be carefully monitored.
- Appreciate the value of Earth science information in improving the quality of life, globally and within the community.

Chapter Timeline

Chapter 2 takes about three weeks to complete, assuming one 45-minute period per day, five days per week. Adjust this guide to suit your school's schedule and standards. Build flexibility into your schedule by manipulating homework and class activities to meet your students' needs.

A sample outline for presenting the chapter is shown on the following page. It assumes that the teacher assigns homework at least three nights a week and assigns **Understanding and Applying What You Have Learned** and **Preparing for the Chapter Challenge** as group work to be completed during class. This outline also assumes that **Inquiring Further** sections are reserved as additional, out-of-class activities. This is only a sample, not a suggested or recommended method of working through the chapter; adjust your daily and weekly plans to meet the needs of your students and your school.

Day	Activity	Homework
1	**Getting Started; Scenario; Chapter Challenge; Assessment Criteria**	
2	**Activity 1 – Investigate**	**Investigate (community survey); Digging Deeper; Check Your Understanding**
3	**Activity 1 – Review; Understanding and Applying; Preparing for the Chapter Challenge**	
4	**Activity 2 – Investigate, Parts A – D**	**Digging Deeper; Check Your Understanding**
5	**Activity 2 – Review; Understanding and Applying; Preparing for the Chapter Challenge**	
6	**Activity 3 – Investigate**	**Digging Deeper; Check Your Understanding**
7	**Activity 3 – Review; Understanding and Applying; Preparing for the Chapter Challenge**	
8	**Activity 4 – Investigate, Parts A – C**	**Digging Deeper; Check Your Understanding**
9	**Activity 4 – Review; Understanding and Applying; Preparing for the Chapter Challenge**	
10	**Activity 5 – Investigate**	**Digging Deeper; Check Your Understanding**
11	**Activity 5 – Review; Understanding and Applying; Preparing for the Chapter Challenge**	
12	**Activity 6 – Investigate, Parts A and B**	**Digging Deeper; Check Your Understanding**
13	**Activity 6 – Review; Understanding and Applying; Preparing for the Chapter Challenge**	
14	**Complete Chapter Report**	**Finalize Chapter Report**
15	**Present Chapter Report**	

Chapter 2

National Science Education Standards

Designing a beverage container using mineral resources sets the stage for the **Chapter Challenge**. Students learn the value of minerals as natural resources. Through a series of activities, students begin to develop the content understandings outlined below.

CONTENT STANDARDS

Unifying Concepts and Processes
- System, order, and organization
- Evidence, models, and explanation
- Constancy, change, and measurement

Science as Inquiry
- Identify questions and concepts that guide scientific investigations
- Design and conduct scientific investigations
- Use technology and mathematics to improve investigations
- Formulate and revise scientific explanations and models using logic and evidence
- Communicate and defend a scientific argument
- Understand scientific inquiry

Earth and Space Science
- Energy in the Earth system
- Geochemical cycles
- Origin and evolution of the Earth system
- Origin and evolution of the universe

Science and Technology
- Communicate the problem, process, and solution
- Understand science and technology

Science in Personal and Social Perspectives
- Personal and community health
- Natural resources
- Environmental quality
- Natural and human-induced hazards
- Science and technology in local, national, and global challenges

History and Nature of Science
- Science as a human endeavor
- Nature of scientific knowledge
- Historical perspectives

Key Science Concepts and Skills

Activities Summaries	Earth Science Principles
Activity 1: Materials Used for Beverage Containers in Your Community Students think about the different kinds of containers used to hold beverages. They survey the kinds of beverage containers found in their school and determine the kinds of materials used to make these containers. Students repeat the survey for the entire community.	• Materials used to make beverage containers
Activity 2: What are Minerals? Students examine a series of mineral samples to develop a list of properties that can be used to identify minerals. They break pieces of calcite and halite to understand the concept of mineral cleavage. Students stack marbles within cardboard enclosures to develop a sense of different atomic arrangements in minerals. Finally, they use a test kit to identify the minerals they examined at the start of the activity.	• Rocks and minerals • Mineral identification • Physical properties of minerals: hardness, luster, streak, specific gravity, cleavage, crystal shape, color
Activity 3: Where are Mineral Resources Found? Students examine a map that shows the distribution of mineral resources in the United States; then they study the mineral resource map for their state. Students use the state map to zoom in and create a mineral resource map for their community. This helps them determine the resources available to make their beverage containers.	• Distribution of mineral resources in the United States • Distribution of mineral resources in the oceans and on the continents • Formation of mineral deposits
Activity 4: How are Minerals Found? Students create a mineral deposit using sand and iron filings. They exchange deposits and use a compass to create a map of the deposit. On the basis of the maps they have produced, students select sites to drill for iron. They drill to extract core samples at different sites, and they monitor the cost of the exploration. Finally, students research the mineral commodities that are found in their community and the techniques used by geologists to find these commodities.	• Geologic maps as exploration tools • Geochemistry in resource exploration • Geophysical resource exploration
Activity 5: What are the Costs and Benefits of Mining Minerals? On the basis of their exploration in Activity 4, students develop a plan for excavating iron from their mineral deposit models. In addition to determining the net income from their excavation, students monitor the cost of the excavation and consider the cost of environmental reclamation following the excavation.	• Surface mining • Underground mining • Environmental reclamation
Activity 6: How are Minerals Turned into Usable Materials? Students separate sulfide from sulfide ore to model the process used to separate a mineral from a rock. Then, they separate copper from copper oxide to model the separation of an element from a mineral.	• Processing mineral ores

Chapter 2

Equipment List for Chapter Two:

Materials needed for each group per activity.

Activity 1 Part A

- A variety of beverage containers, representing a range of sizes, shapes, and materials

Activity 2 Part A

- Set of mineral samples (common minerals like quartz, halite, calcite, muscovite, biotite, feldspar, gypsum, hematite, galena, pyrite, magnetite, and talc)

Activity 2 Part B

- Calcite crystal
- Halite crystal
- Hammer
- Magnifying glass
- Strong light source
- Safety goggles

Activity 2 Part C

- Box of about 200 glass marbles
- Large piece of corrugated cardboard or posterboard
- Duct tape
- Large scissors

Activity 2 Part D

- Mineral identification kit, including copper penny (pre-1982), nail, piece of glass, steel file, streak plate, and magnet (or small object like a paper clip) to test for magnetism
- Copies of a mineral identification table (see sample table on page 341 of this Teacher's Edition)

Activity 3

- Copy of your state mineral resource map*
- Graph paper
- Colored pencils

Activity 4 Part A

- Scissors
- Paper half-gallon (2-L) milk or orange juice carton or a box/tub of similar dimensions
- Marker
- Graph paper
- Dampened sand
- Iron filings or black magnetite sand
- Powdered drink mix (optional)

Activity 4 Part B

- Magnetic compass (or stud finder or pencil magnet)
- Survey flags – toothpicks and masking tape
- Clear plastic straws
- Wooden dowel to push core sample out of straw (optional)

Activity 4 Part C

No additional materials are required.

Activity 5

- Plastic spoons
- Scrap paper
- Stopwatch
- Small containers to store sand and ore
- Magnet
- Balance or scale
- Calculator

Activity 6 Part A

- Clear jar with a lid
- Crushed sulfide ore
- Steel shot

*The *EarthComm* web site provides suggestions for obtaining these resources.

- Water
- Towel
- Piece of wire screen or a sieve
- Plastic cup
- Paper towels
- Liquid bubble-bath solution
- Drinking straw
- Squeeze bulb
- Index card

Activity 6 Part B

- Piece of circular paper or filter paper
- Funnel (about 4 in. diameter at the wide end)
- Crushed copper ore
- Vinegar
- Two bottles for collecting the leaching solution
- Dry-cell battery
- Steel wool
- Alligator clips (four)
- Metal spoon or a dime
- Insulated wire (two pieces)
- Lead fishing weight

Mineral Resources
...and Your Community

Getting Started

The Earth contains a vast array of useful minerals, including metals like gold, silver, and copper, and nonmetallic minerals like quartz and salt. Minerals are a natural resource. Their formation, extraction, and use occur within a complex Earth system. Mineral resources are nonrenewable. Some kinds of mineral deposits can no longer be formed on Earth, even given enough time, because conditions on and within the Earth have changed through geologic time.

- What are minerals?
- What Earth systems are involved in forming a mineral?
- How might the use of minerals change one of the Earth's systems?

What do you think? Look at the diagram of the Earth Systems at the front of this book. In your notebook, draw a picture to show what you know about the Earth systems that are involved in forming a mineral. Draw a second diagram to show how the use of a mineral might change the Earth system. Be sure to label the diagram so that others can understand what you have drawn. Be prepared to discuss your pictures with your small group and the class.

Scenario

A company in your community has developed a new line of beverages, which they claim are the most environmentally friendly drinks on the market. In order to maintain this promise, the company is seeking to design beverage containers that are produced with minimal environmental impacts. Additionally, this relatively new company does not have a large amount of money, and therefore would like the container to be both inexpensive and practical to manufacture.

Getting Started

Uncovering students' conceptions about Mineral Resources and the Earth System

Use **Getting Started** to elicit students' ideas about the main topic. The goal of **Getting Started** is not to seek closure (i.e., the right answer) but to provide you (the teacher) with information about the students' starting point and about the diversity of ideas and beliefs in the classroom. By the end of the chapter, students will have developed a more detailed and accurate understanding of minerals as resources and how they fit into the Earth system.

Students will most likely have heard the term mineral, particularly in association with nutrition. They also may have seen crystals and gems and recognize these as minerals that are mined from the ground (or geosphere). It is most likely that many students will have very little undestanding about what characteristics define a mineral, or how other spheres of the Earth system can be involved in the formation of mineral deposits. Encourage students to think beyond the geosphere and to consider how all of the spheres of the Earth system might relate to mineral resources.

In time, *EarthComm* students will understand that minerals are a component of the geosphere, and that their use and extraction affects the atmosphere, biosphere, cryosphere, and hydrosphere. Changes in the other Earth systems can cause changes in the mode of operation of the geosphere; likewise, changes within the geosphere can affect the other Earth systems. Students will come to appreciate the importance of understanding these changes, inasmuch as they can affect the availability of valuable minerals that we depend on, as well as the environment in which we live.

Chapter Challenge

Your challenge is to use mineral resources, preferably from your community, to design a beverage container that has a minimal environmental impact at all stages, from developing the raw materials to recycling the empty containers. It should also be cost-effective and practical.

You must also produce a report that substantiates your design and addresses these questions:

- What are mineral resources?

- What mineral resources do you plan to use to make your beverage container?

- Where were the mineral resources you plan to use to make your container formed?

- How are the mineral resources that are being used to make the container found?

- How are the mineral resources extracted?

- How can you use these minerals to make a practical container?

- What impact does the use of these mineral resources have on the environment?

- What name will you give the beverage that conveys how unusual and environmentally friendly it is? What is your rationale for the choice of name?

Assessment Criteria

Think about what you have been asked to do. Scan ahead through the chapter activities to see how they might help you to meet the challenge. Work with your classmates and your teacher to define the criteria for assessing your work. Devise a grading sheet for the assessment of the challenge. Your teacher may provide you with a sample rubric to help you get started. Record all this information. Make sure that you understand the criteria and the grading scheme as well as you can before you begin.

R87

Chapter Challenge and Assessment Criteria

Read (or have a student read) the **Chapter Challenge** aloud to the class. Students most likely will not have thought a great deal about mineral resources or where the raw materials for things like a beverage container come from. Students may have difficulty identifying different mineral resources that are likely candidates for making a beverage container. Many may consider mineral resources to be things like crystals and precious gems. Encourage students to consider the different kinds of containers that they have seen and to think about whether or not these things come from mineral resources. Once the students have settled on a few possible mineral resources it will help them to organize their thoughts about the remaining questions in the **Chapter Challenge**. Students may have difficulty answering these questions if they have never seen a mining operation. The mineral resource that students are most likely to have had some contact with is quartz sand, which is used in making glass. Allow students to discuss what they have been asked to do. Have students meet in teams to begin brainstorming what they would like to include in their **Chapter Challenge** reports. Request a brief summary in their own words of what they have been asked to do, and a description of attributes of a high-quality report.

Alternatively, lead a class discussion about the challenge and the expectations. Review the titles of the activities in the Table of Contents. To remind students that the content of the activities corresponds to the content expected for the Chapter Report, ask them to explain how the title of each activity relates to the expectations for the **Chapter Challenge**. Familiarize students with the structure of each activity. When you come to the section titled **Preparing for the Chapter Challenge**, point out that each activity contributes to the challenge in some way.

Guiding questions for discussion include:
- What do the activities have to do with the expectations of the challenge?
- What have you been asked to do?
- What should a good final report contain?

A sample rubric for assessing the **Chapter Challenge** is shown on the following pages. You can copy and distribute the rubric as is, or you can use it as a baseline for developing scoring guidelines and expectations that suit your needs. For example, you might wish to ensure that core concepts and abilities derived from your local or state science frameworks also appear on the rubric. You might also wish to modify the format of the rubric to make it more consistent with your evaluation system. However you decide to evaluate the chapter report, keep in mind that all expectations should be communicated to students and the expectations should be outlined at the start of their work. Please review **Assessment Criteria** (pages xxiv to xxv of this Teacher's Edition for a more detailed explanation of the assessment system developed for the *EarthComm* program.

Assessment Rubric for Chapter Challenge on Mineral Resources

Meets the standard of excellence. **5**	<u>*Significant*</u> information is presented about <u>all</u> of the following: • What mineral resources are • What mineral resources you plan to use to make your beverage container • Where the mineral resources you plan to use were formed • How the mineral resources you plan to use were found • How the mineral resources you plan to use were extracted • How the minerals can be used to make a practical container • What impact the use of these minerals has on the environment • What the name of your beverage is, and how its name conveys that the beverage is unusual and environmentally friendly <u>*All*</u> of the information is accurate and appropriate. The writing is clear and interesting.
Approaches the standard of excellence. **4**	<u>*Significant*</u> information is presented about <u>*most*</u> of the following: • What mineral resources are • What mineral resources you plan to use to make your beverage container • Where the mineral resources you plan to use were formed • How the mineral resources you plan to use were found • How the mineral resources you plan to use were extracted • How the minerals can be used to make a practical container • What impact the use of these minerals has on the environment • What the name of your beverage is, and how its name conveys that the beverage is unusual and environmentally friendly <u>*All*</u> of the information is accurate and appropriate. The writing is clear and interesting.
Meets an acceptable standard. **3**	<u>*Significant*</u> information is presented about <u>*most*</u> of the following: • What mineral resources are • What mineral resources you plan to use to make your beverage container • Where the mineral resources you plan to use were formed • How the mineral resources you plan to use were found • How the mineral resources you plan to use were extracted • How the minerals can be used to make a practical container • What impact the use of these minerals has on the environment • What the name of your beverage is, and how its name conveys that the beverage is unusual and environmentally friendly <u>*Most*</u> of the information is accurate and appropriate. The writing is clear and interesting

Assessment Rubric for Chapter Challenge on Energy Resources

Below acceptable standard and requires remedial help. **2**	_Limited_ information is presented about the following: • What mineral resources you plan to use to make your beverage container • Where the mineral resources you plan to use were formed • How the mineral resources you plan to use were found • How the mineral resources you plan to use were extracted • How the minerals can be used to make a practical container • What impact the use of these minerals has on the environment • What the name of your beverage is, and how its name conveys that the beverage is unusual and environmentally friendly _Most_ of the information is accurate and appropriate. Generally, the writing does not hold the reader's attention.
Basic level that requires remedial help or demonstrates a lack of effort. **I**	_Limited_ information is presented about the following: • What mineral resources are • What mineral resources you plan to use to make your beverage container • Where the mineral resources you plan to use were formed • How the mineral resources you plan to use were found • How the mineral resources you plan to use were extracted • How the minerals can be used to make a practical container • What impact the use of these minerals has on the environment • What the name of your beverage is, and how its name conveys that the beverage is unusual and environmentally friendly _Little_ of the information is accurate and appropriate. The writing is difficult to follow.

Chapter 2

ACTIVITY I—MATERIALS USED FOR BEVERAGE CONTAINERS IN YOUR COMMUNITY

Background Information

Aluminum

Aluminum is a metallic chemical element that occupies the third column of the third row of the periodic table of the elements. It is by far the most abundant metal element in the Earth's crust. Its position in the third row determines its chemical tendency to form positively charged ions with a charge of plus three. In minerals, aluminum tends to form oxides, which usually contain water as well. Aluminum also tends to be a substitute for silicon in the silica tetrahedra of silicate minerals. When rocks containing aluminum-bearing silicate minerals, called aluminosilicates, are weathered, they ultimately leave a residue of hydrous aluminum oxide minerals, which are extremely insoluble and resistant to further weathering.

The main ore of aluminum, called bauxite, is a mixture of hydrous aluminum oxides, usually with some percentage of hydrous iron oxides as well. Bauxite suitable for mining contains more than 50% aluminum oxide. Most bauxite is an earthy and semiconsolidated material. It forms by deep weathering of surface bedrock containing aluminosilicate minerals in a hot and humid climate. The major bauxite deposits in the world are located in Africa, South America, and the Caribbean. In the United States, there are some bauxite deposits in Alabama, Arkansas, and Georgia.

The pure metal, aluminum, is obtained from bauxite by smelting. Smelting involves reduction (removal of oxygen) of the aluminum oxides in the bauxite in high-temperature furnaces in an oxygen-free environment. The process requires a large input of energy, so the smelting operations tend to be located near sources of inexpensive electricity, usually hydropower—it is economically advantageous to bring the ore to the power source, rather than the other way around. Aluminum is a silvery white metal. Pure aluminum is soft, but it is commonly alloyed with small percentages of other metals to make it stronger and harder. There are many alloys of aluminum, designed for various special applications: some aluminum alloys are formulated for ease of machining, and others for enhanced resistance to corrosion. Although aluminum has a lower electrical conductivity than copper, it is often used for high-voltage power transmission lines because it is cheaper and lighter in weight. Although it is not as strong as steel, it is used extensively in the aircraft industry because of its lower density.

Glass

Glass is a hard substance, usually brittle and usually transparent, that consists mainly of silica (SiO_2) and metal elements (mostly sodium, potassium, and calcium) that are melted at high temperatures. The raw material for most glass is pure quartz sand. Sodium and potassium are added to lower the melting point of the quartz, and calcium serves as a stabilizer in the glass-making process. Many kinds of additives are used for special purposes:

• boron, for thermal and electrical resistance
• lead, for brilliance and weight
• barium, to increase the refractive index for optical equipment
• metal oxides, to impart color

Glass is an amorphous material, in the sense that it has no long-range, ordered atomic structure; technically, even naturally occurring glasses are not minerals. Glass forms from the silica melt by rapid cooling, called quenching. In this process, the material effectively becomes a solid by local bonding of silicon and oxygen atoms, without having nearly enough time for the atoms to become organized into a regular three-dimensional array. Because the constituent atoms are not locked into a regular array with strong bonds, however, glass actually has the properties of an extremely viscous liquid. For example, if a long glass rod is supported horizontally at its ends, the rod sags downward over a long period of time to take a permanent set.

Processes of glassmaking have changed little since ancient times. The materials are fused at high temperatures, skimmed, and cooled, and then the viscous molten glass is either poured into molds, blown into hollow objects, or drawn into rods or filaments. The shaped glass is annealed by being held at a high temperature to relieve internal forces set up by the shaping of the glass, and then cooled slowly to room temperature. Today, most hollow containers—like light bulbs or beverage containers—are machine blown, although fine ornamental glass, as well as specialized laboratory glassware, is still blown by hand.

Plastics

In rheology—the science that deals with the flow of materials—a plastic is a material that flows by deformation when an applied deforming force builds up to a certain value, called the yield strength of the material. The materials we know as plastics are rheological plastics only in a certain sense: they are organic materials that have the ability to flow into a desired shape when heat and pressure are applied, and then retain their shape when the temperature and pressure are brought back to room temperature. There are two basic types of plastics:

- thermosetting, which cannot be softened again when the temperature and pressure are raised
- thermoplastic, which can be softened and remolded repeatedly by application of heat and pressure

Chapter 2 deals with thermoplastic containers.

Plastics consist mostly of a binder together with plasticizers, fillers, and pigments for coloring. It is the binder that usually gives the plastic its main characteristics and its name. The binders are mostly synthetic organic materials, most of which are derived from petroleum hydrocarbons. These materials consist of very long, chainlike molecules called polymers. They are derived by polymerization of simpler organic compounds called monomers. When the thermoplastic binder is subjected to heat and pressure, the chainlike molecules can slide past one another, giving the material its plasticity. Plasticizers are added to the binder to increase toughness and flexibility, and fillers are added to improve particular properties, like hardness or shock resistance.

Plastic articles are usually manufactured from raw-material plastics in the form of powders. The powders are made into the desired shapes by molding. Molding is done in various ways. In compression molding, the powder is placed in a mold and compressed into the desired form by application of heat and pressure. In injection molding, an already plastic material is inserted into the mold under pressure. In extrusion molding, the plastic material is squeezed out of a pressurized chamber through an opening of the desired shape.

Chapter 2

Plastic containers for liquids are made by compression molding or injection molding. Most opaque or translucent plastic containers for liquids consist of low-density polyethylene, high-density polyethylene, or polypropylene; most transparent plastic containers for liquids consist of PET (or polyethylene terephthalate, also known as PETE). According to the American Plastics Council, PET is clear, tough, and has good gas and moisture barrier properties. This plastic is commonly used in soft-drink bottles and many other injection-molded consumer product containers. Other applications include strapping, molding compounds, and both food and non-food containers. Cleaned, recycled PET flakes and pellets are in great demand for spinning fiber for carpet yarns and producing fiberfill and geotextiles. Polyester is the nickname for this fiber.

Properties of PET:
• clarity
• strength/toughness
• barrier to gas and moisture
• resistance to heat

Packaging Applications of PET:
• plastic soft drink and water bottles
• beer bottles
• mouthwash bottles
• peanut butter and salad dressing containers
• oven-safe film
• oven-safe pre-prepared food trays

Recycled Products from PET:
• fiber
• tote bags
• bottles
• clothing
• furniture
• carpet

More Information – on the Web
Visit the *EarthComm* web site www.agiweb.org/earthcomm to access a variety of links to web sites that will help you deepen your understanding of content and prepare you to teach this activity. Many of the sites also contain images that you can download.

Goals and Assessment

Clarify that the goals indicate what students should understand and be able to do as a result of the activity. Make sure students understand that Chapter Assessments are based upon these goals.

Goal	Location in Activity	Assessment Opportunity
Investigate the kinds of beverage containers found in your community.	**Investigate**	Students' surveys of beverage containers are adequate; data are recorded correctly in table.
Determine what materials are commonly used to make beverage containers, specifically those found in your community.	**Investigate; Digging Deeper; Check Your Understanding** Questions 1 – 2 **Understanding and Applying What You Have Learned** Questions 1 and 3	Responses to questions are based on findings in the community and in the reading.
Determine why certain materials are used to make the beverage containers.	**Digging Deeper; Check Your Understanding** Question 3 **Understanding and Applying What You Have Learned** Question 3	Responses to questions closely match those given in Teacher's Edition.

Chapter 2

Earth's Natural Resources Mineral Resources

Activity 1

Materials Used for Beverage Containers in Your Community

Goals

In this activity you will:

- Investigate the kinds of beverage containers found in your community.

- Determine what materials are commonly used to make beverage containers, specifically those found in your community.

- Determine why certain materials are used to make the beverage containers.

Think about It

Different materials are sometimes used to perform the same function but with different levels of performance and cost. For example, softball bats can be made of aluminum or wood. The use of materials for a specific function can also vary by geographic location.

- When your community was first founded, what did the settlers use for beverage containers?
- Are those containers commercially practical today?

What do you think? Record your ideas in your *EarthComm* notebook. Be prepared to discuss your responses with your small group and the class.

Activity Overview

Students begin the investigation by brainstorming about different kinds of containers used to hold beverages. They consider the properties that make these containers more or less useful. Using this information, students design a table to collect data about the beverage containers used in their school cafeteria. They complete a similar survey for beverage containers used in the community. Students also begin to think about the materials used to make each container. **Digging Deeper** reviews different materials used for making beverage containers. The reading also looks at the costs and processes involved in the production of different kinds of containers.

Preparation and Materials Needed

You will want to provide your students with some sample empty beverage containers to analyze for the first part of the investigation. Try to find a variety of containers, including those of different sizes, shapes, and materials. Some ideas include glass bottles, cardboard cartons, plastic jugs, wooden containers, ceramic or clay mugs, aluminum cans, etc. The idea is to give students an assortment that will help them determine which properties to look for in the surveys they complete for their school cafeteria and their community.

Students will need to complete the survey of beverage containers in the community as a homework assignment. You may want to give them a couple of days (or a weekend) to do this.

Materials
• A variety of beverage containers, representing a range of sizes, shapes, and materials

Think about It

Student Conceptions

If students seem reluctant to answer this question, provide them with several examples of beverage containers that are in use today and ask them to think about how containers may have differed in the past. Students are likely to say that beverage containers were produced from resources like wood, clay (pottery), or even glass. Most likely, they will say that using these containers is not practical today because we have other containers that can be produced more quickly and at lower cost.

Answers for the Teacher Only

Answers may vary somewhat depending on the location of the community. Most areas of the United States were settled more than a hundred years ago. At that time, and ever since the European settlement of North America, most containers for liquids were made of glass. Beverage containers of the kind the students are likely to think about were made almost entirely of glass. The technology for producing glass containers was well developed long before the time of Columbus. Larger containers like barrels, were made of wood. Metal containers were much less common, although pewter kitchenware and dinnerware were in common use, as were iron and copper cooking pots. Plastic containers made their appearance within the living memory of the students' grandparents!

Assessment Tool

Think about It Evaluation Sheet
Use this evaluation sheet to help students understand and internalize the basic expectations for the warm-up activity.

NOTES

Chapter 2

Investigate

1. Break into small groups and discuss some of the variables that are important in designing a container to hold liquids. Your teacher will provide your group with some empty containers collected from the school or home to analyze.

   a) What are the pros and cons of using each of these containers?

   b) Produce a list of variables that you think are important.

2. As a class, decide which variables are most significant and how they can be measured.

   a) Record the list in your *EarthComm* notebook.

3. On the basis of the variables you selected, design a data table to help in the collection of information on materials used for beverage containers in your school cafeteria. What additional important information should be included on the data sheet (name, location where data are collected, date, etc.)?

   a) Keep a record of your data table in your notebook.

    Obtain permission before entering the school cafeteria. Follow guidelines provided to help maintain sanitary conditions (wash hands, do not touch utensils or cooking surfaces).

4. Use your table to analyze the beverage containers used in your school cafeteria.

   a) Collect and interpret the data.

   b) What hypotheses can you make about the various materials used to make containers?

5. After you have collected information from your school cafeteria, come back to the classroom and have a spokesperson from your group explain what you have learned so far.

6. As a class, discuss what your data means.

   a) Is your cafeteria representative of your whole community when it comes to use of beverage containers? Explain.

   b) How would your choice of location to sample influence your results?

7. Complete a survey of the beverage containers used in your community. As a class, decide where in your community data should be collected. Every student should have a different location. You might consider locations like supermarkets, neighborhood markets, convenience stores, vending machines, discount stores, gas stations that sell snacks, fast-food restaurants, or sports stadiums. Half of the class should collect the information from disposed containers at home or school. You will be using the form you previously developed, or a modified version of it, to collect information on all the kinds of beverages, kinds of containers for each, and frequency of use of each container in your community.

   a) Visit the location decided on in class and fill out your data table.

    Obtain permission from store managers before conducting surveys in the store. Have an adult with you. When collecting data about disposed containers, do not handle the containers or other trash with bare hands.

R 89

Investigate

Assessment Tools

EarthComm Notebook-Entry Checklist
Use this checklist as a quick guide for student self-assessment and/or an opportunity to quickly score student work. Add further criteria specific to your classroom needs or to this particular investigation.

Investigate Notebook-Entry Evaluation Sheet
Point out the criteria listed on this evaluation sheet that are relevant to this particular investigation. Encourage students to internalize the criteria by making them part of your assessment conversations as you circulate around the classroom. For example, while students are working, ask them criteria-driven questions like:
- Is your work thorough and complete?
- Are all of you participating in the activity?
- Do you each have a role to play in solving this problem? And so on.

1. Provide students with the sample beverage containers that you have collected. Encourage them to think of other beverage containers that you may not have supplied.

   a) Answers will vary depending upon the containers students are examining and also upon what they value in a container. For example, students might say that aluminum cans are lightweight, get cold quickly (but warm up quickly as well), are conveniently sized, and can be recycled. Glass is attractive, can also be recycled, but can break if dropped, etc.

   b) Answers will vary. Possible variables (which might better be called properties) might include:
   - size
   - shape
   - appearance
   - composition
   - whether it holds hot or cold liquids
   - thermal insulation properties
   - cost of production
   - possibility of recycling

2. a) Have students write their variables on the blackboard and discuss which variables are more important than others. Have the class choose the most important ones and develop a master list of variables.

3. **a)** As a class, develop a table students can use to collect data about beverage containers. Encourage them to think about additional useful information to include on the data sheet.

4. Arrange to take your students to the school cafeteria, or have them complete their survey on their lunch break. Alternatively, you could collect one of each kind of beverage container from the cafeteria and have the students complete the survey in the classroom. Answers to the questions will vary. In general, common beverage containers will include:
   - cardboard milk cartons
   - aluminum soda cans
   - glass juice bottles
   - plastic drink bottles
   - paper and Styrofoam® cups.

Some potential variables (types of data) that might be collected and put into the table include:
   - The type of material the container is made of.
   - The properties of that material (is it transparent, breakable, heavy, light, insulating, etc.).
   - Where the different containers were recovered from or most often used.
   - The kind of beverage they were intended to hold.

Hypotheses should attempt to relate the use(s) of the containers to some of the properties identified.

5. Student groups should prepare a brief summary of their findings and have one member of the group present the information to the class.

6. **a)** Most likely, the beverage containers in the cafeteria are not representative of the whole community.

   b) No one location is likely to be representative of the entire community, and each will likely favor one or more of the entire range of containers used in your community.

7. Students will need to complete this part of the investigation on their own time. You may want to prepare a list of places that students should visit, and then assign a student to each place.

   a) Students should use the data table designed earlier in the investigation to record their data. They may wish to modify the table slightly.

NOTES

Chapter 2

Earth's Natural Resources Mineral Resources

8. After you have collected your data, share it with the rest of the class and create a large database. On the basis of the entire data set, break into small groups and discuss and answer the following questions:

a) What types of materials are used in your community to hold liquids?

b) From the data you collected, why do you think certain materials are used for certain containers?

c) In your community, what is the most popular material for a new beverage container?

d) In your community, what is the most common material for a recyclable beverage container?

e) What are the raw ingredients necessary to make these containers both directly and indirectly?

f) From where do the raw ingredients come to make the containers?

g) Do you think your community has all the resources necessary to make these containers?

Reflecting on the Activity and the Challenge

In this activity, you discussed the variables that are important to consider when designing a beverage container. You then developed a method to survey the kinds of beverage containers found at your school cafeteria and throughout your community. This will help you to understand what materials are commonly used to make beverage containers and get you thinking about what factors you need to consider in designing your own container.

R 90

8. After the students have completed their surveys, the class will have quite a bit of data to deal with. You may wish to set up a data table using Microsoft Excel or similar spreadsheet software. You can then print out a copy of the table for each student group to discuss. Alternatively, the class could produce a large chart to be posted in the front of the room.

a) Answers will vary.

b) Answers will vary. Possible responses might include:
   • suitability of material for holding hot or cold liquids
   • cost of producing the container relative to the volume of liquid
   • the ability of a container to hold a pressurized liquid (from carbonation)
   • the ability of a container to maintain the flavor of a liquid for an extended shelf life

c) Answers will vary.

d) Answers will vary, but the most common materials will most likely be either aluminum for cans or glass for bottles.

e) Students may have a difficult time answering this question at this point. You may wish to explain to them that the term raw material refers to a material in its natural, unprocessed state. So, for example, the raw material used to produce cardboard containers is wood, and the raw material used to produce aluminum cans is bauxite.

f) Students may also have a difficult time answering this question at this point. Answers depend on what students have cited as the most common materials that are used to make the containers found in the community.

g) Answers will vary. Most communities will not have bauxite for aluminum cans or petroleum hydrocarbons for plastic cans. Some might not have trees for cardboard cartons!

Reflecting on the Activity and the Challenge

Students should now have a good sense of what kinds of beverage containers are common in their community and why these kinds of containers are used. They should also be thinking about what materials are used for each container and why. Have students begin to think about a name for their beverage, how large their container will be, what kind of liquid it will hold, etc.

Digging Deeper

MATERIALS USED FOR BEVERAGE CONTAINERS

Have you ever been on an impossible scavenger hunt? The items on your list just cannot seem to be found anywhere. Well, here is a sample list of items that you probably could not buy or find in your community today: tortoise-shell comb, plywood car, glass insulators, and a hand mirror made of gutta-percha (an early plastic material used in the mid-1800s). These are all real items, but today they can be found in a museum, not a store. What has changed? Although you can buy products that perform the same function, they are made from very different materials.

Figure 1 Glass insulators were first produced in the 1850s for use with telegraph lines.

Cost was formerly one of the main deciding factors in choosing a raw material for the manufacture of a product. Today, many other factors are involved, like manufacturing technology, product performance, and environmental issues. The choice of materials that products are made of affects everyone. The selection of materials influences the choice of resources used, technology development, environmental impacts, and worldwide competitiveness. Everyone has some influence on the choice of material. The manufacturer chooses a material that results in an acceptable product with a low price. The consumer influences a material choice by deciding how much he or she is willing to spend for a product. Society influences material decisions through concern about product standards, safety and environmental regulations, and bans on some material uses.

Digging Deeper

As students read the **Digging Deeper** section, the relevance of the concepts investigated in **Activity 1** will become clearer to them. Assign the reading for homework, along with the questions in **Check Your Understanding** if desired.

Assessment Opportunity

Reword or restructure the questions in **Check Your Understanding** for a brief quiz. Use the quiz (or a class discussion of the questions in the textbook) to assess your students' understanding of the main ideas in the reading and the activity.

Teaching Tip

Glass insulators were originally produced in the 1850s for use with telegraph lines. As technology developed, insulators were needed for a variety of applications, including telephone lines and electric power lines. The insulators covered wires and kept them from coming into contact with each other, and from getting wet and shorting out. Insulators like the one shown in *Figure 1* on page R91 were suspended off the ground on poles. Today, most wires are buried underground, and antique insulators are popular collectibles.

Chapter 2

Earth's Natural Resources Mineral Resources

To a manufacturer the most important performance measure for a soft-drink container is the ability of the container to hold the beverage under pressure while minimizing the amount of escaping carbonation. If too much carbonation is lost, the beverage tastes flat.

Aluminum, steel, glass, and plastic are the main materials for beverage containers. The 12-oz aluminum can dominates the market for reasons of performance and cost. At the 12-oz size plastic (PET — polyethylene terephthalate) bottles cannot compete with aluminum because too much carbonation leaks through the walls of the container, shortening the product's shelf life. About 33 aluminum cans can be made from one pound of aluminum. It is possible to make only 20 steel cans and only 12 glass bottles from one pound of each material. Glass is heavy and thus it costs about 10 times as much to handle and distribute than steel or aluminum.

Figure 2 One pound of aluminum yields about 33 aluminum cans.

Aluminum is preferred over steel largely because the prolific recycling of aluminum makes it much cheaper to produce aluminum cans. Over 65% of all aluminum beverage cans are recycled. The manufacture of aluminum from bauxite ore is extremely energy intensive, but the cost of making aluminum from recycled material is inexpensive. The manufacture of steel from iron ore uses much less energy than aluminum, so the cost savings in recycling steel are not as great.

Teaching Tip

The United States has the world's largest aluminum industry, producing approximately $39.1 billion in aluminum products and exports annually. Each year in the United States, more than 22 billion pounds of metal aluminum are produced, and the aluminum industry is responsible for about 1.8% of all energy consumed by the United States industrial sector. In the year 2000, the aluminum industry employed nearly 150,000 people. Companies in the United States produce more primary aluminum than any other country in the world, with most of the aluminum being used for transportation, beverage cans, and other packaging products. Review these facts with your students, pointing out that regardless of where their community is located, the aluminum industry has some kind of impact on it.

The photograph in *Figure 2* on page R92 shows how many aluminum cans can be made from one pound of aluminum. Technological advances have allowed the number of cans produced from one pound of aluminum to go from 23 in 1975 to about 33 today. Also, aluminum cans are about 52% lighter today than they were 20 years ago. Encourage students to consider how these advances have helped to make aluminum a good choice for producing beverage containers, noting that this helps to conserve resources, including energy.

Chapter 2

Why do you not see one-liter soda cans made from aluminum? Compared to the total amount of carbonation present in a one-liter volume, the amount of leakage is acceptable for a normal shelf life. An aluminum container would need a much thicker wall to withstand the carbonation pressure and still be stackable, and this would make the product too expensive. The raw material cost of plastic is about one-fifth that of aluminum and steel, and half that of glass. Thus, plastic is the material of choice for the larger containers.

The impact on the environment is now a major factor in designing new products and choosing materials for making those products. "Design for the Environment" (DFE) is now an important part of engineering a new product. Government estimates indicate that manufacturing operations in the United States account for 50% of all waste, over 11 billion tons annually, whereas household garbage accounts for only 2% of the waste generated each year. DFE includes keeping track of the environmental cost of components in the product. For example, if a computer has a circuit board that requires use of a chlorinated solvent to manufacture, that circuit board should carry a higher cost, because many chlorinated solvents pollute ground water and can cause cancer if not handled properly. Another aspect of DFE is using recycled materials to manufacture a product, hence reducing use of raw materials. Still another aspect is making a product easier to recycle when its life is over by making the product as simple as possible, using fewer different materials, and making the product easy to disassemble for recycling.

Figure 3 Aluminum cans can be recycled and the aluminum can be reused.

Check Your Understanding

1. What are the main materials used for producing beverage containers?

2. Compare and contrast the use of aluminum and glass for beverage containers.

3. Describe two factors that influence the selection of materials for making products.

Chapter 2

Teaching Tip

Recycling of aluminum cans in the United States (see *Figure 3* on page R93) has increased from a rate of 25% in 1975 to nearly 63% in 2000. Aluminum cans are recycled more than any other recyclable material; they constitute only about 1% of the total volume of waste in landfills today. The aluminum industry has worked to make recycling faster and easier: it can take as little as 60 days for a recycled can to make it back to a store shelf. Although it costs money to recycle cans, it is much less expensive than going through the process of mining and refining aluminum. Every can that is recycled saves energy and natural resources, and is environmentally beneficial in the sense that it reduces solid waste.

Check Your Understanding

1. The main materials used for producing beverage containers are aluminum, steel, glass, and plastic.

2. About 33 aluminum cans can be made from one pound of aluminum, whereas one pound of glass yields about 12 bottles. Aluminum is also preferred over glass because glass is heavy and thus costs about 10 times more to handle.

3. Answers will vary, but possible factors that influence the selection of materials for making products may include:
   - cost
   - manufacturing technology
   - product performance
   - toxicity of materials
   - environmental issues

Assessment Tool

Check Your Understanding Notebook-Entry Evaluation Sheet
Use this sheet to evaluate the extent to which students understand the key concepts explored in **Activity 1** and explained in **Digging Deeper**, and to evaluate the students' clarity of expression.

Teaching Tip

Toxicity of packaging materials, especially those used for edible products, is always a concern. For beverage cans, solubility of the basic material, or of trace substances in the basic material, is never nonzero. Glass and aluminum, the most common containers for edible liquids today, are virtually insoluble in water. Other metals have greater solubilities. Copper and lead have been used in the past as containers for storing and cooking liquids; both present health problems. Some believe that use of leaded wine containers contributed to the downfall of the Roman Empire!

Earth's Natural Resources Mineral Resources

Understanding and Applying What You Have Learned

1. The glass industry produced 10 million tons of glass containers in 1997. The population of the United States is 281 million people. Assume that none of the containers were exported. Answer the following questions. Be sure to show your work and to explain any other assumptions that you make.

   a) How many pounds of glass containers were produced per capita in the United States in 1997?

   b) What is the population of your state? Your community?

   c) How many pounds of glass containers were produced for your state in 1997?

   d) How many pounds of containers were produced for your community in 1997?

2. In 1999, 1.93 billion pounds of aluminum scrap were collected. Assume that one pound of aluminum yields 33.1 beverage containers.

   a) How many aluminum cans were recycled in 1999?

   b) 102.2 billion aluminum cans were produced in 1999. What percentage of cans were recycled?

3. Most glass beverage containers are soda-lime silicate glass, which is made from soda, lime, and silica. A typical soda-lime glass is 15% soda (Na_2O), 10% lime (CaO), and 75% silica (SiO_2). Assume that soda, lime, and silica sand cost $0.171, $0.269, and $0.05, per pound, respectively.

   a) What is the per-pound cost of soda-lime glass?

   b) If 12 glass bottles can be produced per pound of glass, what is the cost per bottle?

   c) What other expenses factor into using glass bottles for beverage containers?

Preparing for the Chapter Challenge

Write a short paper that reviews the common materials used for beverage containers in your community. Which of these materials do you think would be useful in designing your beverage container? Why? Do you think that your community has all of the resources necessary to make your container? You can include this paper as an introduction to your **Chapter Challenge**, to help justify your choice of materials.

R 94

Understanding and Applying What You Have Learned

1. You will want to remind students that there are 2204.62 pounds in a long ton and 2000 pounds in a short ton. When the term ton is used without a qualifying adjective, short ton is usually meant.

    a) To determine the pounds of glass produced per capita in the United States, students should divide the pounds of glass (10 million tons) by the population of the United States (281 million) and convert the resulting number of tons/person to pounds/person. Using short tons, this equates to 71.2 pounds per capita. If the student used the number for long tons, the answer is 78.5 pounds per capita.

    b) Answers will vary. Data on the population of your state can be found from a number of sources. Two possible sources for current census data for each state include the United States Census Bureau or your local state census office. See the *EarthComm* web site for more information regarding the census data of your state.

    c) Answers will vary; students can multiply their state population by the per capita value they calculated in **Question 1(a)**.

    d) Answers will vary; students can multiply their community population by the per capita value they calculated in **Question 1(a)**.

2. a) Approximately 63,883,000,000 cans were recycled in 1999. (1.93 billion x 33.1 cans).

    b) 62.5% of the aluminum cans were recycled in 1999 (63.883 billion cans/ 102.2 billion cans).

3. a) Students are told the composition of soda-lime glass and the cost per pound of each of the constituent materials. To determine the total cost of soda-lime glass, students will multiply the weight of each material (given by the percentage) by the cost.

    > **Soda:** 0.15 x \$0.171 = \$0.0257
    > **Lime:** 0.10 x \$0.269 = \$0.0269
    > **Silica:** 0.75 x \$0.05   = \$0.0375

    Students then sum the values to get the total cost:

    \$0.0257 + \$0.0269 + \$0.0375 = \$0.0901, or about 9 cents/pound

    b) The cost per bottle is equal to the cost per pound divided by the number of bottles produced by one pound:

    $$\$0.09/12 = \$0.0075$$

    c) Answers will vary, but may include responses like the cost of labor, running the machinery, etc.

Chapter 2

Preparing for the Chapter Challenge

This section gives students an opportunity to apply what they have learned to the **Chapter Challenge**. They can work on their paper as a homework assignment, or during class time within groups. **Activity 1** is designed as an introduction to containers that are commonly used to hold beverages and materials that are best suited for this purpose. The information students produce here can be incorporated as an introduction to their **Chapter Report**.

In designing an assessment rubric for this assignment, you may wish to evaluate how information needed to complete this activity is relevant to the example rubric that has been given for the chapter report. (See **Assessment Rubric for Chapter Report on Mineral Resources** on pages 304 and 305.) The following criteria from that example rubric are relevant to this assignment:

- What mineral resources you plan to use to make your beverage container
- Where the mineral resources you plan to use were formed
- How the mineral resources you plan to use were found

NOTES

Inquiring Further

1. **Beverage containers through time**

   How do you think beverage containers have changed over time? How do you think beverage containers have changed from when your parents were your age? From when your grandparents were your age? Visit the *EarthComm* web site to help you get started with your research on how beverage containers have changed through time.

2. **Beverages containing carbon dioxide**

   Strength and flexibility are two important mechanical properties of beverage containers. There are two main reasons for this:

   • Many beverages, like soda, beer, and champagne, are under pressure from dissolved carbon dioxide gas.

   • Many beverage containers need to withstand the force exerted from inside when the liquid freezes to ice.

   Using your school library, community library, or the Internet (visit the *EarthComm* web site to

get started), research the following topics, which are essential to a clear understanding of the two points listed above:

a) What is the maximum concentration of carbon dioxide that can be dissolved in water, and how does that concentration vary with temperature?

b) How much does water expand when it freezes, and how much force can be generated by the expansion?

Inquiring Further

1. Beverage containers through time

Students can interview their parents and grandparents to find out what beverage containers were like in the past. They are likely to discover that the materials used to make beverage containers have not changed much, but that the design (size, shape, etc.) of bottles and cans has varied greatly. Ask students to consider why they think this is so. Have them share their findings with the class. Ask them to consider differences between their findings. For example, students might find that beverage containers differed depending upon where their parents or grandparents grew up. Information about the history of beverage containers is available on the *EarthComm* web site.

2. Beverages containing carbon dioxide

a) The equilibrium concentration of carbon dioxide in water is a function of three factors:

- pressure
- temperature
- concentration of carbon dioxide in the vapor in contact with the water

The equilibrium concentration increases with pressure and with the ambient concentration of carbon dioxide gas, but decreases with increasing temperature. The first part of the question is somewhat misleading, because the answer is not relevant to the intent of this activity. The maximum concentration would be reached when the water temperature is at the freezing point, and the pressure and the concentration of carbon dioxide in the environment of the water are such that the carbon dioxide is at the point of condensation into a liquid.

b) When water freezes it expands its volume by about 9%. (The density of pure water at room temperature and pressure is 1 g/cm^3, whereas the density of ice is 0.917 g/cm^3.) During this expansion considerable pressure can be exerted on a confining vessel, as anybody that's ever seen a bottle full of water break upon freezing can attest! Freezing water can break apart rocks and is also responsible for frost heaves that buckle roadways and lift pilings out of the ground. These forces can be in excess of many thousands (even tens of thousands) of pounds.

ACTIVITY 2— WHAT ARE MINERALS?

Background Information

Minerals

A mineral is a naturally occurring crystalline solid. The definition excludes naturally occurring fluids like petroleum, and it also excludes naturally occurring solids, like amber or glass, that have no crystal structure. The essential characteristic of crystalline solids is that their constituent atoms are arranged in a regular, three-dimensional array, rather than being arranged randomly. Minerals are a very small subset of all the known crystalline solids; most crystalline solids do not occur naturally and are thus known only from being synthesized in the laboratory.

The atoms in minerals are held together by strong forces called bonds. To understand the nature of bonds, you must be somewhat familiar with atomic structure. All atoms consist of a positively charged nucleus around which a number of negatively charged electrons are in orbit. The fundamental nature of the electron orbits is complex. However, to understand how bonds behave in minerals, it is sufficient to know that electron orbits exist in distinctive groups known as shells. Chemical elements like helium, neon, or argon, whose atoms have just enough electrons to fill all the shells, are chemically almost entirely inert. Elements whose atoms have their outermost electron shell almost filled have a strong tendency to take on one or a few electrons to fill the shell; in so doing, they acquire a negative electric charge. Similarly, elements whose atoms have only one or a few electrons in their outermost shell lose those electrons to revert to a filled-shell configuration; in so doing, they acquire a positive electric charge. Such atoms with electric charges—positive or negative—are called ions.

According to Coulomb's Law, unlike electric charges attract one another and like charges repel one another. Certain combinations of positively charged and negatively charged ions can become packed together in a regular three-dimensional array in such a way that the sum of all the forces (both attractive and repulsive) among all the ions in the structure is attractive. This means that the structure is a stable one and can exist as a mineral. Almost all minerals are of this nature, and are called ionic crystals; the bonds in such a crystal are called ionic bonds. Ions can also be formed from two or more atoms that share their electrons in such a way that each atom effectively has a filled-shell configuration. The bonds among such atoms are called covalent bonds. Covalent bonds are much stronger than ionic bonds, and the covalently bonded atoms form a single ion that is then available for ionic bonding with other ions.

Most of the common minerals on Earth are silicate minerals. Silicate minerals are those in which a covalently bonded unit consisting of one silicon atom and four oxygen atoms becomes ionically bonded with various positively charged ions, most commonly calcium, magnesium, potassium, sodium, and/or iron. The distinctive property of silicate minerals is that these covalently bonded units can become combined into rings, chains, sheets, or even three-dimensional networks by sharing oxygen atoms. This phenomenon, called silica polymerization, is almost unique in the universe, and it makes the mineralogy of silicates extremely diverse. Common minerals like quartz, feldspars, micas, pyroxenes, and amphiboles are silicate minerals.

There are several other common groups of minerals in addition to silicates. In oxides, negatively charged oxygen ions are ionically bonded with iron or other positive ions: hematite and magnetite are common iron oxide minerals. In carbonates, a negatively charged carbonate ion consisting of one carbon atom covalently bonded with three oxygen atoms is ionically bonded with calcium, magnesium, and/or iron ions; calcite and dolomite are common carbonate minerals. In sulfides, a negatively charged sulfide ion is ionically bonded with a great variety of positively charged metal ions. Pyrite, an iron sulfide, is one of the most common sulfide minerals. Sulfide minerals serve as the ores of many of the heavy metals.

Mineral Identification

All minerals have distinctive physical properties resulting from the elements present, the arrangement of atoms, and the strength of bonding between atoms or ions. Observation and simple tests of these properties help in the identification of minerals. If you have had little opportunity to test the physical properties of the minerals you intend to use in this activity, you may want to practice beforehand. Use the identification key on pages R100 and R101 of the student text as a guide. The most common diagnostic physical properties are described here:

Specific Gravity is the ratio of mass of a mineral sample to that of an equal volume of water. By definition, water has a specific gravity of 1. The specific gravity of most rock-forming minerals is in the range 2.6 to 3.0. However, the average specific gravity of metallic minerals is about 5. Pure gold has a specific gravity of 19.3.

To compute specific gravity using a spring scale, weigh the sample in air and then suspend it in water and record the difference in weight. The following equation yields the ratio called specific gravity:

specific gravity = weight of sample in air / loss of weight in water

If you are using a balance scale, the mineral sample is weighed in air and then dropped into a container filled to the brim with water. The displaced water is caught in a preweighed container, and the weight of this equal volume of water can be used in the following equation to compute specific gravity:

specific gravity = weight of sample in air / weight of equal volume of water

Although specific gravity can be measured in the laboratory, it is not difficult to get a relative sense by comparing the heft of a nonmetallic mineral like feldspar or quartz (2.6 to 2.7) with a similar-sized piece of a metallic mineral like galena (7.6). Galena is also a good example to compare to a very similar-appearing mineral that can be distinguished by heft—specular hematite (Fe_2O_3), which also has metallic luster and the same dark gray color, but a specific gravity of 5.26.

Hardness is the resistance of the mineral to scratching. The hardness of minerals is expressed in a scale from 1 to 10, called the Mohs scale (named after Mohs, its originator). The following objects are used in mineral test kits to determine hardness:

- fingernail ~ 2.5
- copper penny ~ 3.5
- iron nail ~ 4.5
- glass ~ 5.5
- quartz ~ 7

Hardness is determined by the strength of the bonds between atoms or ions in the crystal structure of the mineral. It is important to note that this is a relative scale based on index minerals, so a hardness of 4 is not necessarily twice as great as a hardness of 2: it is just harder than 3 and not as hard as 5.

Cleavage is the tendency of some minerals to break along smooth, flat, parallel surfaces. Some minerals like mica show cleavage in one direction, whereas others show two or more different directions. Beginners at mineral identification often confuse cleavage planes with crystal growth faces, but broken mineral specimens found in the field or in the laboratory rarely show crystal growth faces.

Streak is the color of the trail of powdered mineral left when a corner of a mineral sample is rubbed on a piece of unglazed porcelain. The streak color may be different from the color of the mineral specimen. Streak is most diagnostic for dark metallic minerals like hematite, magnetite, pyrite, and galena. Although all of these minerals are dark in color, their streak varies from red to gray-green to a thin gray to thick black. Most nonmetallic minerals produce a similar white to tan streak.

An excellent example of streak's diagnostic potential is the differentiation between galena (PbS) and specular hematite (Fe_2O_3). Both have metallic luster and a similar dark gray or black color, making them difficult to distinguish visually. Galena always leaves a dark gray streak. Because specular hematite is a ferric iron mineral, its streak is a rusty brown.

Luster is the appearance of the mineral in reflected light. Most rock-forming minerals have lusters that can be described as glassy, greasy, earthy, dull, or pearly. Metallic and nonmetallic lusters are harder to distinguish, because some nonmetallic minerals, like biotite, are dark in color and very shiny.

Color is not usually diagnostic and should not receive too much attention in the process of mineral identification. The metallic atoms or ions in the crystal lattice of the mineral determine color. There are many white to light tan minerals, and a host of others that appear black to dark gray. In some cases, most notably quartz, one mineral may appear in a variety of colors. For a few minerals, however, color is diagnostic. Cinnabar (an ore of mercury) and malachite are two that can seldom be confused with other minerals. Cinnabar is always a mottled red, and malachite is a very distinctive green. Students may be familiar with turquoise or azurite, in which the blue color is very prominent. In most cases, though, color is the least reliable diagnostic criterion.

More Information – on the Web
Visit the *EarthComm* web site www.agiweb.org/earthcomm to access a variety of links to web sites that will help you deepen your understanding of content and prepare you to teach this activity. Many of the sites also contain images that you can download.

Goals and Assessment

Clarify that the goals indicate what students should understand and be able to do as a result of the activity. Make sure students understand that Chapter Assessments are based upon these goals.

Goal	Location in Activity	Assessment Opportunity
Define the term mineral in your own words.	**Check Your Understanding** Question 1	Definition is written in a complete sentence and in some way addresses the criteria given in the **Digging Deeper** section.
Evaluate the usefulness of various physical properties for describing and identifying different minerals.	**Investigate** Part A **Digging Deeper; Check Your Understanding** Question 3 **Understanding and Applying What You Have Learned** Questions 3 and 6	List of mineral identification properties is useful and thorough.
Explore how mineral crystals are constructed and how the external form of a crystal reflects its ionic structure.	**Investigate** Part B **Investigate** Part C **Digging Deeper; Check Your Understanding** Question 4 **Understanding and Applying What You Have Learned** Question 5	Observations are insightful and are supplemented with sketches. Marbles are accurately aligned.
Identify a variety of mineral specimens according to their physical properties.	**Investigate** Part D **Digging Deeper**	Minerals are correctly identified. Data table is complete, including special properties.

Earth's Natural Resources Mineral Resources

Activity 2   What Are Minerals?

Goals

In this activity you will:

- Define the term "mineral" in your own words.

- Evaluate the usefulness of various physical properties for describing and identifying different minerals.

- Explore how mineral crystals are constructed and how the external form of a crystal reflects its ionic structure.

- Identify a variety of mineral specimens according to their physical properties.

Think about It

There are an estimated 9000 different species of birds around the world.

- What are some ways to identify birds?

What do you think? Even if you are not a birdwatcher, try to list at least four or five different features that could help you identify different kinds of birds. Make sure you consider other features besides purely visual ones. Record your ideas in your *EarthComm* notebook. Be prepared to discuss your responses with your small group and the class.

Activity Overview

Students begin the investigation by examining an array of mineral specimens. From their observations, they develop a list of properties that are useful for identifying different minerals. Students then break samples of halite and calcite. They examine the broken pieces with a magnifying glass to understand that some minerals, like calcite, cleave when broken. Students construct enclosures and stack marbles in a variety of arrangements to understand the various atomic structures of minerals. Finally, they return to the mineral samples they examined at the start of the investigation and attempt to identify them using a mineral identification key. **Digging Deeper** defines the term mineral, reviews common minerals, describes the structure of minerals, and points out common tools for mineral identification.

Preparation and Materials Needed

Part A

Prepare a set of at least eight common minerals for each student group. Try to have mineral samples of approximately the same size. Suggested minerals include quartz, feldspar, muscovite, biotite, calcite, gypsum, fluorite, halite, hematite, galena, pyrite, magnetite, and talc. Include as many samples as possible from the Mohs scale of hardness. Selections should include samples of similar colors and lusters, as well as definite contrasts. Avoid rocks (mixtures of minerals), because students need to be able to examine one mineral at a time. Make sure that the minerals are not named, because students identify the samples in **Part D** of the investigation.

Part B

You will need several pieces of halite and calcite and a hammer. You may wish to carry out this part of the investigation as a class demonstration, for safety reasons. If students will be breaking rocks, provide each student with a pair of safety goggles.

Part C

You may wish to cut the cardboard squares before class, for safety reasons and to save time in class. The dimensions of the squares will depend upon the size of the marbles that you use, but keep in mind that you will need to line up five marbles by five marbles within the final enclosure for **Step 2**, and at least seven marbles in a triangular arrangement for **Step 3**.

Part D

You will need to prepare a mineral test kit for each student group. Each kit should include a pre-1982 copper penny, a nail, a piece of glass, a steel file, a streak plate, and a magnet (or small object like a paper clip to test for magnetism). Make copies of the mineral identification key in the student text. See **Blackline Master Mineral**

Resources 2.1, Mineral Identification Key. You may also wish to prepare a mineral identification table which students can fill in. A sample is provided as **Blackline Master Mineral Resources 2.2, Data Table for Mineral Identification.**

Note: In recent years, pennies have been manufactured from steel-based alloys rather than from copper, and are therefore harder than copper. Try to find pennies that date from before 1982. The alloy used to make pennies remained 95% copper and 5% zinc until 1982, when the composition was changed to 97.5% zinc and 2.5% copper (copper-plated zinc). Pennies of both compositions appeared in 1982. An alternative would be to use heavy-gauge copper electrical wire from which the insulation has been stripped.

Materials
Part A
- Set of mineral samples (common minerals like quartz, halite, calcite, muscovite, biotite, feldspar, gypsum, hematite, galena, pyrite, magnetite, and talc).

Part B
- Calcite crystal
- Halite crystal
- Hammer
- Magnifying glass
- Strong light source
- Safety goggles

Part C
- Box of about 200 glass marbles
- Large piece of corrugated cardboard or posterboard
- Duct tape
- Large scissors

Part D
- Mineral identification kit, including copper penny (pre-1982), nail, piece of glass, steel file, streak plate, and magnet (or small object like a paper clip) to test for magnetism
- Copies of a mineral identification table (see sample table on page 341 of this Teacher's Edition)

Think about It
Student Conceptions

Students are asked to think of various ways to identify birds. The idea is to foster an observational frame of mind, so that students can recognize the physical characteristics that we use to describe and distinguish objects and apply this skill to the identification of minerals. Ask them how they would differentiate between a blue jay and a bluebird, or between a hawk and a crow. It may be helpful to provide pictures of different birds to get them started. Remind your students to consider all characteristics, not just visual ones; these may include color, height, weight, body shape, beak size, song, or migration patterns. As a guiding question, ask students: What is the difference between general physical properties, and specific physical properties that are more characteristic of the given entity? Have them explain their responses.

Answer for the Teacher Only

Professional ornithologists and amateur birdwatchers alike ordinarily identify birds on the basis of external characteristics rather than inherent genetic characteristics. A very large number of bird species are found in the United States There are several good field guides for bird identification. Among these, *The Sibley Guide to Birds*, by David A. Sibley, recently published by the National Audubon Society, is outstanding. In many bird species, there is considerable intraspecific (within-species) variation in appearance, just as there is in our own species. Some of this variation is geographical, and some exists even in the same geographic region. It is natural that such variation exists, because all species are in the process of evolutionary change. The situation with many minerals in analogous: many of the important minerals have the same structure but a range of different compositions (this phenomenon is called isomorphism), which arises from substitutions of certain ions for other ions with about the same effective size and the same electric charge in the crystal structure. These compositional variations are accompanied by differences, often very substantial, in physical appearance.

Assessment Tool

Think about It Evaluation Sheet
Use this evaluation sheet to help students understand and internalize the basic expectations for the warm-up activity.

Teaching Tips

Use the common, easy-to-find objects of known hardness (glass, nail, penny, etc.) listed above to test for hardness instead of specimens of minerals from the Mohs scale (which is given on page R105 of the student text). These are items that working field geologists always keep at hand.

Some of the terms listed in the Student Edition's mineral identification key under the category of "crystal shape" refer to the crystal system to which each mineral belongs. There are six crystal systems, and minerals from each crystal system have some characteristic elements of shape and symmetry that are found in their well formed crystals. Listed from those that have crystals with the least symmetry to those with the most symmetry, the six crystal systems are the: triclinic, monoclinic, orthorhombic, tetragonal, hexagonal, and isometric systems. More information about each of the six crystal systems and the specific symmetry elements that each imparts to the minerals within that system, can be found on the *EarthComm* web site. Alternative descriptions of crystal shape are given in the **Blackline Master** for the mineral identification key. When filling out their mineral identification table, students should simply use descriptive terms to describe the crystal shape (e.g., massive, six-sided prism, cubic, rhombic, etc.)

Blackline Master Mineral Resources 2.1
Mineral Identification Key

Mineral Identification Key							
Mineral Name	**Hardness**	**Streak**	**Specific Gravity**	**Cleavage**	**Crystal Shape**	**Color**	**Other Properties**
Corundum	9	White	4.0 (Med. – High)	None	Six sided (hexagonal) prisms are common	Gray, red, brown, blue	Crystals often have striated, flat ends
Topaz	8	White	3.4 - 3.6 (Med. – High)	One perfect	Commonly prismatic crystals	Colorless, yellow, blue, or brown	Crystal faces often striated
Quartz	7	White	2.7 (Medium)	None	Commonly massive or as hexagonal prisms	Any color to colorless, greasy luster	Conchoidal fracture; crystal faces often striated
Potassium feldspar	6	White	2.6 (Medium)	Two at 90°	Can occur as short prismatic crystals	White, pink, green or brown	Crystals can have a finely veined appearance
Plagioclase feldspar	6	White	2.6 (Medium)	Two at 90°	Can occur as short prismatic crystals	Blue-gray, black, or white	Striations on some cleavage planes
Magnetite	6	Dark gray	5.2 (High)	None	Commonly granular to massive, can exhibit 8-sided (octahedral) and 12-sided (dodecahedral) shapes	Dark gray to black	Magnetic, metallic luster
Pyrite	6	Dark gray	5.0 (High)	None	Cubic crystals are common	Brass yellow, may tarnish brown	Metallic luster, brittle, faces commonly striated
Apatite	5	White	3.2 (Medium)	One poor	Commonly occurs as hexagonal prisms	Brown, green, blue, yellow, or black	Vitreous to subresinous luster
Hematite	5.5 - 6.5	Red to red-brown	5.3 (High)	None	Can take a variety of forms, including aggregates of thin plates or spheres	Red or steel gray	Red form – earthy luster, gray form – metallic luster
Fluorite	4	White	3.2 (Medium)	Octahedral	Crystals are usually cubic	Colorless purple, blue, yellow, green	Some crystals show bands of varying color
Calcite	3	White	2.7 (Medium)	Three perfect	Rhombohedral and prismatic shapes are common	Colorless, white, yellow, gray	Transparent to translucent, reacts with HCl
Muscovite mica	2 - 2.5	White	2.8 - 2.9 (Medium)	One perfect	Commonly occurs as aggregates of scales or thin sheets	Colorless, yellow, light brown	Elastic, flexible sheets
Biotite mica	2.5 - 3	Gray-brown	2.8 - 3.2 (Medium)	One perfect	Commonly occurs as aggregates of scales or thin sheets	Very dark brown to black	Elastic, flexible sheets
Galena	2.5	Gray	7.4 - 7.6 (Very high)	Cubic	Cubic crystals are common	Silvery gray	Metallic luster
Halite	2.5	White	2.2 (Medium)	Cubic	Cubic crystals are common	Colorless, white	Salty taste
Talc	1	White	2.7 - 2.8 (Medium)	One perfect	Generally massive and rare crystals are tabular	White, gray, yellow	Soapy feel, pearly or greasy luster

Chapter 2

Blackline Master Mineral Resources 2.2
Data Table for Mineral Identification

Sample Number	Mineral Name	Hardness	Streak	Specific Gravity	Cleavage	Crystal Shape	Color	Other Properties
1								
2								
3								
4								
5								
6								
7								
8								
9								
10								

NOTES

Investigate

Check your list of properties with your teacher before testing the minerals. Some properties, while useful, should only be tested with teacher direction.

Part A: Properties of Minerals

1. In a small group, study a set of mineral samples.

   a) Make a list of properties you can use to describe the minerals. For example, color may be a property that you would use to describe a particular mineral.

   b) Write a brief description of each mineral sample using the properties that you listed.

   c) Which properties are most useful in describing an individual mineral sample?

   d) Which properties are the least useful?

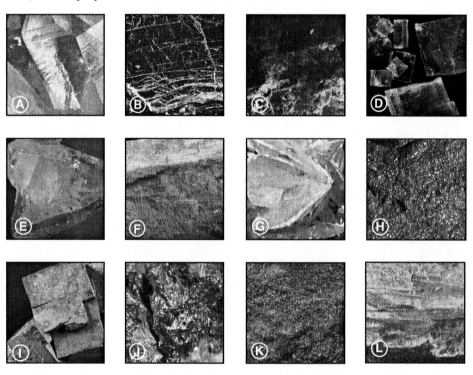

2. Make a class list of all the properties the different groups came up with to describe the minerals. Discuss the usefulness of the properties.

   a) In your notebook keep a record of the class list of the most useful properties in describing minerals.

EarthComm

Investigate

Part A: Properties of Minerals

1. a) If students are having difficulty coming up with properties, ask them to compare different samples (e.g., this piece of calcite has a different shape from this piece of halite—therefore, shape is a distinguishing property). Encourage them to pick up the samples, feel their textures and weights, and examine them closely.

    b) Once students have compiled their lists, they are asked to describe their samples using those properties. This list (along with those from the other class members) will be used as a discussion focal point for the next question and can be revisited (along with the list from **Step 2**) after the mineral-identification exercise in **Part D** of this investigation.

    c) Student responses will vary. Ask students to explain why they think that certain properties are useful.

    d) Student responses will vary. Ask students to explain why they think that certain properties are not as useful.

2. Compile a list of all of the properties and discuss the list as a class. Which properties seem to be most helpful, and why? Which are not helpful, and why not? It may be helpful to combine the sample sets for the entire class and see if the properties they used to describe their sample would work on a different sample of the same mineral.

    a) Lists will vary, but possible properties may include color, hardness, luster, specific gravity (how heavy the sample feels), shape, etc.

Earth's Natural Resources Mineral Resources

Part B: Breaking Minerals

1. Your teacher will supply your small group with a crystal of halite (NaCl) and a crystal of calcite ($CaCO_3$). Put on your safety goggles. Place the halite crystal on a large sheet of paper. Hit it with the hammer, starting very gently and increasing the force of the impact until the crystal breaks.

 Wear goggles.

2. Examine the broken pieces with a magnifying glass under a strong light.

   a) Describe the shape of the fragments, and sketch one or more of them in your *EarthComm* notebook.

3. Select one or a few of the fragments, and break them again with the hammer. Again, examine them with the magnifying glass.

   a) Describe and sketch the fragments in your *EarthComm* notebook.

4. Repeat Steps 1–3 with the calcite crystal.

5. From the results of your investigation, answer the following questions:

   a) How does the characteristic shape of the halite fragments differ from the characteristic shape of the calcite fragments?

   b) Why do you think that the fragments of the crystals have regular and distinctive shapes?

   c) Why do you think that the halite fragments and the calcite fragments have different shapes?

Part C: Stacking Spheres

1. Cut four squares of corrugated cardboard. Make the edges of each piece just long enough so that five marbles can line up along the edges. Tape the pieces together to form a square enclosure with vertical walls, as shown in the diagram below.

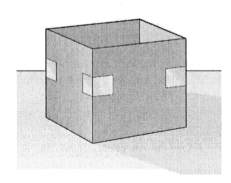

2. Line the bottom of the cardboard enclosure with a single layer of marbles. Next, carefully add a second layer of marbles, with each marble in the second layer directly above a marble of the lowest layer. Continue adding layers until you have filled the enclosure. You have created a regular three-dimensional array of marbles.

   a) In your *EarthComm* notebook, describe the geometry of the array of marbles.

   b) How does the geometry of this array compare to the geometry of the crystal fragments you produced in Part B of the activity?

Part B: Breaking Minerals

1. For safety reasons, you may want to break the mineral samples as a demonstration. You can then distribute fragments to the students for examination. If students break the samples themselves, make sure that they are wearing safety goggles.

2. a) The halite sample will break into fragments with sides that are at right angles to one another. (The fragments will look like cubes—though not necessarily having equal dimensions for all sides. Technically, the shape is a rectangular solid.) Encourage students to describe and sketch several pieces, not just one.

3. a) Again, the halite sample will break into fragments with sides that are at right angles to one another (cubes). Encourage students to describe and sketch several pieces, not just one.

4. Students should observe that the calcite sample breaks along regular planes that are not at right angles to each other (rhombohedral shapes). Have students compare fragments from each step; they should notice the consistency of the planes in each piece. Remind students to describe and sketch several pieces at each step.

5. a) Halite breaks into cubes, where the sides of the fragments are at right angles to one another. The sides of the fragments of calcite are regular, but the planes are not at right angles.

   b) Halite and calcite are minerals that have cleavage, which means that when they break, they tend to break along regularly oriented planes of weakness. Cleavage planes form along planes of weak atomic bonds in the mineral. Because the atomic arrangement of a mineral does not change when it is broken, the mineral breaks along the same planar orientation each time.

   c) The halite and calcite fragments differ in shape because they are different minerals with different atomic structures.

Part C: Stacking Spheres

1. Students can use the illustration on page R98 in the text to help them prepare their square enclosures.

2. a) The marbles are stacked on top of one another in an orderly, cubic array. Each marble after the first layer should rest directly on top of the marble below it. Encourage students to use sketches to describe the geometry of the marbles.

   b) Students should note that the array of marbles resembles the geometry of the halite fragments produced in **Part B**.

Teaching Tip

When adding the second layer of marbles, it will help to tilt the enclosure so that the marbles stack on top of the first layer, not in the pockets between the marbles of the first layer. The type of stacking arrangement illustrated in **Part C, Step 2** is called simple cubic packing and is a fairly inefficient utilization of space—this arrangement actually uses only 52% of the available space (the remaining 48% is void space). The structure illustrated in **Part C, Step 3** is referred to as a hexagonal closest-packed structure and is more efficient, using 74% of the available space. (Visit the *EarthComm* web site for more information on the structure of solids.)

Assessment Tool

EarthComm Notebook-Entry Checklist
Use this checklist as a quick guide for student self-assessment and/or an opportunity to quickly score student work. Add further criteria specific to your classroom needs or to this particular investigation.

Investigate Notebook-Entry Evaluation Sheet
Point out the criteria listed on this evaluation sheet that are relevant to this particular investigation. Encourage students to internalize the criteria by making them part of your assessment conversations as you circulate around the classroom.

NOTES

3. There is another way to pack spheres in a regular three-dimensional array. Build another cardboard enclosure, adjusting its size so that you can arrange the first layer of marbles as shown in the diagram below. Lay in the marbles of the second layer in the "pockets" formed by three adjacent marbles of the first layer. One marble of the second layer is shown in the sketch below. Add layers until you have filled the enclosure.

a) In your *EarthComm* notebook, describe the geometry of this array of marbles.

b) How does the geometry of this array compare to the geometry of the crystal fragments you produced in Part B of the investigation?

c) Can you think of a third way of building a regular three-dimensional array of marbles in a cardboard enclosure? If so, sketch and describe the array in your *EarthComm* notebook.

 Be sure to immediately pick up any marbles that drop on the floor. They can cause accidents.

Part D: Mineral Identification
1. Work in small groups to identify the mineral samples you examined in Part A of this activity. Your teacher will give you a test kit containing:

- a piece of unglazed porcelain tile (called a streak plate),

- a magnet or paper clip, and

- some of the minerals of the Mohs scale to test hardness.

Decide how you could use each piece of equipment to help you identify the minerals. Read the appropriate sections of **Digging Deeper** to help you determine what tests would be appropriate.

Test as many of the properties as you can to describe each mineral as completely as possible.

a) Record your findings in a data table. Be sure to note any special properties exhibited by your samples.

b) Use the Mineral Identification Key on the following pages to assign mineral names to your samples. Compare your list of mineral properties and names to those of other groups. Your teacher will lead a discussion on the correct mineral names and why some teams might have gotten different names for the same mineral.

 Check your plan with your teacher before proceeding.

3. Students can use the illustration in the text as a guide to help them prepare their cardboard enclosures.

   a) The marbles are stacked on top of one another in an orderly array, with marbles above the first layer resting in the "pockets" formed between marbles of the underlying layer. Encourage students to use sketches to describe the geometry of the marbles.

   b) Students should note that the array of marbles resembles the geometry of the calcite fragments produced in **Part B**.

   c) Students answers may vary. One possibility is that the first layer will look like the first layer in **Part C, Step 2** with the second layer filling in the pockets between marbles. In this case, each marble of the second layer will be touching four marbles of the first (as opposed to **Part C, Step 3** where each marble of the second layer touched three marbles of the first). This arrangement (shown in the sketch below) is called body-centered cubic packing.

Part D: Mineral Identification

1. You may want to have your students read the **Digging Deeper** section at this point to learn how various properties are defined. You will need to show your students how to use the materials in their test kit. A good way to do this is to pick a mineral and use the materials provided to identify the sample. Answers to the questions will vary depending upon which samples you provide to the students.

   a) Students should record the properties they listed in **Part A, Step 1(a)** and the descriptions they compiled for each mineral sample in **Part A, Step (b)** in a mineral identification table (see sample data table on page 341 of this Teacher's Edition.) Students can then record the results of their mineral identification tests in the appropriate spaces on the data table.

   b) You may want to explain the terms that appear in the "Crystal Shape" column of the mineral identification key. The words triclinic, monoclinic, orthorhombic, tetragonal, hexagonal, and isometric all refer to the six crystal systems, each of which can impart a distinctive shape and symmetry to minerals within it. More information on the six crystal systems, and the shapes and symmetry that is associated with each, can be found on the *EarthComm* web site. Alternative descriptions of crystal shapes for each of the minerals are given in the **Blackline Master** for the mineral identification key.

Assessment Opportunity

Use a rubric, like the one shown below, to assess student identification of minerals. (You will most likely need to adapt the rubric, depending upon how many mineral samples you provide to your students.)

0 Points	1 Points	2 Points	3 Points
Table was not turned in.	Table is incomplete OR contains a significant number of errors (minerals are incorrectly identified, properties do not match minerals listed).	Table is mostly correct and complete. One or two minerals or mineral properties are misidentified.	Table is complete. Minerals are all identified correctly; there are no more than two errors in mineral property identification.

NOTES

Chapter 2

Earth's Natural Resources Mineral Resources

Mineral Identification Key

Mineral Name	Hardness	Streak	Specific Gravity	Cleavage	Crystal Shape	Color	Other Properties
Corundum	9	White	4 (Med. – High)	None	Commonly six-sided crystals	Gray, red, brown, blue	
Topaz	8	White	3.5 (Med. – High)	One perfect	Orthorhombic or massive	Colorless, yellow, blue, or brown	
Quartz	7	White	2.7 (Medium)	None	Hexagonal or massive	Any color to colorless, greasy luster	Conchoidal fracture
Potassium feldspar	6	White	2.6 (Medium)	Two at 90°	Monoclinic or triclinic	White, pink, or brown	
Plagioclase feldspar	6	White	2.6 (Medium)	Two at 90°	Triclinic (rare)	Blue-gray, black, or white	Striations on some cleavage planes
Magnetite	6	Dark gray	5.2 (High)	None	Massive	Dark gray to black	Magnetic, metallic luster
Pyrite	6	Dark gray	5.0 (High)	None	Cubic crystals common	Brass yellow, may tarnish brown	Metallic luster, brittle
Apatite	5	White	3.1 (Medium)	One poor	Common as six-sided crystals	Brown, green, blue, yellow, or black	
Hematite	5	Red to red-brown	5.0 (High)	None	Hexagonal	Red or steel gray	Red form – earthy luster; gray form – metallic luster
Fluorite	4.5	White	3.0 (Medium)	Octahedral	Crystals usually cubic	Colorless, purple, blue, yellow, green	

NOTES

Mineral Identification Key (continued)

Mineral Name	Hardness	Streak	Specific Gravity	Cleavage	Crystal Shape	Color	Other Properties
Calcite	3	White	2.8 (Medium)	Three perfect	Hexagonal	Colorless, white, yellow, gray	Transparent to translucent, reacts with HCl
Muscovite mica	2.5	White	2.7 (Medium)	One perfect	Monoclinic	Colorless, yellow, light brown	Elastic, flexible sheets
Biotite mica	2.5	Gray-brown	2.7 (Medium)	One perfect	Monoclinic	Very dark brown to black	Elastic, flexible sheets
Galena	2.5	Gray	7.5 (Very high)	Cubic	Cubic crystals common	Silvery gray	Metallic luster
Halite	2.5	White	2.5 (Medium)	Cubic	Cubic crystals	Colorless, white	Salty taste
Talc	1	White	2.7 (Medium)	One perfect	Monoclinic (rare)	White, gray, yellow	Soapy feel, pearly or greasy luster

NOTES

Earth's Natural Resources Mineral Resources

Reflecting on the Activity and the Challenge

In this activity, you saw that there are many different ways to describe minerals. Some properties are more useful than other properties when describing and comparing minerals. Being able to accurately describe minerals helped to identify them. You also learned about how atoms are arranged in mineral crystals, and how the geometry of the arrangement affects the physical properties of the mineral. You must be able to identify minerals and understand their properties to be able to select the best mineral to use for your beverage container.

Geo Words

mineral: a naturally occurring inorganic, solid material that consists of atoms that are arranged in a regular pattern and has characteristic chemical composition, crystal form, and physical properties.

Digging Deeper

MINERALS

Types of Minerals

Minerals have been important to humans for a long time. Early humans used red hematite and black manganese oxide to make cave paintings. People in the Stone Age made tools out of hard, fine-grained rocks. In the Bronze Age, people discovered how to combine copper and tin from minerals into an alloy called bronze. Later, in the Iron Age, people made tools of iron, which is contained in minerals like hematite (Fe_2O_3) and magnetite (Fe_3O_4).

Today, minerals are used in thousands of ways. Feldspar is used to make porcelain. Calcite is used to make cement. Iron and manganese, together with small amounts of several other metals, make steel, which is used to make buildings, trains, cars, and many other things. Gypsum is used to make plaster and wallboard. These are just a few examples of how minerals are used in your daily lives.

Figure 1 The minerals red hematite and black manganese oxide were used by early humans to draw their cave painting.

Reflecting on the Activity and the Challenge

Have students read this brief passage and share their thoughts about the main point of the activity in their own words. Hold a class discussion about how this investigation relates to what is being asked of the students in the **Chapter Challenge**. By understanding the physical properties of different minerals, students will be able to understand why certain mineral resources would be wise choices for beverage containers and why some would not.

Digging Deeper

Assign the reading for homework, along with the questions in **Check Your Understanding** if desired.

Teaching Tip

Use (or rephrase) the questions in **Check Your Understanding** for a brief quiz to check comprehension of key ideas and skills. Use the quiz (or a class discussion) to assess your students' understanding of the main ideas in the reading and the activity. A few sample questions are provided below:

Question: Define mineral.
Answer: A mineral is a naturally occurring, inorganic, solid material that consists of atoms and/or molecules that are arranged in a regular pattern and have characteristic chemical composition, crystal form(s), and physical properties.

Question: Name and define three properties that can be used to identify a mineral.
Answer: Answers can include any three of the following:
- Hardness: the resistance of a mineral to scratching.
- Luster: the reflection of light from the surface of a mineral, described by its quality and intensity.
- Streak: the color of a mineral in its powdered form, usually obtained by rubbing the mineral on a streak plate and observing the mark it leaves.
- Specific Gravity: the ratio of the weight of a given volume of a substance to the weight of an equal volume of water.
- Cleavage: the breaking of a mineral along regularly oriented planes of weakness, thus reflecting crystal structure.
- Conchoidal Fracture: a type of mineral fracture that gives a smoothly curved surface.
- Other properties, like heat conductivity, electrical conductivity, reactions with different acids, radioactivity, magnetic properties, and luminescence (emission of light when subjected to radiation of particular wavelengths).

Chapter 2

To be considered a mineral, a material must meet five criteria:

• Minerals are solid, not gas or liquid.
• Minerals are inorganic, which means they are not alive and never have been.
• Minerals are naturally occurring, not manufactured.
• Minerals have definite chemical compositions, which can be expressed as a formula of elemental symbols (such as SiO_2, Ag, or Fe_2O_3).
• Minerals have a regular three-dimensional arrangement of atoms (called a crystal structure).

Some minerals, called **native-element minerals** consist of only one element (*Figure 2*). A good example is gold (Au), which is often found as nuggets or pieces of pure gold not combined with any other element. Copper (Cu), iron (Fe), and silver (Ag) are other native elements. Most minerals, however, are combinations of elements. Quartz (*Figure 3*), for example, with the formula SiO_2, is made of atoms of silicon and oxygen. Calcite ($CaCO_3$) is made of calcium, carbon, and oxygen.

Geo Words

native-element mineral: a mineral consisting of only one element.

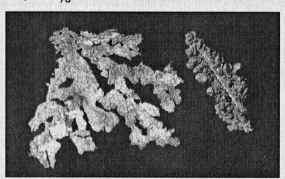

Figure 2 Copper is a native-element mineral.

Figure 3 Quartz is a mineral composed of the elements silicon and oxygen.

EarthComm

Teaching Tips

Copper (*Figure 2* on page R103) is a native-element mineral that is used in electric cables and wires, switches, plumbing, heating, roofing and building construction, and construction of chemical and pharmaceutical machinery. The United States is one of the leading producers of copper in the world.

Quartz (*Figure 3* on page R103) has the chemical formula SiO_2. It can occur in a great variety of colors, but it is most commonly white, clear, purple, or brown. When it can grow in an unbounded fluid medium, quartz forms crystals as hexagonal prisms. The crystals are often large, single, and faultless. They are usually prismatic but often stubby. In rocks, however, quartz usually crystallizes to filling existing cavities—so natural quartz grains are usually irregular in shape. As a crystal, quartz is used as a semiprecious gemstone. Students may be familiar with some gems of quartz, including amethyst, citrine, rose quartz, jasper, and onyx. Quartz has other uses as well. It is used for pressure gauges, oscillators, resonators, and wave stabilizers. It is also used in heat-ray lamps and prism and spectrographic lenses. Relevant to the **Chapter Challenge**, quartz is also the basic raw material in manufacturing glass.

Chapter 2

Earth's Natural Resources Mineral Resources

Geo Words

rocks: naturally occurring aggregates of mineral grains.

ores: rocks that contain valuable minerals.

ions: atoms that have an electric charge because one or more electrons have been added to the atom or removed from the atom.

electrons: particles with a negative electric charge, which orbit around the nucleus of the atom.

Rocks are naturally occurring aggregates of mineral grains. Some rocks consist of only one mineral, but most contain several different kinds of minerals. Sometimes you can see the mineral grains, and sometimes they are too small to see without magnification. Granite consists mostly of large crystals of feldspar, quartz, and mica, as shown in *Figure 4*. Basalt consists mostly of tiny crystals of feldspar and pyroxene.

Figure 4 Granite is a rock composed of several different types of minerals.

Rocks that contain valuable minerals are called **ores**. To remove the valuable minerals, the ore first has to be mined, by digging or blasting. Then the desired mineral is separated from the rest of the ore by processes like crushing, sieving, melting, or settling through a liquid. Most metals and many important nonmetals are refined from ores. Valuable minerals are distributed very unevenly in the Earth's crust. Finding new deposits to meet the needs of industry and technology depends upon understanding the characteristics of the minerals and the ores that contain them.

The Chemistry and Structure of Minerals

As you saw in the activity, minerals are very different from one another. That is because all minerals have a particular chemical composition. Minerals consist of atoms of one or more chemical elements. Each chemical element has different chemical and physical behaviors.

The atoms in minerals are arranged in a regular three-dimensional array (*Figure 5*). The atoms in almost all minerals are in the form of **ions**. Ions are atoms that have an electric charge because one or more **electrons**

Teaching Tip

Igneous rocks form by solidification of molten rock (magma). Granite (*Figure 4* on page R104) is an intrusive igneous rock, which means that it forms when magma cools and crystallizes below the Earth's surface. Granitic rocks are composed of up to about 20% quartz. The dominant minerals in granitic rocks are plagioclase feldspar and potassium feldspar. Granites also contain less than 10% ferromagnesian minerals (minerals like biotite and hornblende that are dark in color and contain magnesium and iron in their structures).

Chapter 2

(particles with a negative electric charge, which orbit around the nucleus of the atom) have been added to the atom or removed from the atom. The ions in a mineral are packed together in an arrangement that brings the ions as close together as possible. The packing arrangement puts positively charged ions in close contact with negatively charged ions. Objects with unlike electric charges are attracted to each other, and it is these attractive forces that hold the mineral together as a solid.

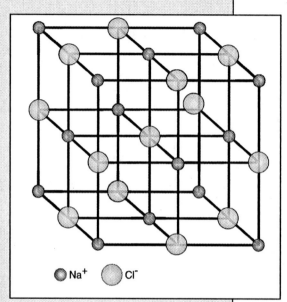

○ Na⁺   ○ Cl⁻

Figure 5 This expanded view of halite shows the orderly three-dimensional arrangement of sodium and chlorine atoms.

Identifying Minerals

The properties of the atoms in a mineral, and also their geometrical arrangement, affect the color, shape, hardness, and other properties of the mineral. Geologists use a variety of tests to describe, compare, and identify minerals. Some of these tests are simple, and can be done with simple equipment. Other tests require special (and expensive!) equipment.

Hardness

Hardness is the resistance of a mineral to scratching. Mineralogists use a relative scale of hardness, called the **Mohs scale**, given below, to test the hardness of a mineral:

1. Talc
2. Gypsum
3. Calcite
4. Fluorite
5. Apatite
6. Orthoclase
7. Quartz
8. Topaz
9. Corundum
10. Diamond

Geo Words

hardness: the resistance of a mineral to scratching.

Mohs scale: a standard of 10 minerals by which the hardness of a mineral may be rated.

EarthComm

Teaching Tip

Make an overhead of *Figure 5* on page R105 using **Blackline Master Mineral Resources 2.3, Expanded View of Halite.** Use this overhead to help students to understand the point of completing **Part C** of **Investigate.** They should understand that when they were stacking the marbles within the square cardboard enclosure, they were essentially creating an atomic arrangement similar to that of halite. Discuss with students how the atomic structure of a mineral influences its physical properties, and how these properties relate to the **Chapter Challenge.**

Teaching Tip

Figure 6 on page R106 shows an assortment of minerals and defines the luster of each mineral. If students did not define the luster of the minerals they examined in the investigation, have them go back and add this information to their data.

Chapter 2

Earth's Natural Resources Mineral Resources

Each mineral in the scale scratches minerals earlier in the scale and is scratched by minerals later in the scale. Diamond (with a hardness of 10) is the hardest natural substance known, and the mineral talc (with a hardness of 1) is one of the softest. The way to test the hardness of a mineral is to scratch an unknown mineral with a material of known hardness. If the mineral is scratched, it is not as hard. If the unknown mineral scratches the known material, then it is harder. Here are the hardnesses, on the Mohs scale, of some common materials:

fingernail: a little more than 2
wire nail: about 4.5
knife blade: a little more than 5
window glass, masonry nail: 5.5
steel file: 6.5

Luster

Geo Words

luster: the reflection of light from the surface of a mineral, described by its quality and intensity.

Luster describes the way a mineral reflects light. Luster is either metallic or nonmetallic. Minerals with metallic luster look like polished metal. Nonmetallic lusters are often further described as glassy (or vitreous), waxy, pearly, earthy, or dull. Pyrite and galena have metallic luster. Quartz and calcite have a glassy luster. Feldspar has a pearly luster. (See *Figure 6A–D*).

Figure 6A Quartz has a vitreous luster.

Figure 6B Feldspar has a pearly luster.

Figure 6C Galena shows a metallic luster.

Figure 6D Pyrite has a metallic luster.

Blackline Master Mineral Resources 2.3
Expanded View of Halite

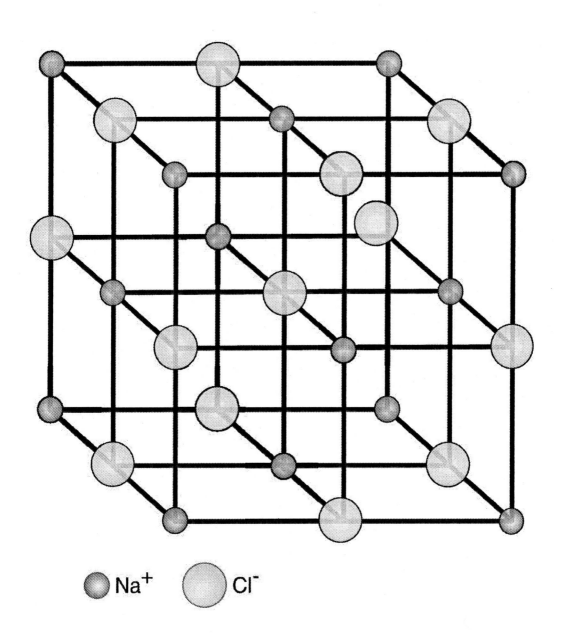

Na⁺   Cl⁻

Streak

Streak is the color of the powdered mineral. To determine the streak of a mineral, scratch the mineral across a piece of unglazed porcelain tile, called a streak plate. Many minerals have a distinctive streak color. Streak color may be different from the color of the mineral sample. For example, hematite is often dark gray in color, but it always has a red streak.

Specific Gravity

Specific gravity is a ratio of the weight of the mineral to the weight of an equal volume of water, which has a specific gravity of 1. Galena, a mineral containing lead, has a specific gravity of approximately 7. That means it is about 7 times as dense as water. Nonmetallic minerals like quartz, feldspar, and calcite mostly have specific gravities less than 3.

Cleavage

Many minerals have **cleavage**, which means that when they break, they tend to break along regularly oriented planes of weakness, as shown in *Figure 7A–D*. Cleavage planes form along planes of weak atomic bonds in the mineral. For example, mica splits easily into sheets, because there are very weak atomic bonds between the layers of atoms in mica. Galena and halite break in cubes; they have cleavage in three directions, all at right angles to one another. Feldspar has cleavage in two directions, at nearly right angles, but it breaks irregularly in other directions. Some minerals, like quartz, have no cleavage. Quartz breaks into irregular shapes and often shows a curved surface called **conchoidal fracture.**

Geo Words

streak: the color of a mineral in its powdered form, usually obtained by scratching the mineral on a streak plate and observing the mark it leaves.

specific gravity: the ratio of the weight of a given volume of a substance to the weight of an equal volume of water.

cleavage: the breaking of a mineral along regularly oriented planes of weakness, thus reflecting crystal structure.

conchoidal fracture: a type of mineral fracture that gives a smoothly curved surface.

Figure 7A Muscovite has one direction of cleavage.

Figure 7B Halite has three directions of cleavage. They are at 90° to each other.

Figure 7C Calcite has three directions of cleavage. They are not at 90° to each other.

Figure 7D Feldspar has two directions of cleavage. ➡

Teaching Tip

Cleavage in minerals is the tendency for a crystal to split or fracture along regular planar surfaces. Cleavage is a manifestation of the geometrical arrangement of ions in the crystal structure. Because of the different bond strengths between different ions, some planar orientations are potentially weaker than others. If certain planar orientations are sufficiently weaker than others, the crystal is likely to split along those planes, rather than irregularly, when broken. If the differences in the strengths of the various planar orientations are insufficient, then the crystal breaks along irregular fracture surfaces. Such minerals are said to have no cleavage.

Halite is a mineral with a pronounced cubic cleavage (three cleavage directions at right angles to one another, reflecting the cubic arrangement of the constituent sodium and chloride ions). At the other extreme, quartz has no cleavage. It's important to make a distinction between the nearly perfect planar fracture surface that results from cleavage, on the one hand, and the nearly perfect planar surface that results from crystal growth, on the other hand. Close examination usually reveals small but significant differences in the small-scale geometry of the two kinds of planar surfaces.

Earth's Natural Resources Mineral Resources

Crystal Shape

When minerals grow in unconfined spaces, they usually develop a regular crystal shape. Quartz crystals grow as six-sided (hexagonal) columns with pointed tops (*Figure 8*). Garnets often grow in regular twelve-sided shapes called dodecahedra (*Figure 9*).

Figure 8 Quartz crystals grow in hexagonal columns.

Figure 9 Garnets often grow in dodecahedral shapes.

Color

Color is usually the first thing you notice about a mineral, but it is the least reliable property in mineral identification. Many minerals have different colors depending on what impurities are present. Corundum (Al_2O_3) is sometimes tinted red by small amounts of chromium; these crystals are known as rubies. The same mineral tinted blue by small amounts of titanium is called a sapphire. Quartz is usually transparent, but it also occurs in a great many colors, depending on what impurities are present. Some minerals tarnish or change color when surfaces are exposed to air. Many minerals have the same color as other minerals. Many prospectors in the gold-rush days were fooled by pyrite (fool's gold), which has a metallic luster and a color similar to gold. Pyrite has a lower specific gravity than gold, is brittle (gold is malleable), and leaves a black streak on a white porcelain tile (gold has a gold-colored streak).

Other Properties

Some minerals have special properties that make them easy to identify. Some of these properties also make these minerals useful for specific purposes.

Teaching Tip

Crystals that show all or most of their crystal faces are referred to as being euhedral, and those that show few of their faces are referred to as anhedral. The degree to which a crystal grows to be a nice euhedral grain depends largely on it having sufficient space to grow unfettered. If given sufficient room to grow, the external shape of a well-formed mineral grain is related to the underlying internal arrangement of atoms within the mineral grain. This internal arrangement helps to determine the crystal class to which a mineral belongs. One of the more diagnostic aspects of crystal shape is the nature of symmetry in a euhedral crystal, and this symmetry is related to the minerals crystal class. The quartz grain shown in *Figure 8* belongs to the hexagonal crystal class, and euhedral quartz grains often occur as pointed hexagonal prisms as shown in the photograph. The garnet grains shown in *Figure 9* commonly occur as twelve-sided polyhedron called dodecahedra. Garnet belongs to the isometric crystal system, which displays the highest degree of symmetry of any of the six crystal systems.

There are many different varieties of both quartz and garnet that display a wide range of colors. Garnets can be (among others) white, green, black, and cinnamon colored, in addition to the typical burgundy color commonly associated with garnet. Quartz can also have a wide range of colors. In each case, the color is typically a result of differences in the chemical composition of the mineral grain.

Chapter 2

- Metals tend to be good conductors of electricity. This makes them useful in the production and distribution of power, and in machinery. Most metals are also malleable (meaning that they can be changed in shape under pressure without breaking) and ductile (meaning that they can be stretched into wire).

- Some carbonate minerals fizz when a drop of weak hydrochloric acid is applied on the mineral. Acid breaks down the chemical bonds in the carbonate and releases CO_2 gas. Acid is a good test to identify the calcium carbonate mineral calcite.

- A few minerals are radioactive, releasing subatomic particles and radiation as the unstable atoms within them decay. For example, uranium minerals can be detected with a Geiger counter.

- Some minerals are magnetic. Magnetite, an important ore of iron, is naturally magnetic.

- Minerals like fluorite that change ultraviolet light to other wavelengths are called fluorescent. A few minerals that store light energy and release it gradually, are called phosphorescence.

Check Your Understanding

1. What is a mineral?
2. Why do different minerals have different properties?
3. Is color a good identifying property of a mineral? Why or why not?
4. What is the difference between cleavage and crystal shape of a mineral?

Understanding and Applying What You Have Learned

1. Refer back to the list of materials used for beverage containers that you compiled in Activity 1.

   a) Which of these materials are made from minerals?
   b) What other materials might be made from minerals and used as beverage containers?

2. Give at least one advantage and one disadvantage to using a native element to produce a product (for example, a beverage container).

3. How might the physical appearance or properties of a mineral be misleading when evaluating its potential for use as a beverage container?

4. Describe at least one possible disadvantage to producing beverage containers from a mineral that has high specific gravity.

5. A student claims that diamond is the hardest mineral because carbon, from which diamond is made, is a very hard element. Use what you have learned about minerals to provide a different explanation as to why diamond is the hardest mineral.

6. Correct the following misconception: "Quartz is always clear or transparent."

R 109

Check Your Understanding

1. A mineral is a naturally occurring, inorganic, solid material that consists of atoms and/or molecules that are arranged in a regular pattern and have characteristic chemical composition, crystal form(s), and physical properties.

2. Minerals have different properties because all minerals have a particular, distinct, combination of chemical composition and internal structure.

3. Although color is usually the first thing that you notice about a mineral, it is probably the least reliable property of the mineral. Color is an unreliable property because a mineral may come in many different colors, depending on the composition and concentration of minor elements and trace elements in the mineral.

4. When a mineral breaks in a regular pattern, it is said to have cleavage. Crystal shape, however, refers to the shape that a mineral takes as it grows.

Teaching Tip

There are different minerals that have the same chemical composition but different internal structure. This phenomenon is called polymorphism (in contrast to isomorphism, whereby different minerals have the same structure but different chemical composition). An example of polymorphism involves graphite and diamond, both of which consist of carbon. Another example is calcite and aragonite, both of which have the same chemical formula, $CaCO_3$ (calcium carbonate).

Understanding and Applying What You Have Learned

1. a) Responses will vary, depending upon the kinds of beverage containers that you examined. Containers made from aluminum, glass, and steel all utilize mineral resources.

   b) As above for **Question 1(a)**.

2. Responses will vary. A sample response might be that an advantage of using native metals is that they tend to be attractive, and a disadvantage is that they are very expensive. Another possible attraction to using a native element is that it is already in a pure form (like copper or gold), so the element does not need to be extracted from an ore.

3. Answers will vary. A sample response might be that a mineral might be attractive to look at but may not be very strong or capable of holding liquid or carbonation. Another possibility is that the mineral resource may not look like the material that would eventually be used to make the container: this would be the case for bauxite being used as the source for aluminum cans.

4. Answers will vary, but might include responses like that a container made from a mineral with a high specific gravity would be more expensive to produce and ship because it is heavy. Another possibility is that consumer satisfaction would be less if the container is heavy, because it would be difficult to carry and tiring to drink from.

5. Student responses should include some mention of the arrangement of atoms in a diamond, and the strength of the bonds between those atoms.

6. Although quartz is usually clear or transparent, it is also found in a variety of colors, depending upon the impurities present in the mineral. Aggregates of many crystals of fine-grained quartz, as in quartz veins, usually appear milky white, even though the individual crystals are transparent; this is an optical effect rather than an inherent property of the quartz crystals.

NOTES

Earth's Natural Resources Mineral Resources

7. Make a concept map that demonstrates your understanding of minerals. Include the following terms: mineral, element, rock, ore, compound.

8. When you broke halite with a hammer, you observed smaller cubic-shaped crystals. What would you observe if you shattered a quartz crystal into many smaller pieces? Why?

Preparing for the Chapter Challenge

Write a paper that will serve to help the president of the beverage company understand what mineral resources are. Be sure to define "mineral" and explain how minerals are identified. Also, explain how the arrangement of atoms in different minerals affects their physical properties and therefore their potential for use as materials to produce beverage containers.

Inquiring Further

1. **Mineral make-up of the Earth**

   Investigate the proportions of various materials in the crust of the Earth. Which minerals are most common? Which elements make up most of the minerals?

2. **Metallic and nonmetallic resources from minerals**

   Which minerals are the source of metals like iron, silver, lead, and copper? What are some nonmetallic resources? Which minerals are the sources of these resources?

7.

Concept Map Showing Relationship between
Rocks, Ores, Minerals, Compounds, and Elements

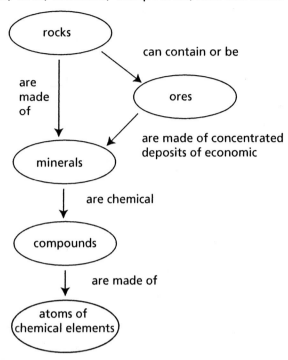

8. Quartz has no cleavage, but rather breaks into irregular shapes and often shows conchoidal fracture.

Assessment Tool

Check Your Understanding Notebook-Entry Evaluation Sheet
Use this sheet to evaluate the extent to which students understand the key concepts explored in **Activity 2** and explained in **Digging Deeper**, and to evaluate the students' clarity of expression.

Preparing for the Chapter Challenge

This section will allow students to pull together their knowledge of minerals. The paper they write can also address the properties that are important when considering which mineral to select for the beverage container. This paper can be used as an introduction, to help the CEO understand student justifications for the mineral resources that they plan to use to make the beverage container.

The goals of the assignment for this section are relevant to the question of what mineral resources are, which is the first bulleted criterion in the example assessment rubric for the chapter report (See **Assessment Rubric for Chapter Report on Mineral Resources** on pages 304 and 305).

Inquiring Further

1. Mineral make-up of the Earth

Ten elements account for 99% of the crust. These are:

- oxygen (46.6%)
- silicon (27.7%)
- aluminum (8.1%)
- iron (5.0%)
- calcium (3.6%)
- sodium (2.8%)
- potassium (2.6%)
- magnesium (2.1%)
- titanium (0.4%)
- hydrogen (0.1%)

The most common rock-forming minerals are the silicates, which are composed (in part) of the two most abundant elements in the Earth's crust: oxygen and silicon. Common silicate minerals include feldspar, quartz, mica, pyroxene, and olivine, with feldspar and quartz being the most common minerals in the crust (olivine and pyroxene are the most common minerals in the Earth's upper mantle). Other major rock-forming minerals include:

- oxides, which are composed of oxygen combined with a metallic element; e.g., hematite (chemical composition: Fe_2O_3)
- carbonates, which are made of calcium and magnesium combined with carbon and oxygen; e.g., calcite (chemical composition: $CaCO_3$)
- sulfides and sulfates, which are based on the element sulfur; e.g., pyrite (chemical composition: FeS_2)

2. Metallic and nonmetallic resources from minerals

Almost all aluminum is produced from bauxite, which is a fine-grained mixture of hydrous aluminum oxides. Iron is derived from several iron-bearing minerals: hematite and magnetite (iron oxides), siderite (iron carbonate), and several iron silicates. Almost all iron comes from vast deposits of iron-bearing rocks, called iron formations, which were deposited during a relatively brief episode in geologic history about two billion years ago (see **Activity 3, Background Information**). Copper is derived mainly from various copper sulfide minerals. Lead and zinc, which tend to occur together, are also derived mainly from sulfide minerals. Silver is found both as the native element and as sulfide minerals. Most gold is obtained in native form in very low concentrations from ancient sedimentary rocks and recent river gravels.

Chapter 2

NOTES

ACTIVITY 3— WHERE ARE MINERAL RESOURCES FOUND?

Background Information

Geologic Setting and Formation of Ore Deposits

Economically valuable mineral deposits are formed in a great variety of geologic settings. Ores of many metals are formed in association with igneous processes. Some of these metal ores are formed directly by precipitation of certain metal-bearing minerals during cooling and crystallization of the magma, and then segregation of the crystals by such processes as settling to the bottom of a magma chamber. Most, however, are precipitated by aqueous solutions or water-rich residua of magmas.

Water is intimately associated with magmas. Many kinds of magma contain as many as a few percent water dissolved in the magma. Compounds with a low boiling point that are dissolved in magma at the great pressures deep in the Earth are called volatiles. If the magma ascends through the crust to reach positions near or at the Earth's surface, the volatiles have a tendency to come out of solution as gases. If, however, the magma remains at depth to crystallize very slowly there, much or most of the water contained in the magma does not fit into the main body of minerals precipitated from the magma. The residual magma (after most has crystallized) then becomes very rich in water, along with a number of chemical components that are concentrated in the residual liquid because they do not fit into the major minerals.

As these residual juices rise buoyantly through the overlying rock toward the surface, they cool. As they cool, a great variety of minerals are precipitated in the form of pegmatite dikes and veins. The most common vein minerals—like quartz, muscovite, and potassium feldspar—are not especially valuable economically, but many other minerals containing such minor elements as lithium, beryllium, boron, or phosphorus can constitute economically valuable deposits.

Another common process that leads to economically valuable metal ore deposits involves convective circulation of deep groundwater occasioned by the presence of hot magma and hot, newly crystallized igneous rock in the crust below. As the groundwaters reach depths at which they are warmed magmatically, their density decreases and they rise back upward. These hot waters tend to be rich in various metal ions contributed by the cooling magma. As the hot waters circulate upward, they cool and deposit hydrothermal minerals. Sulfides and other compounds of metals like gold, silver, lead, zinc, and mercury are commonly formed in this way.

It is convenient to think of hydrothermal settings as having a characteristic kind of plumbing system in which the water circulates slowly through large volumes of rock. The characteristic structure of such plumbing systems is still not well known, because it is impossible to image modern examples of such systems in detail, and their geologic record is largely inaccessible in full three-dimensional geometry.

The ores of two very important metals—iron and aluminum—are formed without association with igneous processes. As noted in **Activity 1, Background Information,** almost the sole ore of aluminum is a material called bauxite, which is formed in near-surface environments by concentration of aluminum in igneous and metamorphic rocks that contain aluminosilicate minerals. Almost all economically valuable ores of iron come from sedimentary rocks called iron formations. These iron-bearing sedimentary rocks were deposited in shallow marine sedimentary environments, interbedded with ordinary sandstones and shales, during a relatively brief time interval in geologic history about two billion years ago. Before that time, the Earth's atmosphere and oceans apparently contained no free oxygen, and enormous quantities of iron in ferrous form was present in solution in the oceans. As the near-surface environment of the Earth gradually became oxygenated, perhaps in association with the proliferation of single-celled photosynthesizing plants, the iron was oxidized to insoluble forms of ferric iron in the form of iron oxides and carbonates. The Proterozoic terranes of all the major continents contain large deposits of iron ores of this kind.

More Information – on the Web
Visit the *EarthComm* web site www.agiweb.org/earthcomm to access a variety of links to web sites that will help you deepen your understanding of content and prepare you to teach this activity. Many of the sites also contain images that you can download.

Chapter 2

Goals and Assessment

Clarify that the goals indicate what students should understand and be able to do as a result of the activity. Make sure students understand that Chapter Assessments are based upon these goals.

Goal	Location in Activity	Assessment Opportunity
Identify the mineral resources and commodities of the United States.	**Investigate; Digging Deeper; Understanding and Applying What You Have Learned** Questions 3 – 4	Mineral resources and commodities are properly identified.
Identify the mineral resources and commodities within your community and state.	**Investigate; Understanding and Applying What You Have Learned** Questions 1 and 5	Mineral resources and commodities are properly identified. Mineral resource map of the community is correctly drafted.
Understand how different minerals are formed and which minerals are best suited for particular tasks.	**Digging Deeper; Check Your Understanding** Questions 2 and 3 **Understanding and Applying What You Have Learned** Questions 2 and 4	Responses to questions closely match those given in Teacher's Edition.
Describe the uses of your state's major mineral resources.	**Investigate; Understanding and Applying What You Have Learned** Questions 1 – 2, and 5	Responses to questions closely match those given in Teacher's Edition.

NOTES

Activity 3   Where Are Mineral Resources Found?

Goals

In this activity you will:

- Identify the mineral resources and commodities of the United States.

- Identify the mineral resources and commodities within your community and state.

- Understand how different minerals are formed and which minerals are best suited for particular tasks.

- Describe the uses of your state's major mineral resources.

Think about It

Suppose that your community was located on a small island in the middle of the ocean.

- Would you find mineral resources on the island?
- Is it possible to find a single place on Earth that has no resources?

What do you think? Record your ideas in your *EarthComm* notebook. Be prepared to discuss your responses with your small group and the class.

Activity Overview

Students begin the investigation by examining a map of the United States that shows the distribution of several mineral resource commodities. Then, they examine a mineral resource map of their state for commodity type and distribution. Students use the state map to produce a mineral resource map for their own community. **Digging Deeper** looks at the distribution of mineral resources in the oceans and on the continents, and explains how deposits form in each location.

Preparation and Materials Needed

Well in advance, you will need to obtain copies of your state mineral resource map. These can be obtained online through the USGS. The *EarthComm* web site contains suggestions to help you find the needed map. Depending on your location, you may need the maps of neighboring states in order to examine the region that encompasses a 100-km radius for your community. If this is the case, you may wish to prepare an appropriate regional state outline map that students can use to complete **Step 3** of this investigation.

Materials
- Copy of your state mineral resource map*
- Graph paper
- Colored pencils

Think about It

Student Conceptions

Student responses will vary. When students think of an island in the middle of the ocean, they most likely will conjure up an image of a tropical island. You may wish to supply them with photographs of islands in different parts of the world (a tropical island, an island in a cold climate, etc.) and ask them to think about what kinds of resources would be found on each. Students may respond to the second question by stating that mineral resources are not found in water.

Answers for the Teacher Only

All islands, large or small, consist of bedrock and/or sediments. (Technically, the world's continents are islands, but they are not usually considered to be so. In any case, they are not small. Most landmasses that are called islands are much smaller than continents, although some, like Madagascar, are large in the absolute sense.) The answer to the first question hinges in part on semantics: what constitutes a mineral resource? Even sand is a valuable mineral resource when it is used for concrete. The limestones that constitute many tropical islands could be ground up and used for agricultural purposes. Some islands, especially those that consist largely of igneous rock, might contain economically valuable metal deposits.

*The *EarthComm* web site has suggestions for obtaining this map.

Chapter 2

It would be difficult, if not impossible, to find anywhere on Earth that does not contain any resources, if resource is taken to mean any naturally occurring material that can be used as a raw material for the manufacture or production of something. Clearly, however, certain areas—e.g., those that contain nothing but dirt, without even vegetation—are much less valuable as a source of resources than other areas.

Assessment Tool

Think about It Evaluation Sheet
Use this evaluation sheet to help students understand and internalize the basic expectations for the warm-up activity.

NOTES

Earth's Natural Resources Mineral Resources

Investigate

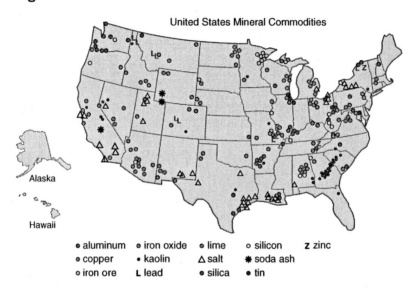

United States Mineral Commodities

Alaska

Hawaii

• aluminum	• iron oxide	• lime	○ silicon	**z** zinc
○ copper	• kaolin	△ salt	✳ soda ash	
○ iron ore	**L** lead	• silica	• tin	

1. State and federal governments compile maps of the locations of mineral deposits. The deposits are usually designated by the mineral commodity name. For example, chalcopyrite is an important copper mineral; the mineral commodity is copper. Examine the map of the United States that shows the distribution of several mineral resource commodities and processing plants.

   a) What mineral commodities are shown on the map?

   b) What minerals would you look for to find those commodities?

   c) What trends do you see in the distribution of the resources? Discuss the possible reasons for this distribution.

   d) If the whole United States were your community, could you find the mineral resources necessary to make all of the beverage containers you use in your community? Explain.

2. Use a mineral resource map of your state (which usually shows resources by commodity, not by mineral name) to answer the following questions:

   a) Are the resources evenly distributed around the state or are they concentrated in some areas?

   b) Which mineral commodities are found in your state?

   c) How do you think geologists decided whether a mineral deposit should be included on the map?

   d) Does a deposit have to be mined to be on the map?

Chapter 2

Investigate

Teaching Tip

The map on page R112 shows the distribution of a selected number of mineral commodities in the United States. For a complete map showing all of the mineral commodities in the United States (and a list indicating the number of each type), visit the *EarthComm* web site. Note that the map in the text does not show the location of mineral deposits, but rather the location of mineral commodities and mineral processing plants. Point this out to students and review the difference with them.

1. a) Many mineral commodities are shown on the map. These include aluminum, copper, iron ore, iron oxide, kaolin, lead, lime, salt, silica, silicon, soda ash, tin, and zinc.

Teaching Tip

You may want to review the mineral commodities on the map with your students, because they may not be familiar with materials like kaolin (a clay mineral). Many of these commodities are discussed further on pages 479–483 of this Teacher Edition. Consider discussing with students why some of the materials would make better choices for beverage containers than others (e.g., why lead would not be a good choice). If you are using the complete mineral commodities map from the USGS, you may ask students to list the top 10 kinds of mining sites, which would include: coal, dimension stone, sulfur, industrial sand and gravel, cement, lime, gold, silver, peat and perlite (combined), and salt.

  b) Students may have a difficult time with this question, because they most likely do not have an extensive knowledge of minerals. You may wish to have them visit the *EarthComm* web site, where they can do some background research about each of the commodities shown on the map.

   • **Aluminum** is derived from bauxite, which is composed mainly of one or more aluminum hydroxide minerals (for example the mineral gibbsite, which is $Al(OH)_3$), plus various mixtures of silica, iron oxide, titania, aluminosilicate, and other impurities in minor or trace amounts.

   • **Copper** is a native-element mineral; before the 20th century, native copper was the main source of copper metal. Since then, more efficient extraction methods have led to the use of a variety of other copper-bearing minerals as the main sources of copper metal. These minerals include (among others) copper-bearing sulfide minerals (like chalcopyrite and chalcocite) and hydrous carbonate and sulfate minerals (like azurite, malachite, and bronchantite).

   • **Iron ore** is recovered from iron oxide minerals, mainly magnetite (Fe_3O_4) and hematite (Fe_2O_3), as well as the iron carbonate siderite ($FeCO_3$) and various iron-bearing silicate minerals.

- **Iron oxides** include a combination of one or more ferrous or ferric oxides, and impurities, like manganese, clay, or organics. Iron oxide materials are used to produce pigments that are nontoxic, nonbleeding, weather resistant, and lightfast.
- **Kaolin** is a clay material, composed of the minerals kaolinite, nacrite, dickite, and anauxite.
- **Lead** is used for a variety of purposes, but perhaps most commonly in batteries. Although lead is a native element, it occurs in its native form only rarely. By far the most common ore of lead is the mineral galena, a lead sulfide. Other ores of lead include (among others) lead sulfates (anglesite) and lead carbonates (cerussite).
- **Quicklime** (calcium oxide) is manufactured by heating high-purity calcitic or dolomitic limestone to high temperature in a process called calcining. Quick lime can also be produced from other calcareous materials like chalk, coral, marble, and shell (all which are composed of the minerals calcite and/or aragonite). Quicklime is used in a wide range of metallurgical, environmental, industrial, and construction applications. Quicklime should not be confused with the lime that is used in gardens and other agricultural applications. Garden lime is simply crushed up limestone that has not been converted into calcium oxide by the calcining process.
- **Salt** is mined from rock salt (halite) deposits.
- **Silica** includes minerals that are composed of silicon and oxygen. For economic purposes, quartz is the most important.
- **Silicon** is an element with the chemical symbol, Si. It is recovered from silica.
- **Soda ash** is the trade name for sodium carbonate, a chemical refined from the mineral trona, or sodium-carbonate-bearing brines (both referred to as natural soda ash), or manufactured from one of several chemical processes (referred to as synthetic soda ash).
- **Tin**, for commercial purposes, is recovered mainly only from the mineral cassiterite (SnO_2). Small amounts of tin are also recovered from the minerals stannite, cylindrite, frankeite, canfieldite, and teallite.
- **Zinc** is recovered from the mineral sphalerite, which is zinc sulfide. It is the 23rd most abundant element in the Earth's crust.

Teaching Tips

Additional information on many of the minerals named above, including their occurrence and chemical composition, can be found on the *EarthComm* web site.

Additional information about the use and recovery of each of these mineral commodities can be found the USGS web site. Visit the *EarthComm* web site to be linked to the appropriate pages.

c) Students should describe the rock types (i.e., general physical characteristics) typically associated with each commodity type and should note if they see any clusters of mineral commodities. For example, they might notice a concentration of kaolin located in Georgia and South Carolina. In investigating the occurrence of kaolin they will find that is mined from near-surface clay deposits, and they may hypothesize that there is a lot of clay in Georgia. Another possibility is that they may notice that lime deposits are widely distributed throughout the country. In investigating lime they will find that is mined from limestone and marble. From these observations they might conclude that limestone is a common and widely distributed rock type.

d) Answers will vary depending upon the materials selected to make their containers, but it is likely that all of the mineral resources could be found in the United States. The United States does have the mineral resource commodities to make beverage containers from the common materials used for such purposes (glass, clay, aluminum, etc.)

2. a) Answers will vary depending upon your state, but it is likely that the resources are concentrated in certain areas.

b) Answers will vary.

c) Generally speaking, only mineral deposits large enough to be profitable, and known reserves of minerals for which there is a developed use and potential market, are included on state or local maps.

d) No, a deposit does not have to be mined to appear on the map.

Chapter 2

3. Depending on where you live, you may need to obtain mineral-resource maps from a nearby state to create your map. Create a new map of the area within a 100-km radius of your community.

   a) Make your map fit on one piece of graph paper. Include your map in your *EarthComm* notebook.

   b) If there are no commodities within 100 km of your community, how far might you have to go to locate some within your state?

Reflecting on the Activity and the Challenge

In this activity, you identified the mineral resources of the United States and those within your state. This will allow you to determine where the materials needed to make your beverage container will be obtained. If your community does not have all the mineral resources necessary to make all the products you use in your community, you must begin to think about relying on an exchange of resources and materials between other communities in order to make all of the products you use.

Digging Deeper

MINERAL RESOURCES ON EARTH

Mineral Resources in the Ocean

Almost all of the Earth's mineral resources are located on the continents. The waters of the ocean contain staggering quantities of many chemical elements, but the concentrations are for the most part extremely small. For most of these chemical elements the costs of extraction make it impractical to use them as mineral resources. Gold is a good example. Its concentration in the ocean is only 0.011 μg (millionths of a gram) per liter, but that adds up to more than ten billion (10^{10}) total kilograms of gold in the ocean! That's far greater than the known reserves of gold in continental ore deposits. The technology for extracting gold from seawater is so difficult and expensive, however, that it is nowhere close to being economically feasible.

Figure 1 Due to the low concentration of most chemicals in ocean water, it is not practical to "mine" minerals from the waters of the ocean.

R 113

3. Student maps will vary depending upon the community. Remind students to include a title and legend on their map.

Assessment Opportunity

The following sample rubric can be used or adapted as a basis for assessing student maps. You should consider handing out the rubric prior to assigning the work so that students are aware of what is expected of them.

Item	Missing	Incomplete/Inaccurate	Complete/Accurate
Mineral resources located within a 100-km radius of the community are represented.			
Mineral-resource locations are correct.			
Map includes key or legend with a scale and a title.			
Map is scaled correctly to fit on one page of graph paper.			

Assessment Tools

EarthComm Notebook-Entry Checklist
Use this checklist as a quick guide for student self-assessment and/or an opportunity to quickly score student work. Add further criteria specific to your classroom needs or to this particular investigation.

Investigate Notebook-Entry Evaluation Sheet
Point out the criteria listed on this evaluation sheet that are relevant to this particular investigation. Encourage students to internalize the criteria by making them part of your assessment conversations as you circulate around the classroom.

Reflecting on the Activity and the Challenge

Students examined the distribution of mineral resources throughout the United States, within their state, and within their community. They should now have a sense of what resources are available for them and where they would need to go to obtain these resources. Discuss with students the consequences of using mineral resources that are found far from the community versus those that are obtained locally. They should recognize that the longer the distance that a resource must be transported, the more expensive it would be. Students should think about how they would justify to the beverage company the need to import resources.

Digging Deeper

Assign the reading for homework, along with the questions in **Check Your Understanding** if desired.

NOTES

Earth's Natural Resources Mineral Resources

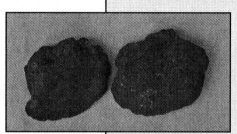

Figure 2 Iron-manganese nodules form slowly on the ocean floor.

Great expanses of the deep-ocean floor, especially in the Pacific Ocean, are covered with dark-colored, rounded masses, called iron–manganese nodules, as shown in *Figure 2.* They range in size from golf balls to large fists. They consist of very fine-grained minerals of iron and manganese, with many other chemical elements in smaller concentrations. Techniques for mining them from the ocean floor have been developed, but there are two problems. The sediment stirred up and suspended in the water during mining would have a harmful effect on the deep ocean environment. Also, the open oceans belong to no one country but to humankind as a whole.

Mineral Resources of the Continents

One of the most basic facts of geology is that the Earth's continents are geologically old. The oldest continental bedrock is known to be four billion years old (only half a billion years younger than the Earth itself), and large areas of the Earth have bedrock that is older than a billion years. On the other hand, geological processes that generate the bedrock record on the continents operate on fairly short geological time scales. That means that there has been plenty of time for the geological record of the continents to become extremely varied and complicated. Just a casual glance at a geologic map of United States like the one in *Figure 3* shows a jumble of irregularly shaped areas, colored in with a great variety of colors and patterns corresponding to rocks of different types and different ages. That should leave you with the impression that a large part of the United States has complex geology.

Figure 3 Geologic map of the continental United States.

R 114

Teaching Tip

Vast areas of the deep ocean floor, at depths of 4 to 5 km, are covered with iron–manganese nodules (*Figure 2* on page R114). The nodules mostly range in size from eggs to large fists. Their spacing on the sea floor is commonly about equal to their diameter, so that the intervening fine sediment is easily visible. They form only in areas where background rates of deposition of very fine-grained marine sediment are extremely small—of the order of a centimeter per thousand years. The nodules consist of fine-grained mixtures of various oxides of manganese and iron, which are precipitated slowly from the seawater. They represent a potential source of metals for humankind, but there are serious environmental problems associated with their mining. Also, no single country has the rights to mine the open ocean.

Figure 3 on page R114 is a geologic map of the continental United States To obtain copies of your state geologic map, and also to find resources on reading and interpreting geologic maps, visit the *EarthComm* web site.

Assessment Opportunity

Reword or restructure the questions in **Check Your Understanding** for a brief quiz. Use the quiz (or a class discussion of the questions in the textbook) to assess your students' understanding of the main ideas in the reading and the activity.

Only a very tiny fraction of the rocks of a continent like North America are ores. Most rocks, like sandstones, shales, limestones, granites, basalts, schists, and gneisses, do not contain economically valuable minerals (although many are used as building stones, and limestones are used in making cement and glass). Ores are formed only in certain very specific and unusual conditions. Metal mining operations in the United States occupy much less than one-tenth of 1% of the total land area. If they blindfolded you, put a parachute on you, put you aboard an airplane, and arranged for you to make your skydive over a random point in the United States, the chance that you would land on or near an ore deposit is extremely small!

Look again at the geologic map of the United States in *Figure 3*. One very striking thing about it is that large areas are occupied by only one or a few colors representing sedimentary rocks. In these areas, the deeper rocks of the Earth's crust are covered by a blanket of nearly horizontally layered sedimentary rocks that in most places have never been buried very deeply or subjected to ore-forming processes. In contrast, most areas of the western United States and some areas of the eastern United States show a complex pattern of rock types. These areas are called **orogenic belts**. The word *orogenic* means "mountain-building." These are areas where collisions between the Earth's **lithospheric plates** have resulted in great uplift of the land surface to make mountains, which in many places have later been worn down from their originally great heights. In those areas, igneous activity (the movement and crystallization of previously melted rock, called **magma**) has led to deposition of various kinds of ores in certain places.

One of the most important ore-forming processes is called **hydrothermal activity**. Magmas contain many economically valuable chemical elements in very small concentrations. They become concentrated in the water-rich "juices" that are left over after most of the magma has crystallized to form ordinary igneous rocks, because these elements tend not to be included in the main minerals that crystallize from the magma. These juices work their way upward toward the surface. As they move upward they cool, causing a great variety of unusual minerals to crystallize. Many deposits such as the one shown in *Figure 4*, called hydrothermal

Figure 4 The backhoe removes silver ore from a hydrothermal deposit in an underground mine.

Geo Words

orogenic belt: a region that has been subjected to folding and deformation during the process of formation of mountains (orogeny).

lithospheric plate: a rigid, thin segment of the outermost layer of the Earth, consisting of the Earth's crust and part of the upper mantle. The plate can be assumed to move horizontally and adjoins other plates.

magma: naturally occurring molten rock material, generated within the Earth, from which igneous rocks are derived through solidification and related processes.

hydrothermal activity: pertaining to hot water, the action of hot water, or to the products of this action, such as a mineral deposit precipitated from a hot aqueous solution.

NOTES

Chapter 2

Earth's Natural Resources Mineral Resources

deposits, are valuable ores. Much of the copper, zinc, tin, lead, mercury, gold, silver, and platinum, among others, come from hydrothermal ore deposits.

Ores of iron and aluminum, the two most-used metals, are a different story. Almost all iron ore comes from special sedimentary rocks, very rich in iron minerals, that were deposited in the oceans far back in geologic time, about two billion years ago. In the United States, these rocks are mined for iron ore in northern Minnesota and northern Michigan. In contrast, aluminum ore, called bauxite, consists of aluminum oxides that are formed when rocks containing aluminum are weathered at the Earth's surface in warm and humid climates. Some bauxite is mined in the United States, but most comes from other countries.

Mineral Resources around the World

Some countries of the world are richer in mineral deposits than others. Countries that include large areas of very old continental rocks are especially rich in mineral resources. Canada, Russia, Congo, South Africa, Brazil, and Australia are especially rich in mineral deposits.

The United States has abundant energy reserves (coal, oil, natural gas) but is not as rich in most mineral deposits as many other large countries. The United States has fairly abundant iron, copper, and tin deposits, but almost all of the aluminum, as well as most of the ores of several special metals that are important in making steel, like nickel, cobalt, or chromium, must be imported. The United States government maintains stockpiles of important metals, in case supplies from other countries become reduced or cut off in the future.

Check Your Understanding

1. What mineral resources are retrieved from the oceans?
2. What mineral resources are produced through hydrothermal activity?
3. What are the two most commonly used metals and where do they come from?

Understanding and Applying What You Have Learned

1. Look up additional information about the mineral resources found in your state.

   a) How are your state's mineral resources used?
   b) What products are manufactured from them?

2. Refer back to the mineral resources map of your community that you produced. Could any of the minerals or commodities near your community be used to make a beverage container? If yes, which ones? If no, how far would you need to go to locate minerals for making beverage containers?

3. Why is such a small proportion of the area of the United States underlain by ore deposits?

Check Your Understanding

1. Iron–manganese nodules are the main mineral resource recovered from the oceans.

2. Hydrothermal activity produces valuable ores of metals like copper, zinc, tin, lead, mercury, gold, silver, and platinum.

3. Two of the most commonly used metals are iron and aluminum. Iron ore comes from sedimentary rocks that are very rich in iron minerals and were deposited in the oceans very far back in geologic time. Aluminum ore, called bauxite, consists of aluminum hydroxides that are formed when rocks containing aluminum are weathered at the Earth's surface in warm and humid climates.

> **Assessment Tool**
>
> **Check Your Understanding Notebook-Entry Evaluation Sheet**
> Use this sheet to evaluate the extent to which students understand the key concepts explored in **Activity 3** and explained in **Digging Deeper**, and to evaluate the students' clarity of expression.

Understanding and Applying What You Have Learned

1. Answers to these questions will vary depending upon where you live. Your state geological survey or department of natural resources should be able to provide you with the necessary information.

2. Answers will vary depending upon your community.

3. Ore deposits are formed only in certain, very specific, unusual conditions. Most common rocks in the United States—like sandstones, shales, limestones, granites, basalts, schists, and gneisses—do not contain sufficient quantities of economically valuable ore minerals to make mining worthwhile.

4. Why are many ores associated with orogenic belts? Describe an example of an important ore in the United States that is not associated with mountain building.

5. If producing a beverage container depended primarily upon how close your community was to the source of the primary mineral resource, would your community produce containers made of glass (silica sand), steel (iron ore), or aluminum? Explain.

Preparing for the Chapter Challenge

Prepare a short paper in which you address the following questions:

- Which of the minerals in or near your community could be used to make any container that could hold a liquid?

- Which minerals in or near your community could contribute to the manufacture of a beverage container? What could those minerals contribute?

- If your community were not allowed to import a mineral commodity like aluminum, what impact would the lack of aluminum have? What minerals could be used instead?

Inquiring Further

1. **Worldwide distribution of mineral resources**

   On a map showing worldwide distribution of mineral deposits, find the possible sources of minerals to make beverage containers.

2. **Communities and mineral resources**

   Describe the locations of communities in relation to the mineral resources of your state. What relationships seem to exist? What industries grew because of the raw materials available locally?

3. **The Hall process**

   Research the Hall process used to produce aluminum from aluminum ore. How does the process work? Why did this process impact the aluminum industry? Who was Hall, and why did he become famous and wealthy?

4. Orogenic belts are areas where collisions between the Earth's lithospheric plates have resulted in great uplift of the land surface to make mountains. In these areas, igneous activity has led to the deposition of various kinds of ores. The iron ores mined in the United States, however, are an example of an important ore that is extracted from sedimentary rocks, not the igneous rocks associated with orogenic belts.

5. Answers will vary depending upon your community. Students should pick the material for which the needed resources are closest.

Preparing for the Chapter Challenge

Students should address all of the bulleted points in a short paper. Answers to the three questions will vary depending upon the area in which you live. These questions will help students think more about what materials are available for them to make their beverage containers, and what costs might be associated with acquiring resources from far away.

Relevant criteria for assessing this section (see **Assessment Rubric for Chapter Report on Mineral Resources** on pages 304 and 305) include a discussion on:
- what mineral resources you plan to use to make your beverage container
- where the mineral resources you plan to use to make your container were formed
- how the mineral resources you plan to use to make your container were formed

Review these criteria with your students so that they can be certain to include the appropriate information in their short papers. From a health perspective, many minerals are unsuitable for use as beverage containers. For example, although the mineral galena (lead sulfide) may be abundant in your state and can be used to make a container, lead is harmful to human health. You might wish to visit the *EarthComm* web site for further information about minerals that are not suitable for making a product that will hold liquids for human consumption.

Inquiring Further

1. Worldwide distribution of mineral resources

The *EarthComm* web site provides links to useful sources of information that will help your students to explore this question. Key minerals used to produce the raw materials for beverage containers include:

- **Bauxite** is a naturally occurring, heterogeneous material composed mainly of one or more aluminum hydroxide minerals, plus various mixtures of silica, iron oxide, titania, aluminosilicate, and other impurities in minor or trace amounts. The principal aluminum hydroxide minerals found in varying proportions with bauxites are gibbsite and its polymorphs boehmite and diaspore. Bauxites are typically classified according to their intended commercial application: abrasive, cement, chemical, metallurgical, refractory, etc. The bulk of world bauxite

Chapter 2

production (approximately 85%) is used as feed for the manufacture of aluminum. Leading producers include Guinea, Australia, Brazil, Jamaica, India, and China.

- **Iron ore** includes hematite (Fe_2O_3) and magnetite (Fe_3O_4) together with carbonates and silicates. Iron ore is mined in about 50 countries. The seven largest of these producing countries account for about three-quarters of total world production. Australia and Brazil together dominate the world's iron-ore exports, each having about one-third of total exports.

2. Communities and mineral resources

Students should look at the mineral resources map they examined in the investigation. Answers will vary depending upon the state, but in general students should find that communities tend to cluster around mineral deposits.

3. The Hall process

The Hall process is the term used for the modern industrial electrolytic method of aluminum production. The process was discovered simultaneously and independently by Charles Martin Hall in the United States and Paul-Louis Héroult in France in 1886. The basic concept of the process is still relevant, although there have been various technical improvements since then. Alumina (the term for aluminum oxide) derived from the bauxite ore is dissolved in a fused (molten) electrolyte (the term for an electrically conducting medium) that is derived from cryolite, an aluminum-bearing mineral with the chemical formula, Na_3AlF_6. The alumina is electrolyzed by a DC current that is passed through the electrolyte. CO_2 is liberated at the anode, and aluminum metal is deposited on the carbon lining of the vessel, which acts as the cathode. The operation is semicontinuous; alumina is added periodically, and eventually the aluminum is removed from the vessel. The operating temperature is 960 to 970°C. The energy efficiency is about 40%; most of the inefficiency is a consequence of the waste heat that is produced as the current flows through the cell.

NOTES

ACTIVITY 4 — HOW ARE MINERALS FOUND?

Background Information

Mineral Exploration

Early mineral exploration was conducted by prospectors. Armed with pack animals, camping supplies, pick, shovel, a gold pan, hammer, drill bits, blasting powder, and very little scientific knowledge of geology, they searched for gold and silver exposed in rock outcrops along mountainsides and in canyons. Finding a usable deposit was largely a matter of luck. Later, as the demand for copper, lead, and iron increased, they used many of the same techniques to locate and remove these exposed metallic ores.

Today, most mineral resources are hidden below the surface or trapped deep within rock formations, calling for more sophisticated exploration methods. Today's prospector is a team of well-trained specialists using the latest in technological sensors and a thoroughly planned systematic search based on evidence. Team members include specialists in geology, geophysics, geochemistry, computer technology, drilling, metallurgy, and mineral economics. Finding new mineral resources is no longer a matter of chance.

Geologic Maps and Cross Sections

Geologic maps show the distribution of bedrock that is exposed at the Earth's surface or buried beneath a thin layer of surface soil or sediment. A geologic map is more than just a map of rock types: most geologic maps show the locations and relationships of rock units.

Each rock unit is identified on the map by a symbol of some kind, which is explained in a legend or key, and is often given a distinctive color as well. Part of the legend of a geologic map consists of one or more columns of little rectangles, with appropriate colors and symbols, identifying the various rock units shown on the map. There is often a very brief description of the units in this part of the legend. The rectangles for the units are arranged in order of decreasing age upward. Usually, the ages of the units, in terms of the standard relative geologic time scale, is shown as well.

All geologic maps convey certain other information as well. They show the symbols that are used to represent such features as folds, faults, and attitudes of planar features like stratification or foliation. They have information about latitude and longitude, and/or location relative to some standard geographic grid system. They always have a scale, expressed both as a labeled scale bar and as what is called a representative fraction—1:25,000, for example—whose first number is a unit of distance on the map and whose second number is the corresponding distance on the actual land surface.

All geologic maps (except perhaps special-purpose maps that show all the details of an area that might be the size of a small room!) involve some degree of generalization. Such generalization is the responsibility of the geologist doing the mapping. Obviously, it is not practical to represent features as small as a few meters wide on a map that covers many square miles: the line depicting the feature on the map would be far finer than the finest possible ink line. The degree of generalization necessarily increases as the area covered by the map increases. You could easily see this for yourself if you had access to a geologic map of some small area together with the corresponding geologic map of the entire state: the state map would show far less detail of the small area than the full map of that same small area.

Most geologic maps are accompanied by one or more vertical cross sections, which are views of what the geology would look like in an imaginary vertical plane downward from some line on the land surface. The geologist constructs these cross sections after the map is completed. Their locations are selected so as to best reveal the three-dimensional nature of the geology. Cross sections are constructed by projecting downward the geologic features and relationships that are observed at the surface. The degree of certainty about the geology shown on the cross section decreases downward with depth below the surface.

Constructing cross sections requires the geologist to be able to visualize the geology in his or her mind. This kind of visualization is difficult for some people; they need extra instruction and help with maps and cross sections.

Activity 4 could be enhanced by introducing students to geologic maps of mineral resources, and some of the standard symbols used on them, before the investigation begins. If you do not already have a mineral resource map for your state, try contacting your local state geologic survey. You can get the telephone number of the state geological survey from directory assistance or find their contact information through links on the *EarthComm* web site. Students can conduct the investigation without previous experience with maps, but be aware that formulating their own codes takes additional time. They may also struggle with the concept of

mapping depth in addition to surface features. Emphasize drawing cross-sectional diagrams of at least three sides of their models to help them acquire this perception.

Seismic Surveys

Seismic reflection profiling is the most important technique for imaging rock structures in the subsurface. Seismic profiling is based on the propagation of seismic waves—elastic waves that travel through the Earth after being generated by an earthquake or an artificial explosion. When a body of rock is subjected to an impact, a vibration, or an explosion, seismic waves are generated, and these waves propagate at high speeds in all directions away from the region of the original disturbance.

A good way of visualizing the generation and propagation of elastic waves is to imagine striking one end of a long cylindrical rod of rock with a hammer. The blow of the hammer compresses the rock at the face of the rod. As the rock reexpands elastically, it causes compression of the next layer of rock along the rod, which in turn reexpands, and so on. In that way, an elastic wave propagates along the rod. The speed of the wave is of the order of several kilometers per second. In reality, a train of many waves is generated rather than a single wave front.

When elastic waves reach a surface of discontinuity in wave velocity within the body of rock, some of the wave energy is transmitted through the discontinuity but some is reflected. The reflected wave propagates back in a direction opposite to that of the original wave. The velocity of elastic waves is mainly a function of rock density and rock composition, so any change in these rock properties from layer to layer causes seismic waves to be reflected.

In a seismic survey, seismic waves are created at a large number of points at the Earth's surface along a line that typically stretches for thousands of meters. The vibrations may be caused by explosions or by pounding of the land surface by a special vehicle that travels along the line. The reflections are detected by special instruments called geophones. The times of the explosions and the times when the geophones receive the reflections must be recorded extremely accurately. Powerful computers then process the data to produce an image of the positions of the reflectors in a cross section vertically downward from the seismic line.

In recent times, computers have become sufficiently fast and powerful to allow processing of three-dimensional arrays as well as two-dimensional arrays. The image cubes can then be sectioned by computer in any desired orientation. Resolution of reflectors is now as fine as several meters. There is, however, an inherent lower limit to resolution, set by the wavelength of the seismic waves and the fact that shorter-wavelength seismic waves are attenuated more strongly than longer-wavelength seismic waves and thus cannot penetrate sufficiently deep into the Earth.

More Information – on the Web
Visit the *EarthComm* web site www.agiweb.org/earthcomm to access a variety of links to web sites that will help you deepen your understanding of content and prepare you to teach this activity. Many of the sites also contain images that you can download.

Goals and Assessment

Clarify that the goals indicate what students should understand and be able to do as a result of the activity. Make sure students understand that Chapter Assessments are based upon these goals.

Goal	Location in Activity	Assessment Opportunity
Construct a model of a mineral deposit and map the relative positions of the deposit.	**Investigate** Part A **Investigate** Part B	Models are constructed following instructions. Map notes compass deflection at several locations and includes a legend.
Understand how geologists explore for mineral resources by conducting a survey to locate ore deposits in another group's model.	**Investigate** Part B **Digging Deeper; Check Your Understanding** Questions 2 and 3 **Understanding and Applying What You Have Learned** Questions 2 – 3	Survey is completed properly. Responses to questions closely match those given in Teacher's Edition.
Use your survey results to drill for the ore.	**Investigate** Part B	Final cross section is drawn correctly to reflect data collected from well logs. Less than $100,000 is spent.
Understand the necessity and benefits of exploratory surveys in locating minerals.	**Investigate** Part B **Digging Deeper; Check Your Understanding** Questions 1 and 3 **Understanding and Applying What You Have Learned** Questions 2 – 3	Responses to questions are reasonable, based on available data.

Chapter 2

Earth's Natural Resources Mineral Resources

Activity 4   How Are Minerals Found?

Goals

In this activity you will:

- Construct a model of a mineral deposit and map the relative positions of the deposits.

- Understand how geologists explore for mineral resources by conducting a survey to locate ore deposits in another group's model.

- Use your survey results to drill for the ore.

- Understand the necessity and benefits of exploratory surveys in locating minerals.

Think about It

Suppose that your neighbor's dog buried your family car keys somewhere in your backyard.

- What tools or information would you want to help you to locate your keys?

What do you think? Record your ideas in your *EarthComm* notebook. Be prepared to discuss your suggestions with your small group and the class.

Activity Overview

Students begin the investigation by constructing a model of a mineral deposit using a paper milk carton, sand, and iron filings. They produce a map to indicate where they placed the iron filings in the sand. Students then exchange models. Using compasses, they survey the "unknown" model to produce a cross section indicating how they think the deposits are distributed in the sand. On the basis of their cross sections, students drill to recover core samples. Then they compare their peers' findings of the model they created to the maps they produced when the model was constructed. Finally, students investigate the different kinds of mineral deposits found in their community and the techniques that geologists use to find these deposits. **Digging Deeper** reviews techniques of resource exploration, including mapping, geochemistry, and geophysics.

Preparation and Materials Needed

Part A

You may want to ask students to bring the paper milk or orange juice cartons into class. Be sure to clean the cartons thoroughly before using them. If you are running low on time, you may wish to prepare the carton for the students by cutting off one side of the container and drawing the grid on the other side. This way, students can begin building their models right away.

> **Teaching Tip**
>
> You can have students use magnetite sand or iron filings to represent the ore in their models. An alternative is to have students use a mixture of small magnets and powdered drink mix. The magnets will be sufficient to deflect the compass needle during the survey, and the powdered drink mix is easily visible during drilling.
>
> Students will be using these models again in the next activity to excavate the deposits. At the end of **Activity 5**, you can then simply rinse the sand to remove the drink mix, and pick the magnets out by hand.

Part B

This part of the investigation builds upon the models produced in **Part A**, so there is little advance preparation that you can do. You may want to prepare survey flags using toothpicks and masking tape.

Teaching Tip

In **Part B** of the investigation, students use compasses to survey the models. You may find that you have better success with a stud finder. You can also obtain an instrument called a pencil magnet, which is a tool used by geologists in the field to detect magnetic deposits. These instruments are available at low cost and are highly sensitive. Information about obtaining pencil magnets is available on the *EarthComm* web site.

Part C

Students can use the mineral resource maps they examined and produced in **Activity 3** to complete part of this investigation. They will need to do additional research to answer the questions regarding exploration techniques. You may want to provide them with resources to answer these questions. Your state geological survey will most likely be able to provide you with helpful information. You can access your state survey's web page through the *EarthComm* web site. If time and computer availability is not an issue, students can also use the *EarthComm* web site to complete their research.

Materials

Part A
- Scissors
- Paper half-gallon (2-L) milk or orange juice carton or a box/tub of similar dimensions
- Marker
- Graph paper
- Dampened sand
- Iron filings or black magnetite sand
- Powdered drink mix (optional)

Part B
- Magnetic compass (or stud finder or pencil magnet)
- Survey flags – toothpicks and masking tape
- Clear plastic straws
- Wooden dowel to push core sample out of straw (optional)

Part C
- No additional materials are required.

Think about It

Student Conceptions

Responses will vary. Students may say that they would use a metal detector to find the keys, or look for areas where the ground has been obviously disturbed.

Answer for the Teacher Only

There are no right or wrong answers for this particular question. Accept all reasonable responses.

Assessment Tool

Think about It Evaluation Sheet
Use this evaluation sheet to help students understand and internalize the basic expectations for the warm-up activity.

Investigate

Part A: Creating a Model of a Mineral Deposit

1. Draw a grid of 1-cm squares on one side of a 2-L (half-gallon) paper orange juice or milk carton. On graph paper draw the same number of squares in the same configuration. Do the same for the bottom of the carton.
   Label the drawings and the carton with corresponding letters.
   Also draw a graph of the top of the carton (what will be the exposed sand surface). Label the drawing.

   a) Write your group's name on the graph paper and on the carton.

2. Use scissors to remove one side of the carton. Place the carton so that the open side is up, as shown.

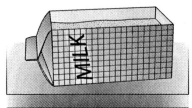

3. Next, fill up the carton to within 2 cm of the top using dampened sand to represent sediment and black sand (magnetite) or iron filings to represent ore bodies. You will need to distribute the "ore" throughout the sand in lines (called veins or ore shoots), in concentrated layers, or in pods. Do not bury the filings too deeply, and make sure that the deposits you produce are large enough to deflect a compass needle.

   ⚠ Wear goggles.

   a) As you fill the carton, sketch ("map") the position of the "ore" onto the graph paper representations of the side, the end, and the top of the carton.

4. When your model is complete, give it your teacher. Do not let other groups see your model or your sketches, because later each group will try to find the ore in another group's model. Keep your map for later use.

Part B: Exploring a Model of a Mineral Deposit

1. Obtain another group's model to explore. You will have a limited amount of money (an imaginary $100,000) with which to conduct your exploration. You will use a magnetic compass to survey the model. The buried magnetite or iron filings will cause the compass needle to be deflected from north. Each measurement made with the compass will cost you $500.

2. First, run your compass quickly over the model to see the maximum amount of deflection. This will help you scale your measurements. Design symbols to represent no needle deflection, maximum needle deflection, and one or two magnitudes in between. Then plan where you intend to make measurements.

   a) Draft a legend to explain the symbols.

   b) Using a piece of graph paper, make a map showing all the locations on the model where you will measure the deflection of the compass needle.

Investigate

Part A: Creating a Model of a Mineral Deposit

1 – 2. Students can use the illustration on page R119 in the text to help them to build their models.

3. The sand used to construct the model should be damp. Remind them not to bury the iron filings too deeply in the sand, and tell them to make fewer, large deposits rather than many smaller ones. The number of deposits should be limited to three or fewer. This is both to make sure that the deposits are large enough to deflect a compass needle, and so that they have the potential to find the deposits in **Part B** with only a limited number of drilling sites (they have enough "money" to only drill three sites). Tell them to run their compass over the top of each deposit to test that the deposits are large enough to be detected.

   a) Remind students to sketch the position of the ore deposit as they are creating their model. It is very important that students create accurate maps of the deposits, to serve as the key to the model at the end of the investigation.

4. Students should exchange models.

Part B: Exploring a Model of a Mineral Deposit

1. Review the parameters set forth for the mineral exploration and discuss what students are expected to do. Throughout the exercise, students will want to keep in mind that they have a total of $100,000, that each compass reading will cost them $500, and that each drill hole will cost $30,000.

2. a) Remind students to include a legend to explain the various symbols they place on their map.

   b) Students should run the compass over the model and sketch a map of where they think the deposits are located. Remind them that they must pay $500 for each compass reading.

Teaching Tip

This initial survey should be used as a tool for determining which areas the students will examine more closely in the next step. Discuss with students why they do not complete an intensive survey of the entire model. They should note that an intensive survey would be time consuming; also, because they are being assessed $500 for each compass reading, it would be quite expensive. The quick survey saves time and allows students to focus their studies.

Earth's Natural Resources Mineral Resources

3. Now it's time to complete the magnetic survey. Pick one person to operate the compass, one to record the measurements for each location in a data table, and a third person to plot the data on your map.

a) Measure at each location marked on your map. On the map and in the data table, write the symbol for the strength of the measurement based on the movement of the compass needle.

b) After a few measurements, you may decide to alter your plan or make more or fewer measurements. Record any changes you make in your survey plan, along with the reasons for them.

c) When you are satisfied that you have as many measurements as you want (or can afford), draw a cross section of what you think your deposit looks like.

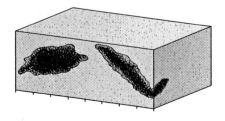

4. To confirm your hypothesis of the location of the deposit, select sites to "drill" holes with a soda straw. Each drill hole will cost your group $30,000.

a) Mark the locations of your planned holes on your map. Be sure to total your survey account, because you cannot drill more holes than your company can pay for.

5. Mark the locations of the drill holes on the surface of the sand with "survey flags" (toothpicks and masking tape). Carefully push a straw all the way down into a location marked by a survey flag. (It may be necessary to dampen the sand again if enough time has passed for the model to dry out. The sand should stick together as a column inside the straw.)

6. Carefully pull out the straw and inspect your core through the clear plastic, or insert a wooden dowel and push the core out onto scrap paper for viewing.

a) Estimate the percent recovery of the "rocks" in the drill core.

b) Create a vertical description (called a log) of each hole by recording the type of material at regular intervals (for example, every centimeter). Supplement your descriptions with sketches.

"Hole A"	Depth	Description
	0-1 cm	100% light tan sand
	1-2 cm	mostly sand (75%), small amount of black iron deposit
	2-3 cm	100% black iron deposit
	3-4 cm	50-50 light tan sand and black iron deposit
	4-5 cm	mostly sand (80%), small amount of black iron deposit
	5-6 cm	100% sand
	6-7 cm	100% light tan sand
	7-8 cm	100% light tan sand

3. Students should take readings at the locations indicated on their maps from **Step 2**. Cross sections should be drawn to resemble the illustration in the left-hand column of page R120 in the student text. Have students tally the cost of the survey (number of readings x $500) when they are done.

Teaching Tip

Students will likely be concerned about spending too much money on compass readings. However, remind them that the compass readings will help them create a more accurate cross section of the model, which will in turn help them decide where it would be appropriate for them to drill. A $500 compass reading is much less expensive than a $30,000 drill hole that yields no ore.

4. a) Students should use their cross sections to determine where they will drill for ore. Remind students that each drill hole will cost them $30,000, and that they have available $100,000 minus the cost of the survey (**Step 3**). Students will not be able to drill more than three holes.

5. Students may need to dampen the sand if it has dried out. Dampening the sand will make the extraction of the core sample much easier.

6. a) Answers will vary. Students may be confused by the concept of 100% recovery. This simply means that drill core sample was recovered over the entire depth of the drill hole.

   b) Student well logs should resemble the illustration in the right-hand column of page R120 in the text.

Chapter 2

7. Complete "drilling" at all locations.

a) When all the holes have been drilled and logged, draw a cross section of the ore deposit. On a piece of graph paper plot the drill holes and the drilling results. Connect the plotted points to show where the ore is located based on your observations.

b) Record how much money your group spent on your project, and return the models to your teacher for use in the next activity.

8. Report your exploration venture to the class, and discuss the various groups' outcomes and experiences. Obtain a copy of the cross section produced by the group that explored your model. However, do not yet reveal the original sketch (map) of your model to the other group.

a) In your *EarthComm* notebook summarize your class discussions.

 Dispose of the straws after use or mark them as lab equipment. Wash your hands after you have completed the activity.

Part C: Mineral Commodities in Your Community

1. In the previous activity you researched the mineral commodities located near your community. Use the information you collected to answer the following questions:

a) What types of deposits do the minerals found in your community typically form?

b) Which of the minerals have been found by geochemical or geophysical techniques?

c) Which of the mineral resources are located by geologic mapping?

d) Are techniques of seismic wave propagation and gravity surveys used in your state? For which mineral deposits are these methods used?

Reflecting on the Activity and the Challenge

Mineral-bearing ores are difficult to locate from the surface. This activity illustrates that exploration is expensive, and not always accurate. However, the costs of your project would be much higher if the drilling had been done by random sample without any survey. Although the ore body shown on your cross-section and exploration map probably differs in size or shape from the actual ore body, you would not have found it at all without the survey to guide your test drilling. Think about how you can use what you have learned to address an important aspect of your **Chapter Challenge**—explaining how mineral resources used to make beverage containers are found.

7. **a)** On the basis of their well logs and the survey, students should construct a final cross section to indicate the location of ore deposits in the model.

   b) The amount of money spent by each group will vary, but it should not exceed $100,000. Before you collect the models from the students at the end of the exercise, make sure that they have noted which model they surveyed, because they will use the same model in **Activity 5**.

8. Students should obtain the cross section from the group that explored their model. They should not trade maps. Student experiences will vary. As a class, discuss the importance of the initial survey, the intensive survey, and the development of a preliminary cross section. You may also want to encourage students to discuss how these steps will help them in **Activity 5**, when they actually excavate the mineral deposits.

Assessment Opportunity

The following sample rubric can be used or adapted as a basis for assessing student work. You should consider handing out the rubric prior to assigning the work, so students are aware of what is expected of them.

Item	Missing	Incomplete/Inaccurate	Complete/Accurate
Survey Map and Date Table			
Includes key or legend explaining all symbols used.			
Map and table are consistent.			
Compass readings are properly recorded.			
Drilling locations are represented.			
Well Logs			
Descriptions are useful, detailed.			
Information is recorded at regular, appropriately spaced intervals.			
Includes illustration.			
Final Cross Section			
Consistent with survey and drilling results.			

Part C: Mineral Commodities in Your Community

1. Students should use the maps and information assembled in **Activity 3** to answer these questions. Answers to these questions will vary depending upon your community. Contact your state geological survey for information to help students answer the questions.

Assessment Tools

EarthComm Notebook-Entry Checklist
Use this checklist as a quick guide for student self-assessment and/or an opportunity to quickly score student work. Add further criteria specific to your classroom needs or to this particular investigation.

Investigate Notebook-Entry Evaluation Sheet
Point out the criteria listed on this evaluation sheet that are relevant to this particular investigation. Encourage students to internalize the criteria by making them part of your assessment conversations as you circulate around the classroom.

Reflecting on the Activity and the Challenge

Have a student read this section aloud to the class, and discuss the major points raised. Students should now realize the cost associated with mineral exploration, but they should also realize that cost could be reduced through careful surveying and planning. Discuss how this investigation fits into the **Chapter Challenge**, and explore why the beverage company would be concerned about how minerals are found (cost of exploration and extraction affects cost of materials).

Blackline Master Mineral Resources 4.1

Activity 4: Understanding and Applying, Question 4

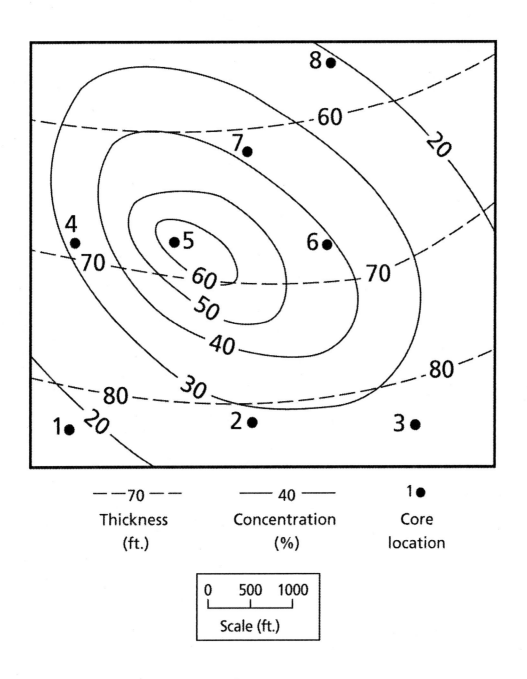

- - 70 - -   Thickness (ft.)

—— 40 ——   Concentration (%)

1●   Core location

0   500   1000
Scale (ft.)

Earth's Natural Resources Mineral Resources

Digging Deeper

RESOURCE EXPLORATION

Finding Minerals

Exploration for mineral resources relies heavily on experience, observation, data, analysis, inference, and a good understanding of geology, geochemistry, and geophysics. Even with all that, it still involves quite a lot of uncertainty. There is never a guarantee that minerals will be in the place prospectors expect them to be. Also, even if the minerals are located, there is no guarantee that the concentration will be high enough for the mineral to be extracted profitably.

Exploration is very expensive. Drilling is the most expensive way to explore, but it is the only sure way to confirm the type and amount of minerals present. In mineral exploration, drill holes are often at least 300 m deep, and they may cost $150 per meter. In oil exploration, a single drill hole may cost millions of dollars. To avoid spending money on "dry holes," geologists use other techniques to find out more about what is below the surface and to eliminate areas with low exploration potential.

Mapping

Geologists map the surface by examining rock types in the field. Geologists know that all mineral deposits are associated with specific kinds of rocks. Therefore, they can look for those specific rocks as "guides to ore," before doing any other kind of work. They then look for folds, faults, and fractures in the rocks, any unusual colors, and rock formations that seem inconsistent with the surroundings. They also may take rock samples to analyze in the laboratory. Then they record all the information on a map. **Geologic maps** like the one shown in *Figure 1* help geologists infer what lies below the surface.

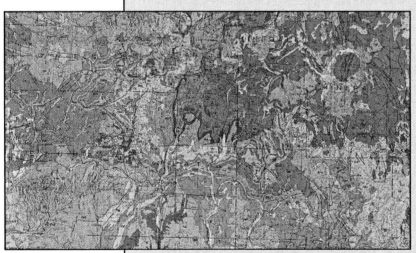

Figure 1 Geologic map of Southern Colorado.

Geo Words

geologic map: a map on which the distribution, nature, and age relationships of rock units are recorded.

geochemist: a geologist who studies the distribution and amounts of chemical elements in minerals, ores, rocks, soils, water, and the atmosphere, and their circulation in nature.

geophysicist: a geologist who studies the physical properties of the Earth, or applies physical measurement to geological problems.

Digging Deeper

Assign the reading for homework, along with the questions in **Check Your Understanding** if desired.

Assessment Opportunity

Use a quiz to assess student understanding of the concepts presented in **Activity 4.** Some sample questions are listed below:

Question: How can geochemistry be useful in mineral exploration? What is a major limitation associated with using geochemistry in mineral exploration?
Answer: Geochemistry can be used to determine in what abundance different elements are present in a rock or other material. This can help to determine whether or not a mineral deposit exists below the surface, based upon the chemistry of rock and mineral samples taken from the ground. Geochemistry does not tell a geologist where, or how far below the surface, a mineral deposit is located.

Question: Why are geologic maps important in mineral exploration?
Answer: Geologic maps help geologists infer what is below the surface of the Earth, which can give them a sense of where mineral deposits might potentially be located.

Question: How could a geophysicist use measurements of the local strength of the Earth's gravity field to find an ore body?
Answer: By using a very sensitive instrument called a gravimeter, a geophysicist can measure small differences in the Earth's gravity field. Because ore bodies tend to be denser than common rocks, the acceleration of gravity is slightly greater above the ore bodies. So, the geophysicist can look for places with a greater acceleration of gravity to indicate where ore bodies might be.

Teaching Tip

Figure 1 shows an example of a geologic map. Geologic maps represent information in a number of different ways. The different colored regions that are visible in *Figure 1* each represent the occurrence of a different geologic formation (a distinct unit that can be mapped). The colors are usually related to the age of that unit, and there will generally be an text abbreviation and sometimes a fill pattern that further identifies each different unit. Geologic structures like folds and faults are also noted on these maps. Many such features can be seen in *Figure 1* as heavy-ruled lines, some of which have small arrows emanating from them. More detail on the symbols used in geologic maps and how the maps are read can be found on the *EarthComm* web site.

Chapter 2

Geochemistry

There is no guarantee that what is on the surface matches what is deeper underground. **Geochemists** are geologists who specialize in analyzing the chemistry of rocks and minerals. Samples taken at the surface, from pits dug in the ground, or from a few drilled holes, can be analyzed to determine all the chemical elements present. These elements may be present in very small concentrations, often just a few parts per million (ppm), as halos above or around valuable mineral deposits. In addition to sampling rocks and soil, geochemists may also sample vegetation and water, because elements from the deposit can be absorbed by plant roots or dissolved in water. Their presence can help geochemists get an idea about whether a mineral deposit might exist below the surface. However, the geochemist cannot tell how far below the surface a mineral deposit might exist.

Geo Words

seismic wave: a general term for all elastic waves in the Earth, produced by earthquakes or generated artificially by explosions.

geophone: a seismic detector placed on or in the ground that responds to ground motion at its point of location.

Geophysics

The **geophysicist** is another important part of the exploration team. Oil companies and mineral companies employ thousands of geophysicists worldwide. Geophysics involves measuring properties of the Earth and of specific rocks and minerals. In oil exploration, the most common geophysical technique is **seismic-wave** reflection. A vibration is created by a small explosion or by striking the ground with a steel plate. The vibrations travel through the crust until they are reflected back to the surface by a rock layer. The returning vibration is recorded on a sensitive instrument called a **geophone**, as shown in *Figure 3*. The time it takes for the vibrations to return to the surface tells the geophysicist about the depth to the reflecting layer. Using this information, the geophysicist can more accurately determine if the subsurface is folded, faulted, or has other features where oil, gas, or mineral deposits may accumulate.

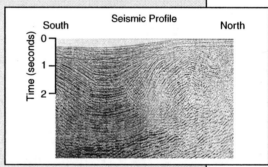

Figure 2 Seismic section.

Figure 3 A geophone stuck into the Earth and connected to the geophone cable with two clips.

Teaching Tip

The seismic section shown in *Figure 2* on page R123 comes from the Santa Barbara channel (offshore California). The sedimentary layers filling in the channel can be seen at the left of the figure.

Earth's Natural Resources Mineral Resources

Geophysicists also measure the local strength of the Earth's gravity field, which is affected by the density of the rocks below the Earth's surface. Dense rocks cause a slightly greater gravitational attraction than the less dense rocks around them, so gravity at the surface is slightly greater. Most ores are denser than the common rocks in the Earth's crust, like granite, basalt, sandstone, or limestone. The gravity field over most ore bodies is therefore slightly stronger than over nearby areas. An instrument called a **gravimeter** as shown in *Figure 4*, which measures the acceleration due to gravity, can detect these small differences in the Earth's gravity field. You would weigh slightly more over such an ore body than elsewhere, although not enough to feel it.

Other Techniques

An instrument called a **magnetometer** is used to detect changes in the Earth's magnetic field. Rocks that contain abundant iron-bearing minerals affect the local magnetic field. Geologists can also measure how well rocks conduct electricity. Most of the electrical current that geologists put in the ground for exploration flows through water in the pore spaces of rock. If a rock conducts electricity very well, it is likely to contain either a lot of water or a lot of metallic minerals.

The Bottom Line

No exploration technique can give a complete picture of the subsurface. The exploration geologist must be very good at reading the story of the rocks and discovering all the clues to successfully find minerals. Millions of dollars are spent on a typical exploration venture, but only a very small number of such ventures become economically profitable mines.

Geo Words

gravimeter: an instrument for measuring variations in Earth's gravitational field.

magnetometer: an instrument for measuring variations in Earth's magnetic field.

Check Your Understanding

1. What are the benefits of mineral exploration?

2. What geophysical technique did you use to survey the model in class?

3. How do maps help geologists be more successful in exploration?

Figure 4 Geologists gather data using gravimeters. The silver plate at the lower left is a gravimeter.

EarthComm

Teaching Tip

Figure 4 on page R124 shows one of the instruments used by geophysicists. Point out to your students that exploration techniques like those described in **Digging Deeper** have relatively little impact on the environment. Therefore, not only is mineral exploration a way to save money, but it is also much more environmentally friendly than digging up ground in hopes of finding minerals.

Check Your Understanding

1. Mineral exploration helps geologists find out more about what is below the surface of the Earth and eliminate areas with low mineral yield, thus saving time and money in the long run.

2. In the investigation, students measured changes in magnetic field produced by the iron filings. This technique is similar to the use of a magnetometer.

3. Geologic maps help geologists infer what is below the surface of the Earth, which can give them a sense of where mineral deposits might potentially be located.

Assessment Tool

Check Your Understanding Notebook-Entry Evaluation Sheet
Use this sheet to evaluate the extent to which students understand the key concepts explored in **Activity 4** and explained in **Digging Deeper**, and to evaluate the students' clarity of expression.

Understanding and Applying What You Have Learned

1. Compare your original diagram of the model your group built to the map made by the group that surveyed it.

   a) What differences exist between the two maps? Why do they not match?

   b) What other exploration techniques might you have used to gather more information about your model? What instruments would have been needed?

2. a) How could you have reduced the cost of your exploration?

   b) How do you think mineral explorers decide how much money and time to spend on exploration?

3. How do exploratory surveys help to reduce the cost of obtaining mineral resources?

4. The map on the right shows the thickness of a rock formation made of limestone and dolostone. The data for the map came from drilling into the Earth, taking core samples, and measuring the thickness of the formation in each core. The map also shows the percentage of dolostone in the formation. The dolostone contains the highest percentage of lead-zinc ore. Use a copy of the map to complete the following questions:

   a) Areas where dolostone exceeds 50% have profitable ore. Color these areas green.

   b) What is the thickness range of dolostone in the profitable area?

   c) What is the surface area (in square feet) of dolostone in the profitable area?

   d) If the average profits on the dolostone is $0.30 per cubic yard, what is the approximate company gain?

   e) Assume that the top of the formation is level. Draw a cross section between core holes 1, 5, 7 and 8. Indicate the profitable area on the cross section.

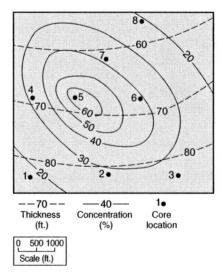

Understanding and Applying What You Have Learned

1. a) Because mineral exploration is not an exact science, the original maps, created when the model was put together, will likely differ from the maps created by the students during their surveys. The differences between the two maps will vary, but some attempt at explaining the differences should be made.

    b) Students may cite one of the other techniques discussed, like gravity or seismic surveys, or other techniques discussed at the beginning of **Activity 4** in the **Think about It** exercise.

2. a) Answers will vary depending upon how much initial magnetic surveying was done and whether or not any of the drill holes failed to produce ore. One way to reduce the cost of exploration would be to take a few compass readings over a broad area (to identify potential deposits) and then focus in on the potential deposit by taking additional readings in a specific area. Another way to reduce the cost would be to do more work with the compass (which costs only $500 per reading) prior to drilling (which costs $30,000 per drill hole). In general, the relative cost of drilling is such that a failure of even one drill hole to produce ore should lead to the conclusion that more detailed initial surveys would have yielded a net savings in exploration expense.

    b) The time and money mineral explorers decide to spend on exploration depends on how much net financial gain will be made with the discovery of a deposit.

3. Exploratory surveys help geologists determine what lies below the surface and where exploratory drilling would be profitable. Without these surveys, drilling sites would be selected through guesswork, which could be quite expensive.

4. a)

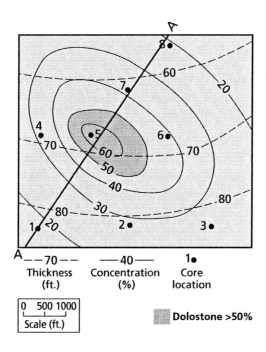

Chapter 2

b) The dolostone in the profitable area is between about 65 and 73 ft thick.

c) Review with students how to calculate the **surface area of an ellipse**:

$$A = \pi\, r_1 r_2,$$ where r_1 is the radius of the long axis and r_2 is the radius of the short axis

Students should use the scale to measure the radii. Their measurements may be slightly different, and therefore their calculations may yield results slightly different from those that follow. If $r_1 = 2000$ ft and $r_2 = 1150$ ft, then the surface area of the dolostone is equal to 7,225,663 ft^2 (rounded to the nearest whole number).

d) Students will need to assume that both the upper and lower surfaces of the deposit are flat and horizontal, and they will then need to calculate the volume of the deposit. To do this they must know the formula to calculate the **volume of a cylinder**:

$$V = 4/3(\pi\, r_1 r_2 r_3),$$ where r_1 is the radius of the long axis, r_2 is the radius of the short axis, and r_3 is the height, or depth, of the cylinder

Students can either convert the r_1, r_2, and r_3 values from feet to yards **before** calculating the volume, or they will also need to know that there are 27 cubic feet in one cubic yard. Rounding errors are minimized if the conversion to cubic yards is done after the calculation of the volume. If $r_1 = 2000$ ft, $r_2 = 1150$ ft, and $r_3 = 70$ ft, then the volume of the cylinder of the dolostone is equal to 674,395,223 ft^3. Dividing by 27 cubic feet per cubic yard yields 24,977,601 yd^3. Therefore, for this cylinder of dolostone, the company would gain approximately $7,493,280 (24,977,601 yd^3 x $0.30).

e)

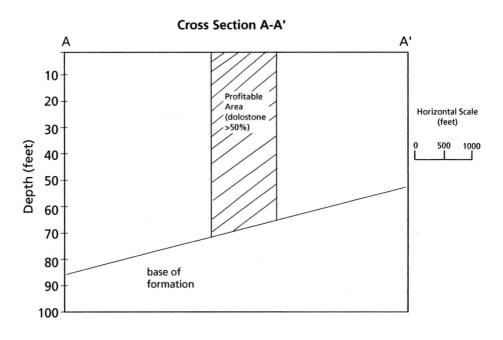

NOTES

_____

_____

_____

_____

_____

_____

_____

_____

_____

_____

_____

_____

_____

_____

_____

_____

_____

_____

Chapter 2

Earth's Natural Resources Mineral Resources

Preparing for the Chapter Challenge

On the basis of the mineral commodities located near your community and the types of deposits such minerals might form, discuss how you could explore for the type of deposit most economically feasible for making a beverage container. Answer the following questions:

• How big do you expect the deposit to be?

• Is the deposit visible from the surface?

• How deep might the deposit be?

• Does a model exist for the deposit? How does the model affect the way you would explore for it?

• What laws apply in your community to mineral exploration and staking claims?

Inquiring Further

1. **Mineral supply and demand**

   Are common minerals and materials likely to have a high or low economic value? Are minerals and materials that have a low economic value usually produced in large or small volumes? How far are minerals and materials with low economic unit value typically transported? What about minerals and materials with high economic unit value? Explain why this is so.

2. **Maps and mineral deposits**

   How do geologists decide which mineral deposits should be included on survey maps?

Preparing for the Chapter Challenge

Students should address all of the bulleted points in this section. By this time, they should have a good idea of what materials they are going to use to construct their beverage container. This section will help them begin thinking about where they will need to go to find any needed resources and what costs are associated with the location of the necessary deposits. Students will need to do a bit of extra research to address all of the questions raised.

Relevant criteria for assessing this section (see **Assessment Rubric for Chapter Report on Mineral Resources** on pages 304 and 305) include discussions of:

- how the mineral resources are found
- how the mineral resources are extracted
- what impact the use of these minerals has on the environment

Inquiring Further

1. Mineral supply and demand

This question reviews the concept of supply and demand. If a mineral is common, and therefore abundant and easy to locate, it is likely that the mineral can be mined in sufficient quantity to meet or exceed whatever demand there is for it (provided that it can be extracted at a reasonable cost). The lower cost associated with exploration will tend to keep the cost of extraction low. Accordingly, the economic unit value for that mineral will be kept low and, given sufficient demand, it can be produced in large volumes. Mineral commodities that are common and have a low economic value tend to not be transported for too great a distance, in order to keep the transportation costs low relative to the value of the mineral.

Minerals that are difficult to locate (i.e., have a low supply) will have a higher economic value if the demand for them is also high. If more of the mineral is needed than can be supplied from the exploration and mining activities (i.e., demand is greater than supply), then people will compete for what is available, driving the value up.

2. Maps and mineral deposits

Generally speaking, only mineral deposits large enough to be profitable, as well as known reserves of minerals for which there is a developed use and potential market, are included on state or local maps.

ACTIVITY 5 — WHAT ARE THE COSTS AND BENEFITS OF MINING MINERALS?

Background Information

Open-Pit Mining

Surface mining is advantageous where the desired material lies at or near the Earth's surface. Where the top of a large but areally limited body of ore or other resource is exposed at the surface, open-pit mining is used. Open-pit mining creates an increasingly deeper and wider hole, but the area of the mining operation remains local. The volume of overburden (unwanted material that covers the body of the resource) is usually small compared to the volume of the ore or other resource that is eventually recovered. The depth of the pit is limited either by the depth of the resource or the increasing cost of recovery. Ores of iron, copper, and other metals are recovered by open-pit mining, as are nonmetallic bulk resources like sand, gravel, and stone. Stone quarries represent one form of open-pit mining.

Reclamation of open-pit operations is usually not possible because of the large volume of the resource that is removed relative to the overburden available for replacement. In pits that are not too deep, various imaginative uses have been devised: tank storage of chemicals, underground parking structures, archive storage, and even shopping centers and office complexes that are partly above ground and partly below ground. In all of these cases, the groundwater table must lie well below the original ground surface. If the groundwater table is near the surface, the pit is flooded. If there are no problems of toxicity, as with sand and gravel operations or stone quarrying, the resulting lake can be used for recreation or wildlife habitats.

Dredging

Certain resources can be mined by dredging, which is usually carried out along watercourses. Dredging involves either a large-scale, powerful, underwater suction system or mechanized devices like mechanical buckets. Sand and gravel are commonly mined by dredging. Certain metal ores in the form of placer deposits are also mined by dredging. A placer is a sedimentary deposit in which some large percentage of particles of a valuable metal or mineral with density much greater than ordinary sediment has become concentrated by processes of fluid flow and sediment deposition. Gold, tin, and certain other materials are mined as placer deposits.

Strip Mining

Resources that are areally extensive but do not extend deep into the Earth are commonly mined using a technique of surface mining called strip mining. With heavy earth-moving machinery, first the overburden and then the resource is removed. The overburden is piled in mounds or ridges as the mining proceeds across the landscape. Without reclamation, such surface mining reshapes the land surface drastically and typically alters the pre-existing patterns of surface and subsurface drainage. Reclamation involves replacement of the overburden to restore in an approximate way the original topography of the area, together with revegetation to hinder erosion of the unprotected surficial material.

More Information – on the Web
Visit the *EarthComm* web site
www.agiweb.org/earthcomm to access a
variety of links to web sites that will help
you deepen your understanding of content
and prepare you to teach this activity.
Many of the sites also contain images that
you can download.

The *EarthComm* web site also provides
information about how to obtain *Metal,
Mining, and the Environment* (AGI
Environmental Awareness Series, Volume 3),
a 64-page booklet that focuses on why we
need metals, how metals are recovered
from the Earth, and how we address the
major environmental concerns related to
producing metals.

Chapter 2

Goals and Assessment

Clarify that the goals indicate what students should understand and be able to do as a result of the activity. Make sure students understand that Chapter Assessments are based upon these goals.

Goal	Location in Activity	Assessment Opportunity
Design a procedure for mining the model used in **Activity 4** that optimizes time, cost, and environmental impact.	Investigate	Plan is reasonable and is drafted within the rules of the excavation
Analyze the outcomes of mining the model to understand demand and market value, and the costs of labor, refining, transportation, and environmental reclamation.	**Investigate; Digging Deeper; Understanding and Applying What You Have Learned** Questions 1 – 2, and 4	Responses are reasonable and accurately reflect mineral deposit models.
Understand the benefits and drawbacks to mineral exploration and mining.	**Investigate; Digging Deeper; Check Your Understanding** Question 3 **Understanding and Applying What You Have Learned** Questions 3 – 4	Responses are reasonable, and closely match those given in Teacher's Edition.
Analyze the economic importance of mining in your state.	**Digging Deeper; Understanding and Applying What You Have Learned** Question 5	Students develop questions, if they are role playing the part of the government or community member, to ask the company representatives that address the question of economic impact, or if they are representing the company, students develop a plan that addresses the economic impact on the state, not just the benefits for the company.

NOTES

Activity 5

What Are the Costs and Benefits of Mining Minerals?

Goals

In this activity you will:

- Design a procedure for mining the model used in Activity 4 that optimizes time and cost, and reduces environmental impact.

- Analyze the outcomes of mining the model to understand demand and market value, and the costs of labor, refining, transportation, and environmental reclamation.

- Understand the benefits and drawbacks to mineral exploration and mining.

- Analyze the economic importance of mining in your state.

Think about It

Minerals are excavated from the Earth through mining.

- What is a mine, and what does it look like?
- What does the process of mining involve?

What do you think? Have you ever seen a working or abandoned mine? In your *EarthComm* notebook, sketch a picture of a mine that you have seen or one that you imagine. Describe what processes take place in the mine. Be prepared to discuss your ideas with your small group and the class.

Activity Overview

Students use the information obtained in the previous activity to formulate a plan for excavating iron ore from the model of the mineral deposit. In developing their plans, students must stay within the constraints set forth by the rules of the excavation. They must be concerned with time, cost of sand removal, and cost of environmental reclamation. Students then carry out their plan and determine how much money they made (or lost!) from the excavation. **Digging Deeper** reviews the different methods of mining and points out the kinds of resources that are typically extracted using each method. The reading also looks at the environmental consequences of mining and traces the steps that must be taken to restore the environment after an excavation.

Preparation and Materials Needed

Beyond collecting the necessary materials, no special preparation is required for this investigation.

Materials
- Plastic spoons
- Scrap paper
- Stopwatch
- Small containers to store sand and ore
- Magnet
- Balance or scale
- Calculator

Think about It

Student Conceptions

Some students may have seen mines, but probably most have not. It is likely that when students think of mines, they conjure up images of shafts that go deep below the surface of the Earth. Encourage them to use illustrations to describe how a mine looks. Students may not have a sense of some of the processes involved in mining. They are likely to think of miners digging materials from the Earth, or heavy machinery plowing deep into the ground, but many students do not realize that mining also involves restoring or reclaiming the land.

Answer for the Teacher Only

A mine is an underground excavation for the extraction of mineral deposits, in contrast to surficial excavations like quarries. The term is also applied to various types of open-pit workings. In another sense, a mine is also the area or property of a mineral deposit that is being excavated. Mines vary considerably in depth and areal extent. Key facets to the process of mining involve excavation, removal, and restoration (reclamation). Reclamation involves soil treatment, water treatment, and/or acid rock drainage.

Earth's Natural Resources Mineral Resources

Investigate

1. Using the same model that you surveyed in Activity 4, make a plan for excavating the ore. Here are the rules for excavation:

   • Your mining company has $200,000 to spend on this project.

   • You have 15 minutes to complete mining.

   • It costs $1000 to remove each gram of sand from the model.

   • It costs $1000 for every minute spent mining.

   • It costs $500 to replace each gram of sand for environmental restoration after the mine is closed.

   • If your company disrupts more than 25% of the surface of the model, you must pay a $5000 restoration fee.

   • Your company will receive $4000 for each gram of ore recovered.

   Using the cross section you drew in the previous activity, formulate a plan for removing the ore. Consider the factors of ore concentration, depth, and surrounding material in your plan. Discuss the ways you will keep your excavation localized to avoid the $5000 restoration fee. Be sure every member of your group understands his or her role in the mining operation before you begin, because time is worth money.

   a) Outline your plan in your *EarthComm* notebook.

2. You will use plastic spoons to remove the sand and ore. A timekeeper will start the clock on your mining. As you remove material from the model, separate the sand from the ore by placing a sheet of scrap paper over the ore and touching the "unrefined" ore with a magnet through the paper. Transfer the magnet, the paper, and whatever ore sticks to it to another container. When you lift the magnet away from the paper, the released ore will be caught in the container. Do this several times to remove all the ore from the sand. Save all materials to take to the "refinery."

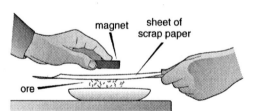

3. When you bring the containers of sand and ore and the remains of the model to the refinery (weigh-in station), the clock stops for your group.

   a) Write down the number of minutes you used in mining.

    Wear goggles throughout the activity. Wash your hands when you are done.

Investigate

Assessment Tool

Think about It Evaluation Sheet
Use this evaluation sheet to help students understand and internalize the basic
expectations for the warm-up activity.

1. Students should use the same model for this investigation that they surveyed in
   Activity 4. Review the rules of the excavation to make sure that they understand
   all the associated costs. Have students read through all the steps of the
   investigation as they design their excavation plan.

   a) Check student plans before they begin the actual excavation process to make
      certain that they are feasible, given the rules of the excavation.

2. Remind students to start the clock once they begin mining. Remind them to be
   careful when removing sand, because they will need to try to remove as little sand
   as possible and to disturb as little of the sand surface as possible. Students can use
   the illustration on page R128 in the text to help them separate the sand and ore.
   Given the rules of the excavation, students are allowed no more than 15 minutes
   for mining. However, they do not have to use the entire time, and they will save
   money if they do not use all of the 15 minutes.

Teaching Tip

Round times to the nearest minute, and be sure that students record elapsed
mining time as a factor of their mining operation. Remind students that the
surface area of the models must be measured later in see **Step 4 (c)**, and DO NOT
ALLOW THEM TO DESTROY THE MODELS.

3. a) Students should record the total time they spend mining, from the first spoonful
      of sand that is removed until they have separated the sand and ore and arrived
      at the weigh-in station. Students are charged $1000 for each minute they have
      spent mining.

4. Calculate the costs for environmental reclamation.

   a) Find the mass of the sand and record it in your notebook.

   b) Record in your notebook the total cost for returning the sand to the model.

   c) Measure the surface area of your mine and compare that to the total surface area of the model. Did you disturb more than 25% of the surface? If so, charge your company $5000 for environmental reclamation.

5. Calculate your income.

   a) Find the mass of the ore and calculate your company's income determined by the current market value of the ore ($4000 per gram).

   b) Write the information in your *EarthComm* notebook.

Reflecting on the Activity and the Challenge

Your model mine illustrated some of the factors involved in recovering mineral deposits from the Earth. The depth of the deposit, the type of material around it, and the concentration of the ore are all geological factors that influence mining. These factors influence the decision to mine and the costs of the mineral resources used to produce goods, such as containers.

4. a) Students should record the mass of the sand. They are charged $1000 for each gram of sand they remove from the model.

    b) Students are charged $500 to replace each gram of sand.

    c) If students have disturbed more than 25% of the model surface, they are charged a $5000 restoration fee. You may need to help them determine whether or not 25% of the surface has been disturbed.

5. a) Students should weigh the ore that they have collected. They receive $4000 for each gram.

    b) Students can take the difference between the money they have spent on the excavation and the money they netted from the recovered ore. Ask them to note whether or not they consider their excavation a success (i.e., did they turn a profit, or did they spend more money than they made?).

Assessment Tool

EarthComm Notebook-Entry Checklist
Use this checklist as a quick guide for student self-assessment and/or an opportunity to quickly score student work. Add further criteria specific to your classroom needs or to this particular investigation.

Investigate Notebook-Entry Evaluation Sheet
Point out the criteria listed on this evaluation sheet that are relevant to this particular investigation. Encourage students to internalize the criteria by making them part of your assessment conversations as you circulate around the classroom.

Assessment Opportunity

The final step in this mining simulation is to weigh and sell the ore recovered from the model. The arbitrary market value of the ore ($4000/g) could be manipulated to change outcomes and stimulate discussion. It is important that students record all information for their evaluation of the economic success of their mining venture and consider the factors that created their situation. Check notebooks for complete information. A profit/loss statement of the group's mine could be used as an embedded assessment.

Reflecting on the Activity and the Challenge

Developing and defending the mining plan, looking at its results, and analyzing the outcomes play an important role in mineral use. Review mining methods and safety in a class discussion, comparing and analyzing techniques used to mine the model in the investigation. Discuss with students the implications of the exercise they have completed with respect to the **Chapter Challenge**. Again, ask students to consider why it is important for them to be able to explain to the beverage company how mineral resources are extracted from the ground (i.e., the cost of extraction will affect the cost of producing the container).

NOTES

Chapter 2

Earth's Natural Resources Mineral Resources

Digging Deeper

REMOVING MINERAL RESOURCES FROM THE GROUND

Types of Mining

There are two main ways to get minerals out of the ground: surface mining and underground mining. The method used depends on the location, size, depth, and grade or quality of the ore. Both surface mining and underground mining require planning and money to ensure that the ore is removed in the most cost-effective, safe, and environmentally sound way possible.

Surface Mining

Geo Words

spoil: overburden and other waste material removed in mining, quarrying, dredging, or excavating.

In the United States, most minerals are removed from the ground by a kind of surface mining called open-pit mining. This method of mining has advantages when the ore body is very large and close to the surface. Large earth-moving equipment removes the soil and rocks above the ore. The original topsoil is stored in large piles for later use in reclamation, and the underlying rock material that covers the ore body, called **spoil**, is stored elsewhere. The operation continues from a series of large benches and roads that wind their way down into a large, deep pit. The benches serve as both working areas and as roads to haul the ore to the surface. Ore and waste in the pit is drilled, blasted, and loaded into huge trucks that carry it to crushers or waste piles. The crushed ore is then transported to facilities for storage and refining. The Bingham Canyon copper mine, shown in *Figure 1*, near Salt Lake City, Utah, is the largest open-pit in the world. It is approximately two miles in diameter and more than a half mile deep.

Figure 1 The Bingham Canyon copper mine in Utah is the largest open-pit mine in the world.

Digging Deeper

Assign the reading for homework, along with the questions in **Check Your Understanding** if desired.

Assessment Opportunity

Use a brief quiz (or a class discussion) to assess your students' understanding of the main ideas in the reading and the activity. A few sample questions are provided below:

Question: Name and describe the two types of mining.
Answer:
- Surface Mining – involves the removal of soil and rocks from the Earth's surface to extract minerals that are not deeply buried. Open-pit mining is one kind of surface mining.
- Underground Mining – used when the mineral resources lie deep beneath the surface of the Earth or when an ore body has irregular geometry.

Question: What are the environmental implications associated with surface mining?
Answer: Surface mining disturbs the land surface and can therefore destroy an ecosystem and change patterns of surface-water runoff. Drainage from surface mining can also pollute surface water and groundwater.

Question: What risks to humans are associated with mining?
Answer: Mining involves working with heavy equipment. Underground mines can collapse. Long-term exposure to many minerals causes health problems for the miners and processors.

Teaching Tip

Bingham Canyon Mine in Utah, pictured in *Figure 1* on page R130, is about 2.5 mi. wide and more than 0.5 mi. deep (and still growing!). It is the largest human-made excavation in the world and, along with the Great Wall of China, is one of only two human-made structures visible from space. Open-pit mining began in Bingham Canyon Mine in 1906, and to date the mine has produced nearly 16 million tons of copper. Estimates suggest that levels of ore will allow open-pit mining to continue through the year 2012. After that, underground mining could continue for up to 20 years.

Resources like sand, gravel, coal, phosphate, as well as iron ore and copper ore, are extracted from surface mines. Much of the coal in the western United States comes from open-pit mines called strip mines. Strip mining is used where the deposit is in the form of thin but widespread sheet-like layers near the surface.

As world reserves of petroleum become scarcer in the coming decades, the great reserves of oil shale in Colorado and Wyoming are likely to be mined from strip mines, for conversion to petroleum. In Canada, deposits called tar sands, which contain enormous reserves of petroleum, are already being mined from strip mines, and are likely to be a major source of petroleum for both Canada and the United States in the future.

Figure 2 Strip-mined land not yet reclaimed.

Dimension stone quarries are special types of open-pit mines. Large blocks of granite, marble, limestone, or sandstone are removed intact (not crushed). Quarrying is used when the desired final product is large blocks of the rock itself.

Underground Mining

Underground mining is used when the mineral resources lie deep beneath the surface, or when an ore body has irregular geometry. Shafts and

R 131

Teaching Tip

After the desired material is removed during the strip-mining process, the resulting pit is backfilled with the overburden (seen as the low ridge on the right in *Figure 2* on page R131) that had to be removed to gain access to the desired material.

Chapter 2

Earth's Natural Resources Mineral Resources

tunnels provide access and ventilation. Depending upon the orientation of the ore body, the mineshaft may be vertical, horizontal, inclined, or shaped like a corkscrew. Tunnels, on the other hand, are close to being horizontal. After the ore body is reached, additional tunnels are excavated through the body and the ore is hauled to the surface by trains, loaders, trucks, or elevators. Many mining operations use trackless semi-automated equipment to speed extraction and reduce the number of people exposed to the hazards of working underground. In most underground mines, large "rooms" are formed when the ore is removed. Pillars of ore and/or waste rock material are left to support the mine roof. The grade of the ore and the strength of the overlying rock help determine the size of the rooms and the support pillars. Modern techniques include back-filling tunnels to recover ore pillars and prevent cave-ins and environmental problems like acid mine drainage.

Figure 3 Silver ore is retrieved by blasting tunnels in underground mines.

Environmental and Safety Concerns

A surface mine disturbs the land in many ways. Removing **overburden** (rock material that overlies a mineral deposit) and vegetation destroys established ecosystems. Patterns of surface water runoff are changed. Dissolved chemicals from some of the rocks and minerals exposed at the surface during mining can contaminate surface water and ground water.

Geo Words

overburden: rock material that overlies a mineral deposit and must be removed prior to mining.

Teaching Tip

Removal of ore or coal in deep underground mines, as pictured in *Figure 3* on page R132, is nowadays highly automated, in contrast to earlier times, when excavation was by hand. The maximum depth of underground mines is limited partly by the strength of the rock in the walls and ceiling of the shafts and tunnels and partly by the capability of air conditioning to counteract the high rock temperatures at depth.

Activity 5 What Are the Costs and Benefits of Mining Minerals?

All states require mining operations to file environmental reclamation plans before any ground is disturbed. During reclamation, the mining company attempts to stabilize the ecosystem, returning the land to productivity unrelated to mining. The first step is to spread the spoil back over the strip-mined area to restore the original topography of the land. Then any available topsoil from the original stripping is spread over the spoil, and native trees and shrubs are replanted.

Figure 4 After mining operations finish at a site, native vegetation is replanted.

Mining involves heavy equipment in both surface mines and underground mines, and worker safety is an ongoing concern. Underground mine disasters involving the collapse of tunnels are much less frequent in the United States than in many other countries, but disasters do occasionally happen. Most mining deaths in the United States are related to improperly maintained equipment, failure to wear or use safety equipment, or working in unsafe locations within the mine.

Many mining operations can also expose workers to hazardous materials. Long-term exposure to certain minerals causes health problems for miners and people who process mined ore. Silica dust, heavy metals, asbestos fibers, and other materials can cause serious illnesses. The presence of these materials and the levels of exposure must be constantly monitored to protect mine workers. This raises the cost of mineral excavation.

Check Your Understanding

1. What are the two main kinds of mining? Which kind did your model simulate?

2. What types of mineral resources are extracted from surface mines?

3. What are some of the hazards associated with underground mining?

Teaching Tip

Environmental reclamation requires significant investment of money and careful planning. Millions of dollars are spent each year in the United States to restore old mine sites, as depicted in *Figure 4* on page R133. For example, in 2000, the United States Office of Surface Mining devoted $195 million dollars (nearly one-half of its annual budget) to the reclamation of abandoned mines.

Vegetation in the reclaimed area is usually not the same as the original vegetation. However, if the reclamation is done well, however, environmental differences are not great. In most areas, however, the time that would be needed for complete reversion to the natural state would be very long. The major problem is that it is difficult to duplicate the nature of the original soil structure.

Check Your Understanding

1. The two main kinds of mining are surface mining and underground mining. Student models simulated surface (open-pit) mining.

2. Resources like sand, gravel, coal, and phosphate, as well as iron ore and copper ore, are extracted from surface mines.

3. Cave-ins and environmental problems like acid mine drainage are associated with underground mining, as well as exposure of workers to hazardous materials.

Assessment Tool

Check Your Understanding Notebook-Entry Evaluation Sheet
Use this sheet to evaluate the extent to which students understand the key concepts explored in **Activity 5** and explained in **Digging Deeper**, and to evaluate the students' clarity of expression.

Earth's Natural Resources Mineral Resources

Understanding and Applying What You Have Learned

1. In your simulation, what real-life mining cost did the fee of $1000 per minute represent? What real-life mining cost was represented by the fee of $500 per gram of sand to be replaced?

2. Analyze your group's mining venture.

   a) Did your company make a profit or a loss?
   b) What factors affected your profitability?
   c) What was the largest expense associated with your mining venture?

3. Imagine your model enlarged to a depth of one meter.

   a) What changes would you incorporate in the plan to recover ore?
   b) What risks would be involved?

4. How might delays like equipment repair, environmental claims, labor strikes, etc., affect your profit?

5. A company has approached your community leaders, seeking approval to open a new mine nearby. Both the mine and the associated refining plant will employ people in your community. In class, designate some members as government representatives, some as company representatives, and the rest as members of the community.

   • If you are designated a government representative, come up with at least three questions to ask the company representatives about the mine.
   • If you are a company representative, decide on the commodity to be mined and the method by which it will be extracted. Prepare a brief presentation on the benefits of the mine and be prepared to answer questions.
   • If you are a member of the community, come up with at least three questions for the government council and company representatives.

   a) Write your questions or your presentation points in your notebook.

   b) With your class members role-playing the parts outlined, conduct a town hall meeting to discuss the pros and cons of opening the mine. Write the questions and answers given in your notebook. After the council, discuss as a class how the meeting went, and summarize the outcome in writing.

Understanding and Applying What You Have Learned

1. The $1000/min. fee would represent things like labor, cost of running the equipment needed in the mine, insurance, etc. The $500/g fee would cover costs associated with environmental reclamation.

2. Answers to these questions will vary. Most likely, the time spent will be the largest expenditure. Factors that influence status include:
    - size, depth, and orientation of the mineral deposit
    - concentration of the ore within the surrounding material
    - limits on surface disturbance
    - working capital
    - market price of the ore

3. Changes in the plan should address:
    - time
    - amount of material removed
    - amount of surface area that can be disturbed
    - equipment used to excavate a hole 1 meter deep

    Risks should include the possibilities of cave-ins and collapse in sand, and some sort of preventive measures would have to be taken. Costs of mining and removal might exceed income from the sale of the mineral. Price of the ore could drop, decreasing the profit to be made. Environmental impact and safety risks could be too great to justify, etc.

4. Delays like equipment repair, environmental claims, and labor strikes will increase the time involved in extraction and decrease overall profit.

5. Outcomes will vary. Assign roles or ask for volunteers. Provide sufficient, yet brief, planning time. Alternatively, assign the preparation as homework, and then conduct the activity in class. Ask students to ground their analysis of the town hall meeting in the model mining activity and the reading on mining methods and safety just completed.

    Facilitate and monitor the classroom discussion in which students summarize and critique the town council meeting. Keep the learning objectives in mind as the discussion proceeds. Use the discussion to informally assess the students' understanding of the major ideas presented in the chapter. Have the council vote on the evidence presented in the meeting, and be sure students record the outcome and the council's justification for their decision in their notebooks.

Chapter 2

Activity 5 What Are the Costs and Benefits of Mining Minerals?

Preparing for the Chapter Challenge

Choose a mineral commodity in your community that could be used to make a beverage container and research how it is mined. You may want to take a class field trip to the mine. Use the questions below to conduct an interview with an engineer or geologist at the mine. Write an article about the mine. Your article should include the following information plus other information you have learned in previous activities:

- Is the deposit mined from the surface or underground?
- How large an area does the mine affect on the surface?
- How deep is the mine?
- How many people work at the mine?
- What happens to the minerals that are removed from the ground?
- What effect, if any, does the mine have on the environment?
- What laws apply to mining in your community?

Inquiring Further

1. **Mineral resources and state revenue**

   Visit the *EarthComm* web site, or use encyclopedias, almanacs, or tourist information, to investigate the contribution of mining to your state's economic structure. How much of the state's revenue is generated by mining-related operations?

2. **Environmental impacts of mining**

   What kinds of environmental impact result from surface mining? From underground mining? What kinds of environmental impact regulations for mining are in place for your state? Visit the *EarthComm* web site to get you started with your research.

3. **Sources of minerals**

   Investigate the sources of minerals used in local manufacturing. Why are many minerals found in the United States imported from other countries? Why are most construction materials obtained from local sources?

R 135

Preparing for the Chapter Challenge

If possible, arrange for a field trip to a local mining operation or invite a worker from the operation to come to your classroom. Students can pose their questions to the mine staff and use the responses to write a short article. Encourage them to think of additional questions that they can ask the visitor. If a field trip or class visitor is not an option, students can research the answers to the questions in the library or online.

For assessment of this section (see **Assessment Rubric for Chapter Report on Mineral Resources** on pages 304 and 305) students would need to include a discussion on:
  • how the mineral resources are extracted
  • what impact the use of these minerals has on the environment

Inquiring Further

1. Mineral resources and state revenue
Answers to the question will vary. This is an opportunity for students to add to their knowledge of the resources available in their state.

2. Environmental impacts of mining
Surface mining creates the greatest observable environmental impact. Strip mining completely re-sculpts the landscape. Quarrying leaves permanent scars on the land. Open-pit mines must dispose of the waste removed from around the ore in some way. In the United States, mines are required to restore the surface after the mining venture closes, and that restoration contributes to the cost of extracting the minerals.

The impacts of underground mining are less visible but no less important. Underground mines must also dispose of the material removed from around the ore. Previously, this material was dumped outside the mine entrance in piles of tailings, but it is now often replaced as backfill. Ventilation shafts, the main mine opening, and the buildings and support structures necessary to the operation also affect the area. Closed mines are subject to chemical leaching, disruption of the water table, and contamination of circulating groundwater.

3. Sources of minerals
Assessment of the sources of minerals often involves a cost comparison factor. Many of the environmental protection standards required in the United States are not followed in other countries. Consequently, many minerals available here are produced more cheaply in other areas and are imported from countries where their extraction causes much more environmental damage.

In the case of construction materials (like sand, gravel, and building stone, for instance), the cost of transporting large quantities of these materials makes it much more cost efficient to use local sources to supply these commodities.

ACTIVITY 6 — HOW ARE MINERALS TURNED INTO USABLE MATERIALS?

Background Information

Ore Dressing

Ore dressing is the term used for the treatment of ores to concentrate the valuable constituents (the desired mineral or minerals) into smaller bulk and collect the worthless materials, called gangue, into discardable waste. Ore dressing involves two basic steps:
- severance (mechanical disaggregation of the ore)
- beneficiation (separation of the disaggregated components)

Severance involves coarse crushing, further fine crushing, grinding, and then screening through a series of finer and finer screens to separate size fractions suitable for beneficiation. Beneficiation involves selection, to separate the desired material from the gangue, together with physical movement of the mineral away from the gangue. This selection is based on one or more physical or chemical properties that differ between the mineral and the gangue.

The oldest method of beneficiation is simply hand picking. Other useful properties include specific gravity, magnetic and electrical properties, and surficial or bulk chemical properties. Separation by specific gravity is carried out using a variety of geometrical configurations, but it is fundamentally based on a difference in settling velocity between the mineral and the gangue. In some cases, centrifuging is used to accentuate the difference in fall velocities; in essence, centrifuging increases the effective force of gravity, thereby speeding up the process.

One particularly common technique of separation by specific gravity, which is especially useful for separation of sulfide minerals from gangue, is froth flotation. Froth flotation is the process used to separate particulate solids suspended in a fluid by bringing about selective attachment of the desired mineral particles to a less dense fluid, commonly air bubbles, in a denser fluid. The mineralized fluid aggregate rises to the surface of the denser fluid, where it can easily be removed from the separation apparatus. Froth flotation is used to recover sulfide minerals that are too fine to be recovered by simple gravity methods. The working principle is to treat the surfaces of the particles in such a way that the particles readily become attached to the bubbles. The bubbles, along with the attached mineral particles, then float up to the surface of the vessel.

Two components are in common use in the froth flotation process: collectors and frothers. Collectors are organic molecules of certain kinds that form a thin coating on the particles; frothers are chemical compounds that help to form a stable and manageable froth structure, to aid in removal. The air bubbles are usually much larger than the fine particles; several particles become attached to a single bubble. In modern practice, differential flotation is also possible. For example, by adjusting the chemical nature of the collectors, the process can be arranged so that only the valuable sulfide minerals are retrieved by flotation, leaving valueless particles like pyrite (fool's gold) behind in the liquid.

More Information – on the Web
Visit the *EarthComm* web site
www.agiweb.org/earthcomm to access a
variety of links to web sites that will help
you deepen your understanding of content
and prepare you to teach this activity.
Many of the sites also contain images that
you can download.

Chapter 2

Goals and Assessment

Clarify that the goals indicate what students should understand and be able to do as a result of the activity. Make sure students understand that Chapter Assessments are based upon these goals.

Goal	Location in Activity	Assessment Opportunity
Model the separation of a mineral from a rock.	**Investigate** Part A **Understanding and Applying What You Have Learned** Question 1	Recorded observations are detailed and accurate.
Model the separation of an element from a mineral.	**Investigate** Part B **Understanding and Applying What You Have Learned** Question 2	Recorded observations are detailed and accurate.
Understand some of the techniques involved in mineral processing and the problems associated with them.	**Digging Deeper; Check Your Understanding** Questions 1, 2, and 3 **Understanding and Applying What You Have Learned** Questions 1 – 2	Responses to questions are reasonable, and are based on reading and observations made in investigation.

NOTES

Chapter 2

Earth's Natural Resources Mineral Resources

Activity 6

How Are Minerals Turned into Usable Materials?

Goals

In this activity you will:

- Model the separation of a mineral from a rock.

- Model the separation of an element from a mineral.

- Understand some of the techniques involved in mineral processing and the problems associated with them.

Think about It

When making your favorite cookies, you mix the ingredients in the right proportions and in the right order. You heat the cookie dough to create a chemical reaction that will result in cookies. Now imagine taking one of the cookies and trying to get the original ingredients back out of it. Imagine if the only way to get flour was to extract it from cookies.

- How would that change the cost of flour?
- How would it change how careful you are about the way you use flour?

What do you think? Record your ideas in your *EarthComm* notebook and be prepared to discuss them with the class.

R 136

Activity Overview

In the first part of the investigation, students model the separation of a mineral from a rock by agitating a mixture of sulfide ore and steel shot and pouring off the slurry. They clean the slurry using liquid soap. In the second part of the investigation, students model the separation of an element from a mineral by first using vinegar to produce a leaching solution from crushed copper oxide ore. Then, they separate the copper from the solution using a dry-cell battery, a spoon, and a lead fishing weight. **Digging Deeper** takes a closer look at the processes that are used to refine minerals and prepare them for use.

Preparation and Materials Needed

You will need to collect the materials for the investigation before class. The amounts of ores and solutions needed will vary somewhat depending on the specific materials you have gathered. Additionally, the time required for **Part B** will depend on the strength of the battery used and the concentration of the leaching solution. Accordingly, you may wish to test these investigations before class.

Materials

Part A
- Clear jar with a lid
- Crushed sulfide ore
- Steel shot
- Water
- Towel
- Piece of wire screen or sieve
- Plastic cup
- Paper towels
- Bubble-bath solution
- Drinking straw
- Squeeze bulb
- Index card

Part B
- Piece of circular paper or filter paper
- Funnel (3 to 4 in. diameter)
- Crushed copper ore
- Vinegar
- Two bottles for collecting the leaching solution
- Dry-cell battery
- Steel wool
- Alligator clips (four)
- Metal spoon or a dime
- Insulated wire (two pieces)
- Lead fishing weight

Think about It

Student Conceptions

Give students a few minutes to think about and respond to the questions. Hold a brief discussion. Students will likely recognize that the cost of flour would greatly increase if it had to be extracted from cookies. Students are also likely to say that they would be more careful about its use.

Answers for the Teacher Only

Some chemical changes involving plant and animal tissue are effectively irreversible. Flour is a substance that owes its origin to the specific development of plant tissue in the form of seeds of a particular kind of grass we call wheat. Once that material reacts with other components of the cookie dough during baking, there is no turning back. An inorganic constituent like salt (sodium chloride), on the other hand, could be extracted from the cookies by dissolving the soluble materials from the cookies and reprecipitating the salt. The latter process is more relevant to industrial extraction of valuable minerals from ores.

Assessment Tool

Think about It Evaluation Sheet
Use this evaluation sheet to help students understand and internalize the basic expectations for the warm-up activity.

NOTES

Investigate

Part A: Separating a Mineral from a Rock

1. Fill a clear-plastic jar one-third full of crushed sulfide ore and steel shot. Keep a few pieces of ore and shot aside for comparison purposes during the investigation.

2. Add enough water to cover the ore and shot with 1 cm of water. Cover the jar securely with a lid.

3. Wrap the jar in a towel and shake the jar for two minutes.

4. Screen the mixture by pouring it over a wire screen into a plastic cup to collect the slurry.

5. Observe the mixture you poured off (the slurry) and the ore and shot.

   a) Record your observations in your *EarthComm* notebook.

6. Return the oversize ore pieces and steel shot to the jar and recycle the water by pouring the captured water back into the container. Repeat the process three to five times.

7. After the last repetition of the process, do not recycle the materials. Add about a tablespoon of water to the jar to retrieve the settled ore by sloshing it around and pouring it into the plastic cup.

8. Save the slurry (the screened liquid and ore) by pouring it into a plastic cup.

9. Separate the large pieces of ore and steel shot and put them on a paper towel to dry.

10. Add 4-6 teaspoons of liquid bubble-bath solution to the plastic cup of slurry and stir thoroughly. Place a long straw with a squeeze bulb in the mixture and blow gently.

    a) Make observations about what happens and record this in your notebook.

11. Scrape off the bubbles with an index card and place them on a paper towel to dry. Repeat the process three times.

    a) Record your observations each time.

 Wear goggles and lab apron. Dispose of the straw after the activity.

Investigate

Part A: Separating a Mineral from a Rock

1 – 4. Students should follow the instructions in the text, using the illustration on page R137 as a reference. Remind them to secure the jar lid tightly and to be careful when shaking the jar.

The purpose of the steel shot is to facilitate crushing of the ore. The amount of steel shot and ore needed are not great, and depending on the size of your plastic jar, it may not need to be one-third full. Generally speaking, 10–20 mL of shot and 5 mL of ore should be sufficient if a small jar (50–150 mL) is used.

5. a) The slurry should appear to be a mixture of water and finer-grained particles suspended in the water (and settling out from the water). It will include all of the particles finer than the wire screen mesh. The ore and shot now consist only of particles coarser than the screen mesh.

6. – 9. Students should repeat the process until they have broken down most of the large pieces of sulfide ore. They should save the ore pieces and slurry, setting the ore pieces and steel shot on a paper towel to dry and collecting the slurry in the same plastic cup they have been using all along (they should rinse out the jar with a little water to be sure to get all of the crushed sulfide ore).

10. a) Some material should be entrained in the bubbles, while other material sinks to the bottom of the cup.

11. a) The ore material should be entrained in the bubbles, while the waste material sinks to the bottom of the cup. Repeating this step should separate the ore from the waste material.

Chapter 2

Earth's Natural Resources Mineral Resources

Part B: Separating an Element from a Mineral

1. Fold a piece of circular paper as shown in the diagram. Open the paper so that it forms a cone. Place the paper cone in a funnel.

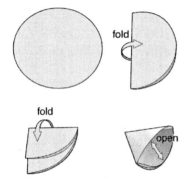

2. Fill the cone three-quarters full with crushed copper oxide ore.

3. Hold the funnel over a collection bottle. Gently pour vinegar through the funnel. The material you collect is the "leaching solution."

4. Put the funnel over a second bottle. Pour the leaching solution from the first bottle back through the funnel.

5. Repeat Step 4 several times to make the leaching solution as strong as possible. A greener solution indicates increasing concentration of copper.

   a) Record your observations after each pour.

6. Clean the battery contacts of a dry-cell battery by rubbing them with steel wool. Use alligator clips to clip insulated wire to connect a spoon to the negative terminal of the battery. Use the insulated wire and clips to connect a lead fishing weight to the positive terminal of the battery.

7. Suspend the lead weight and the spoon in the leached solution so they do not touch each other, as shown in the diagram. Let them sit until you see a change on the spoon.

   a) Keep track of the time. Record the changes on the spoon.

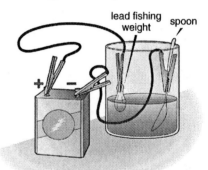

8. Switch the clips to the battery so that the spoon is attached to the positive terminal and the lead weight is attached to the negative terminal. Let them sit until you see a change.

   a) Record your observations.

    Wear goggles, gloves, and a lab apron. Do not leave the battery unattended while it is connected to the objects in the solution. Mark the spoon as lab equipment.

Reflecting on the Activity and the Challenge

In the first part of the investigation, you modeled the separation of minerals from rocks. In the second part of the investigation, you examined how elements are removed from minerals. These investigations should help you to think about the processes needed to prepare a mineral resource for actual use.

Part B: Separating an Element from a Mineral

1. – 4. Students can use the illustration in the left-hand column of page R138 to help them make the funnel. They should pour the vinegar slowly over the copper oxide ore and allow the liquid to drain before adding more vinegar.

Teaching Tip

The "mineral" chrysocolla [$Cu_4H_4Si_4O_{10}(OH)_8$] works well for this investigation. It should be crushed to a grain size roughly similar to aquarium gravel. Chrysocolla is a naturally occurring partially crystalline material (so it is not a mineral in the strict sense) that is found associated with the oxidized zones of copper deposits. It is sometimes used as a minor ore of copper and is sometimes cut as a gemstone. It is found, among other places, in the copper districts of Arizona and New Mexico.

5. a) After each pour, students should observe that the leaching solution becomes a darker blue-green color, indicating that the concentration of copper in the solution is increasing.

6. Students can use the illustration in the right-hand column of page R138 to help them with the setup for this step.

Teaching Tip

Steel wool is an electrical conductor, so make sure the students take care not to touch the steel wool to both terminals at the same time!

7. a) The time required to see changes in the spoon will depend largely on the concentration of the solution. The effects should become visible within less than about 15 minutes, but it could take longer depending on the concentration of the leaching solution and the strength of the battery. The most obvious copper plating will probably occur close to the alligator clips.

8. a) Students should observe the plating on the spoon disappearing and should see the lead weight becoming plated.

Assessment Tools

EarthComm **Notebook-Entry Checklist**
Use this checklist as a quick guide for student self-assessment and/or an opportunity to quickly score student work. Add further criteria specific to your classroom needs or to this particular investigation.

Investigate Notebook-Entry Evaluation Sheet
Point out the criteria listed on this evaluation sheet that are relevant to this particular investigation. Encourage students to internalize the criteria by making them part of your assessment conversations as you circulate around the classroom.

Chapter 2

Reflecting on the Activity and the Challenge

In this investigation, students modeled the separation of a mineral from a rock and the separation of an element from a mineral. Discuss with students how this ties into the **Chapter Challenge**. Ask them to consider the costs associated with processing a mineral resource: would it be more, or less, expensive to use resources that require a great deal of refining for use? They can also consider environmental implications of the refining process, with a focus on the waste products that are produced during processing.

NOTES

Chapter 2

Activity 6 How Are Minerals Turned into Usable Materials?

Digging Deeper

PROCESSING MINERAL ORES

Geo Words

smelting: melting ores to separate impurities from pure metal.

Gold is one of the few elements that can be found in its native state as a mineral. If gold occurs as placer deposits (nuggets in stream gravel) or in a vein with quartz, it does not take much processing other than to separate it from the host rock and then collect enough to melt into a bullion bar. Most of the gold produced in the United States, however, does not come from placer or vein deposits; most comes from ore deposits in the western United States that have such fine flakes of gold disseminated throughout the rock that the gold cannot be seen with the unaided eye. Deposits like these represent a special challenge in separating the gold from the rest of the rock.

The rock containing the gold is first crushed to a very small size. The rock is mixed with a cyanide solution, and the cyanide dissolves the gold out of the rock. The solution containing the gold is passed through a charcoal filter and the gold is precipitated onto the charcoal. The cyanide can be reused. The gold is then separated from the charcoal.

Figure 1 Gold ore.

Scientists have found that some naturally occurring bacteria can help to separate minerals from rocks, and elements from minerals, without using so many potentially dangerous chemicals. One such bacterium is called *Thiobacillus fero-oxidans*. This bacterium was discovered in coal-mine drainage waters in 1947. It prefers a warm environment with a pH of 2 to 3.5, lots of sulfur, and a little nitrogen and carbon dioxide. *Thiobacillus* generates its energy by catalyzing the oxidation of iron. Because many copper minerals contain iron and sulfur, *Thiobacillus* can be used to help separate copper from some minerals. Other bacteria are being experimented with to help process gold without the use of cyanide.

Some minerals cannot easily be leached with a chemical or eaten by bacteria. Iron and some copper minerals have to be **smelted** in order to separate the useful elements from the less useful elements. In the case of copper, the copper is wanted, and the iron and sulfur are waste. In the case of iron, the iron is wanted, but not the oxygen or sulfur. Smelting involves the use of high temperatures in order to break the chemical bonds holding the atoms together. In the past, smelters emitted a lot of pollution, especially sulfuric acid, which tends to cause acid rain. Modern smelters use filters to

Figure 2 Native copper.

Digging Deeper

Assign the reading for homework, along with the questions in **Check Your Understanding** if desired.

Assessment Opportunity

Use a brief quiz (or a class discussion) to assess your students' understanding of the main ideas in the reading and the activity. A few sample questions are provided below.

Question: What is the difference between leaching and smelting?
Answer: Smelting uses high temperatures to break the chemical bonds holding different atoms together (separating different elements from one another), whereas leaching uses chemicals to extract the desired parts of an ore (as in **Part B** of the investigation, where vinegar was used to extract copper from the ore).

Question: What other means besides smelting and leaching can be used to help separate an element from an ore?
Answer: The bacterium *Thiobacillus* is used to help extract copper from some minerals.

Assessment Tool

Check Your Understanding Notebook-Entry Evaluation Sheet
Use this sheet to evaluate the extent to which students understand the key concepts explored in **Activity 6** and explained in **Digging Deeper**, and to evaluate the students' clarity of expression.

Teaching Tip

Figures 1 and 2 on page R139 picture gold and native copper respectively. In certain mineralizing environments, native copper (the pure metal) is precipitated directly from hot, circulating solutions containing copper. These masses are called nuggets. Gold is also common as nuggets. The shapes of nuggets are usually very irregular, because the metal is precipitated in the interstices within the host rock.

Chapter 2

Earth's Natural Resources Mineral Resources

collect the harmful gases before they escape into the atmosphere. Copper smelters produce liquid sulfuric acid, which is sold to companies that use it to produce fertilizer. Both copper and iron smelters have a solid waste product called **slag**. Slag is recycled as a construction material and is used as ballast on railroad tracks.

A consequence of all mineral processing is the production of enormous quantities of waste. In a typical copper mine, for every 1 kg of copper that is produced there is 200 kg of waste rock to be disposed of. For a typical gold mine, for every 1 kg of gold produced, there is approximately 1600 kg of waste. For a typical bauxite mine, for every 1 kg of aluminum that is produced, there is 3 kg of waste. This waste is one of the biggest environmental problems faced by the mining industry, because it must be stored on the mine property or at the processing site. Although uses have been found for some waste products, many more are waiting for a clever person to come along and find a use for them.

Figure 3 Agricultural lime makes use of waste from zinc mines.

Geo Words

slag: solid waste product that comes from the smelting of metal ore or when making steel.

Check Your Understanding

1. Explain how gold is separated from rock.
2. What is smelting?
3. Identify a problem associated with mineral processing.

Understanding and Applying What You Have Learned

1. Answer the following questions about Part A of the investigation:

   a) What is the purpose of the jar, and what does it represent in a mineral processing plant?

   b) What is the purpose of the steel shot, and what does it represent in a mineral processing plant?

   c) What is the purpose of the screen?

   d) Why is this process necessary?

   e) Froth flotation is a method of separating the metal-bearing mineral from unwanted material in an ore. Ore is ground to a powder and mixed with water and frothing agents. Air is blown through the mixture. The mineral particles cling to the bubbles, which rise to form a froth on the surface. The waste material settles to the bottom. The froth is skimmed off and the water and chemicals are removed, leaving concentrated metal. Explain how you modeled froth flotation in the investigation.

2. Answer the following questions about Part B of the investigation:

   a) Is all of the copper mineral collected after a few trials?

   b) What happened to the vinegar solution during the leaching experiment?

Check Your Understanding

1. The rock containing the gold is first crushed to a very small size. Then, the rock is mixed with a cyanide solution, and the cyanide dissolves the gold out of the rock. The solution containing the gold is passed through a charcoal filter and the gold is precipitated onto the charcoal. Finally, the gold is separated from the charcoal.

2. Smelting involves the use of high temperatures in order to break the chemical bonds holding the atoms together.

3. A problem associated with mineral processing is that it results in the production of enormous quantities of waste, which must be disposed of or stored.

Understanding and Applying What You Have Learned

1. **a)** The process used in **Part A** of **Investigate** is a froth flotation process. This process requires that the ore material to be processed must first be ground to a powder. The jar serves as the grinding vessel.

    b) The steel shot is used to help in pulverizing the ore to a fine grain size (powder). Because it is dense and hard and resistant to breaking up, the shot keeps its integrity as it bashes the ore material to finer and finer grain size as the jar is shaken.

    c) The screen separates the fine-grained material (powdered sulfide ore) from the coarser-grained steel shot and uncrushed ore.

    d) The froth flotation separation method used in **Investigate, Steps 10** and **11** requires that the ore material be crushed (powdered) to a fine grain size in order for the process to work effectively.

    e) The froth flotation separation method was modeled by using liquid bubble-bath solution as the frothing agent and bubbling air into the ore–water–soap solution with a straw. The ore was then separated by skimming the ore–froth combination with an index card.

2. **a)** No, not all of the copper is collected after a few trials. This can be deduced from the fact that the solution continued to change color with each successive leaching.

    b) As copper was leached out of the copper oxide ore, the vinegar became green. There are several blue to green copper salts which are combinations of the metal copper and various other chemicals. Examples include copper sulfate and copper acetate. Vinegar is a solution of acetic acid, and as it reacts with the copper oxide ore, it produces copper acetate, which colors the solution green.

c) Were copper ions attracted to the anode or cathode during the first observation? How about the second time?

d) Why didn't the copper ore stay at the bottom?

e) What happens to the waste?

f) How could the waste be used?

g) Where does the copper mineral go from here?

h) Why did the changes take place in the vinegar solution?

i) How long did you leave the spoon in the solution?

j) Would a longer time make a thicker coating?

k) What would happen if you added more solution?

l) What happened when the wires were switched?

3. Explain how the properties of your state's minerals allow them to be used in their present states for various applications. What processes must your state's mineral resources go through to be refined prior to use?

Preparing for the Chapter Challenge

Pick a mineral deposit in your area that could contribute to the manufacture of a beverage container (the same deposit you have been studying) and research how the mineral from that deposit is processed to produce a final material. Create a flow chart that illustrates the process from mining to finished material. Write a report in your notebook based on your research.

Consider what impact mineral processing has on the environment and how scientists and engineers minimize that impact. Examine the mineral deposit you researched in the context of how the deposit and its use (past or present) has affected the environment. Consider all aspects of the environment, including air, water, land, plants, animals, and the related impact of transporting, using, and disposing of the related products.

Inquiring Further

1. **Minerals used to make common materials**

   Using data sources from reference books and the Internet, make a table that gives information on several common materials used in daily life that are derived from mineral resources.

   Materials to investigate: aluminum, copper, gold, gypsum, magnesium, silver, steel, titanium, zinc.

   Information: major ores, percentage produced in the United States, where produced in the United States, methods of production, major uses, annual consumption in United States, percent recycled, known world reserves, estimate of how long reserves will last, given annual world consumption.

Teaching Tip

The green color of the Statue of Liberty is the result of the exposure of copper to the environment. Weathering of the copper causes oxidation and the production of green salts. An interesting sideline is that in olden days, around the time of Washington, for example, paint pigments were generally not available. One way people formed their own paint pigments was to suspend copper metal in a container over a pool of acetic acid. The copper would corrode and the copper salt on the surface of the copper (copper acetate) could be scraped, ground up, and used as a pigment in paint.

c) The positively charged copper ions were attracted to the cathode (the negatively charged terminal) during both observations.

d) Copper salts leached from the mineral ore were dissolved in the vinegar solution, so they did not collect on the bottom.

e) In this case, the copper was not fully removed from the mineral, so it might be possible to put the ore through another, more effective extraction process. For many waste products, alternative uses have been found (like for construction material, for example). If no use is found for the waste product, then the waste would need to be disposed of (stored) in a safe and responsible way.

f) Answers will vary depending on the ore used.

g) Useful waste byproducts can be sold to other companies. The waste ore material that cannot be used must be stored on the mine property or at the processing site.

h) Changes took place in the vinegar solution because the acidity of the vinegar reacted with the copper ore and leached some of the copper from the ore in the form of copper salts. These copper salts colored the vinegar solution.

i) Answers will vary.

j) Students should note that the coating did apparently become thicker as time passed.

k) This depends on the concentration of the solution added. If more pure vinegar were simply added to the copper-rich solution, the concentration of copper would be reduced and the electroplating process would take longer. Addition of more copper-rich solution would increase the amount of copper available for electroplating so that more items could be plated.

l) When the wires were switched, the electroplating was reversed and the spoon lost its plating while the lead weight became plated.

3. Answers will vary depending upon the state in which you live.

Chapter 2

Preparing for the Chapter Challenge

Students are asked to investigate the steps required to process the minerals they plan to use in their beverage container. The flow chart that they produce can be included in their final report. In addition to considering the steps required to refine the mineral, students should consider the environmental implications of the mineral processing, and how this would affect the cost of producing the beverage container.

For assessment (see **Assessment Rubric for Chapter Report on Mineral Resources** on pages 304 and 305), remind students that their reports and flow chart should reflect:

- how the minerals can be used to make a practical container
- what impact the use of these minerals has on the environment

Inquiring Further

1. Minerals used to make common materials

Direct students to the *EarthComm* web site to find the information that they need to make the data table. Information is provided below, but your students may find additional information.

Aluminum

Major ores:	Bauxite
% produced in United States:	In 2000, the United States produced ~ 15.5% of total world Al, and ~ 45% of the Al used in the United States was also produced in the United States (some U.S.-produced aluminum was exported).
Methods of production:	Smelting, from bauxite via the Hall-Héroult process, from scrap Al (recycling).
Major uses:	Transportation (automobiles, airplanes, trucks, railcars, marine vessels, etc.), packaging (cans, foil, etc.), construction (windows, doors, siding, etc.), consumer durables (appliances, cooking utensils, etc.), electrical transmission lines, machinery, and many other applications.
Annual United States consumption:	In 2000, the United States consumed 7900 metric tons of aluminum. The United States recovered 4 million metric tons of scrap aluminum, of which 40% was from old scrap (discarded aluminum products) and 60% was new scrap (manufacturing waste). The recovered old scrap represents 20% of consumption. Relative to aluminum production, approximately 30% of the United States-produced aluminum came from recovered old scrap.
% recycled:	In 2000, the United States produced approximately 5300 metric tons of Al, 1600 of which were produced from recycled Al (old scrap).
Known world reserves:	
Estimate of how long reserves will last:	The world reserve base for bauxite is sufficient to meet world demands for metal well into the 21st century.

Copper

Major ores:	Native copper
% produced in United States:	In 2000, the United States produced ~ 11% of total world Cu, and ~ 25% of the Cu used in the United States was also produced in the United States.
Methods of production:	Pure copper metal is generally produced from a multi-stage process, beginning with the mining and concentration of low-grade ores containing copper sulfide minerals, followed by smelting and electrolytic refining to produce a pure copper cathode. An increasing share of copper is produced from acid leaching of oxidized ores.
Major uses:	Electrical uses of copper, including power transmission and generation, building wiring, telecommunications, and electrical and electronic products. Also used for building construction, electronics and electronic products, transportation, industrial machinery, and consumer and general products.
Annual United States consumption	In 2000, the United States consumed 6120 metric tons of Cu.
% recycled:	In 2000, ~ 12% of copper was derived from old scrap.
Known world reserves:	As of 2000, land-based resources are estimated at ~ 1.6 billion tons and deep-sea nodules are estimated to contain ~ 700 million tons.

Gold

Major ores:	Native-element mineral (Au).
% produced in United States:	In 2000, the United States produced ~ 10% of total world Au, and ~ 75% of the Au used in the United States was also produced in the United States.
Methods of production:	Three types of gold mining today: underground mining of high-grade lode and placer deposits, dredging of surface placer deposits, and open-pit mining. Heap leaching is often used to remove finely disseminated gold from low-grade ore. In this process, mounds of crushed ore are placed on an impermeable pad and sprayed with a dilute cyanide solution. The cyanide solution percolates through the ore and dissolves fine gold particles. The gold is then electrolytically recovered from solution and poured into ingots. Gold is also a by-product of sand and gravel production and base metal (copper, lead, and zinc) mining.
Major uses:	Primary use is for the manufacture of jewelry. Also used as an industrial metal, in computers, communications equipment, spacecraft, jet aircraft engines, and a host of other products.
Annual United States consumption	In 2000, the United States consumed 250,000 metric tons of Au.
% recycled:	In 2000, ~ 56% of reported consumption of gold was recycled.
Known world reserves:	As of 2000, the total world resources of gold were estimated at 100,000 tons. Approximately half of these resources are found in South Africa, and the United States and Brazil each have about 9%.

Gypsum

Major ores:	Gypsum is itself a mineral that is directly mined, so it does not need to be extracted from an ore.
% produced in United States:	In 2000, the United States produced ~ 23% of total world gypsum, and produced an excess of 14,900 metric tons. Approximately 78% of the gypsum used in the United States was produced in the United States.
Methods of production:	Most gypsum is mined as crude gypsum, some of which is then calcined (partially dehydrated). In 2000, approximately 20% of the United States gypsum production was synthetically produced gypsum. Synthetic gypsum is produced as an industrial by-product, largely from coal-fired electric plants.
Major uses:	Used to make wallboard for homes, offices, and commercial buildings, in concrete for highways, bridges, buildings, and many other structures that are part of our everyday life. Also used as a soil conditioner.
Annual United States consumption:	In 2000, the United States consumed 39,900 metric tons of gypsum.
% recycled:	Only a very small amount of gypsum is ever recycled.
Known world reserves:	As of 2000, the United States and Canada had an estimated 1,150,000 metric tons of gypsum in reserve between them (700,000 being in the United States).

Silver

Major ores:	Native-element mineral (Ag).
% produced in United States:	In 2000, the United States produced ~ 12% of total world Ag, and ~ 27% of the Ag used in the United States was also produced in the United States.
Methods of production:	Extracted during the refining process as byproduct of nickel processing. Also a by-product from gold, and copper, lead and zinc mining. Substantial amounts are recovered along with gold during cyanidation.
Major uses:	Industrial applications, as in mirrors, electrical and electronic products, and photography. Also used as a catalyst in oxidation reactions.
Annual United States consumption:	In 2000, the United States consumed 7700 metric tons of Ag.
% recycled:	In 2000, ~ 1600 tons of Ag were recovered from old and new scrap.
Known world reserves:	In 2000, world reserves of Ag were estimated at 280,000 metric tons.

Magnesium

Major ores:	Although magnesium is found in over 60 minerals, only dolomite, magnesite, brucite, carnallite, and olivine are of commercial importance.
% produced in United States:	For the year 2000, estimates of the world primary magnesium production capacity were 497,000 metric tons and the United States production capacity was 83,000 metric tons (~17% of the world production capacity). In 2000, an additional 82,300 metric tons of secondary magnesium were recovered from processed scrap.
Methods of production:	Mg metals are recovered from seawater.
Major uses:	As refractory material in furnace linings for producing iron and steel, nonferrous metals, glass, and cement; in agricultural, chemical, and construction industries. The principal use of magnesium metal is as an alloying addition to aluminum, and these aluminum–magnesium alloys are used for beverage cans and as structural components of automobiles and machinery. Magnesium also is used to remove sulfur from iron and steel.
Annual United States consumption:	In 2000, the United States consumed 730,000 metric tons of Mg compounds and 175,000 metric tons of Mg metal. In 2000 a total of 91,400 metric tons of magnesium in the forms of waste and scrap, metal, alloys, and other forms were imported for consumption.
% recycled:	In the United States, the recent production of primary magnesium metal has been declining. In the year 2000, nearly as much secondary magnesium metal was recovered from processed scrap (82,300 metric tons) as our total primary magnesium metal production capacity (83,000 metric tons).
Known world reserves:	As of 2000, there were an estimated 12 billion tons of magnesium compounds. Because Mg metal is recovered from seawater, the reserves are extensive.

Titanium

Major ores:	Minerals of economic importance include ilmenite, leucoxene, and rutile.
% produced in United States:	In 2000, the United States produced ~ 12% of total world Ag, and ~ 27% of the Ag used in the United States was also produced in the United States.
Methods of production:	Usually recovered through surface mining methods. Production is a two-step process. The first step is a refinement step to purify the ore, and is achieved by the sulfate process, which uses sulfuric acid as a liberating agent, or the chloride process, which uses chlorine as the liberating agent. Once refined, and developed to the appropriate particle size, pigment is surface treated with inorganic oxides or an organic material to give each grade its unique characteristics.
Major uses:	Used mainly as a white pigment in paints, paper, and plastics.
Annual United States consumption:	In 2000, the United States consumed 1430 metric tons of TiO_2.
% recycled:	None.
Annual world consumption:	In 2000, world consumption of titanium was ~ 56,000 tons.
Known world reserves:	Ilmenite supplies ~ 90% of the world's titanium demand. World ilmenite resources are estimated to contain 1 billion tons of TiO_2. Rutile resources are estimated to contain 230 million tons of TiO_2.

Zinc

Major ores:	Sphalerite (zinc sulfide).
% produced in United States:	In 2000, the United States produced ~ 11% of total world zinc, and ~ 52% of the zinc used in the United States was also produced in the United States.
Methods of production:	In the United States, about two-thirds of zinc is produced from ores (primary zinc) and the remaining one-third from scrap and residues (secondary zinc).
Major uses:	Used as a coating to protect iron and steel from corrosion (galvanized metal), as alloying metal to make bronze and brass, as zinc-based die casting alloy, and as rolled zinc. Also used by the rubber, chemical, paint, and agricultural industries. Also a necessary element for proper growth and development of humans, animals, and plants.
Annual United States consumption:	In 2000, the United States consumed ~ 1670 metric tons of zinc (1400 refined).
% recycled:	In 2000, 410,000 tons were recovered.
Known world reserves:	Total world reserves, as of 2000, estimated at 1.9 billion tons.

Chapter 2

Earth Science at Work

ATMOSPHERE: *Automotive Engineer*
New design techniques and the use of new materials can make cars lighter and more aerodynamic. This translates into a car that is more energy efficient.

BIOSPHERE: *Bioengineer*
The human body is a complex machine that can break down. Medical technology has developed a variety of artificial body parts to help people continue to enjoy an active lifestyle.

CRYOSPHERE: *Groundskeeper*
The removal of ice from sidewalks and roads often involves the use of chemicals. The sources of these chemicals and their effect on the environment are important considerations for a groundskeeper.

GEOSPHERE: *Commodities Broker*
The value of shares in a mining company is very dependent on the quantity and types of minerals that have been discovered, as well as the potential outcome of future explorations.

HYDROSPHERE: *Plumber*
The mineral content in water can cause damage to home plumbing. Water softeners can be installed to avoid some of the problems associated with "hard" water.

How is each person's work related to the Earth system, and to mineral resources?

NOTES

Mineral Resources and Your Community End-of-Chapter Assessment

1. Which of the following is an advantage to using aluminum for beverage containers?
    a) Manufacturing aluminum from aluminum ore requires little energy.
    b) It takes less energy to manufacture aluminum than it does to manufacture steel.
    c) Aluminum is the only metal that can be recycled.
    d) The cost of making aluminum from recycled material is inexpensive.

2. Considering how the manufacture of a product might affect the biosphere is known as
    a) environmental protection compliance.
    b) design for the environment.
    c) ethical product design.
    d) protocol for environmental manufacture.

3. What is the major disadvantage to using glass for beverage containers?
    a) The expense of handling and shipping heavy materials.
    b) The loss of carbon dioxide through the walls of the glass container.
    c) The lack of deposits of silica, a rare mineral resource.
    d) The expense of silica used to make glass containers.

4. Which of the following is NOT part of the definition of a mineral? A mineral must:
    a) be a solid.
    b) have a definite chemical composition.
    c) be naturally occurring.
    d) be made of two or more elements.

5. Which response shows the relationship between rocks and minerals? A rock is to a mineral as
    a) a parking lot is to a stadium.
    b) a table is to chairs.
    c) darkness is to light.
    d) a paragraph is to words.

6. If you observe a variety of samples of quartz, which property is likely to vary the most?
    a) color.
    b) streak.
    c) hardness.
    d) cleavage.
    e) density.

7. Atoms that have an electric charge because one or more electrons have been added to or removed from the atoms are known as
   a) ores.
   b) protons.
   c) ions.
   d) neutrons.

8. To determine the hardness of a mineral, you should
   a) crush it in a vise using steadily applied pressure.
   b) break it with a hammer and count the number of pieces that break off.
   c) scratch it against other minerals and objects of known hardness.
   d) divide its mass by its volume to calculate how compact it is.

9. The relationship between a rock and an ore is that
   a) ores are concentrations of valuable rocks.
   b) ores are rocks that contain valuable minerals.
   c) rocks are made of minerals, and ore is of made of crystals.
   d) rocks are made of one or more minerals, but ores are made of one mineral.

10. What holds a mineral together as a solid?
    a) attractions between ions.
    b) cement that builds up between elements.
    c) compaction of elements into a regular three-dimensional pattern.
    d) a lattice of connecting rods between elements.

11. Why don't we actively extract valuable elements like gold from ocean waters?
    a) Ocean waters do not contain large amounts of gold.
    b) We do not have the technology to remove gold from ocean water.
    c) Gold is located at too great a depth to bring to the surface.
    d) The extraction of gold from ocean water is not profitable.

12. Ores are _____ rocks of the North American continent.
    a) evenly distributed throughout the
    b) a very small fraction of the
    c) usually near the Earth's surface in the
    d) most often found in sedimentary

13. The formation of ores through the action of superheated mineral-rich water associated with magma is known as
    a) orogenic belt mineralization.
    b) cascade volcanism.
    c) lithospheric plate magmatism.
    d) hydrothermal activity.

14. What are the two most commonly used metals?
    a) copper and zinc.
    b) aluminum and iron.
    c) lead and brass.
    d) platinum and gold.

15. When exploring for minerals, what is the advantage to using geophysical techniques before drilling?
    a) it eliminates areas with low exploration potential.
    b) it confirms exactly where minerals are located.
    c) unlike drilling, there are no costs involved in using geophysics.
    d) mineral deposits cannot be located through drilling.

16. How do magnetometers work?
    a) By pointing in the direction of magnetic minerals.
    b) By detecting local changes in the Earth's magnetic field.
    c) By calculating the quantity of a magnetic mineral deposits.
    d) By measuring how well rocks conduct electricity.

17. How do geologic maps aid in mineral exploration?
    a) They show the depth of mineral deposits.
    b) They help geologists infer what lies beneath the surface.
    c) They show geologists where to drill for minerals.
    d) They show the amount of mineral in each kind of rock.

18. What method would be most appropriate for a horizontally layered mineral deposit located 15 meters below the Earth's surface?
    a) strip mining.
    b) underground mining.
    c) closed pit mining.
    d) spoil-pile mining.

19. How do mining companies ensure worker safety in underground mines?
    a) By removing ore slowly by hand.
    b) By filling in the open areas after ore is removed.
    c) By leaving pillars of ore to support the mine roof.
    d) By removing no more than 15% of the ore.

20. Why do states require environmental reclamation plans for mining operations?
    a) The government wants to know how much ore will be removed.
    b) Removing overburden and vegetation destroys established ecosystems.
    c) By law, the land must look better after mining than it did prior to mining.
    d) By law animals and plants cannot be disturbed by mining.

21. Smelting separates useful elements from waste material in an ore, but has the disadvantage of
    a) being too expensive to be practical.
    b) requiring high temperatures.
    c) producing pollutants that must be treated.
    d) requiring the addition of chemicals.

Answer Key

1. d	8. c	15. a
2. b	9. b	16. b
3. a	10. a	17. b
4. d	11. d	18. a
5. d	12. b	19. c
6. a	13. d	20. b
7. c	14. b	21. c

Teacher Review

Use this section to reflect on and review the investigation. Keep in mind that your notes here are likely to be especially helpful when you teach this investigation again. Questions listed here are examples only.

Student Achievement

What evidence do you have that all students have met the science content objectives?

_____

Are there any students who need more help in reaching these objectives? If so, how can you provide this?_____

_____

What evidence do you have that all students have demonstrated their understanding of the inquiry processes?_____

_____

Which of these inquiry objectives do your students need to improve upon in future investigations? _____

_____

What evidence do the journal entries contain about what your students learned from this investigation? _____

_____

Planning

How well did this investigation fit into your class time?_____

_____

What changes can you make to improve your planning next time? _____

_____

Guiding and Facilitating Learning

How well did you focus and support inquiry while interacting with students?

_____

What changes can you make to improve classroom management for the next investigation or the next time you teach this investigation? _____

_____

How successful were you in encouraging all students to participate fully in science learning?_____

_____

How did you encourage and model the skills values, and attitudes of scientific inquiry? _____

_____

How did you nurture collaboration among students?_____

_____

Materials and Resources

What challenges did you encounter obtaining or using materials and/or resources needed for the activity? _____

_____

What changes can you make to better obtain and better manage materials and resources next time? _____

_____

Student Evaluation

Describe how you evaluated student progress. What worked well? What needs to be improved? _____

_____

How will you adapt your evaluation methods for next time?_____

_____

Describe how you guided students in self-assessment. _____

_____

Self Evaluation

How would you rate your teaching of this investigation? _____

_____

What advice would you give to a colleague who is planning to teach this investigation? _____

_____

NOTES

Earth's Natural Resources – Mineral Resources and Your Community

Water Resources ...and Your Community

EARTH'S NATURAL RESOURCES CHAPTER 3

WATER RESOURCES... AND YOUR COMMUNITY

Chapter Overview

The **Chapter Challenge** for **Water Resources and Your Community** is for students to prepare a report that will help their community leaders determine how the water resource needs of the community would change with the development of an industrial park, a mini-mall, and a residential area in the community.

Students use an Earth systems approach to investigate the hydrologic cycle and global biogeochemical cycles (the nitrogen cycle) in the context of community water resources and development. Students identify the human and natural factors that determine the income and expenditure of water resources. They determine how to measure domestic water use, and they obtain information on the quantity of water used by industry and agriculture. These activities help them identify methods used to conserve water. They determine how rainfall, temperature, and other natural factors affect proper management and usage. They investigate the vulnerability of water resources to pollution by both human use and natural cycles or processes, and they make models of water-treatment processes. By the end of the chapter, students are better prepared to understand potential water-quality problems and efficient management of this precious and vital resource.

Chapter Goals for Students

- Understand how the hydrosphere is a part of Earth systems.
- Participate in scientific inquiry and construct logical conclusions based on evidence.
- Recognize that water is an indispensable natural resource whose use and quality needs to be monitored carefully.
- Appreciate the value of Earth science information in improving the quality of lives, globally and within the community.

Chapter Timeline

Chapter 3 takes about three weeks to complete, assuming one 45-minute period per day, five days per week. Adjust this guide to suit your school's schedule and standards. Build flexibility into your schedule by manipulating homework and class activities to meet your students' needs.

A sample outline for presenting the chapter is shown on the next page. This plan assumes that the teacher assigns homework at least three nights a week and assigns **Understanding and Applying What You Have Learned** and **Preparing for the Chapter Challenge** as group work to be completed during class. This outline also assumes that **Inquiring Further** sections are reserved as additional, out-of-class activities. This is only a sample, not a suggested or recommended method of working through the chapter; adjust your daily and weekly plans to meet the needs of your students and your school.

Day	Activity	Homework
1	Getting Started; Scenario; Chapter Challenge; Assessment Criteria	
2	Activity 1 – Investigate, Parts A – B	
3	Activity 1 – Investigate, Parts C – D	Digging Deeper; Check Your Understanding
4	Activity 1 – Review; Understanding and Applying; Preparing for the Chapter Challenge	
5	Activity 2 – Investigate; Parts A, B and C	Digging Deeper; Check Your Understanding
6	Activity 2 – Review; Understanding and Applying; Preparing for the Chapter Challenge	
7	Activity 3 – Investigate Parts A and B	Digging Deeper; Check Your Understanding
8	Activity 3 – Review; Understanding and Applying; Preparing for the Chapter Challenge	
9	Activity 4 – Investigate, Parts A and B	Digging Deeper; Check Your Understanding
10	Activity 4 – Review; Understanding and Applying; Preparing for the Chapter Challenge	
11	Activity 5 – Parts A and B	Digging Deeper; Check Your Understanding
12	Activity 5 – Investigate, Part C; Review; Understanding and Applying; Preparing for the Chapter Challenge	Digging Deeper; Check Your Understanding
13	Activity 6 – Investigate, Parts A and B	Digging Deeper; Check Your Understanding
14	Activity 6 – Review; Understanding and Applying; Preparing for the Chapter Challenge	
15	Complete Chapter Report	Finalize Chapter Report
16	Present Chapter Report	

Chapter 3

National Science Education Standards

Producing a report to help community leaders assess how the water needs of the community would change with the construction of an industrial park, a mini-mall, and a residential area sets the stage for the **Chapter Challenge.** Students learn the value of water as a natural resource. Through a series of activities, students begin to develop the content understandings outlined below.

CONTENT STANDARDS

Unifying Concepts and Processes
- Systems, order, and organization
- Evidence, models, and explanation
- Constancy, change, and measurement
- Evolution and equilibrium

Science as Inquiry
- Identify questions and concepts that guide scientific investigations
- Design and conduct scientific investigations
- Use technology and mathematics to improve investigations
- Formulate and revise scientific explanations and models using logic and evidence
- Communicate and defend a scientific argument
- Understand scientific inquiry

Earth and Space Science
- Energy in the Earth system
- Geochemical cycles
- Origin and evolution of the Earth system

Science and Technology
- Identify a problem or design an opportunity
- Propose designs and choose between alternative solutions
- Implement a proposed solution
- Communicate the problem, process, and solution
- Understand science and technology

Science in Personal and Social Perspectives
- Personal and community health
- Natural resources
- Environmental quality
- Natural and human-induced hazards
- Science and technology in local, national, and global challenges

History and Nature of Science
- Science as a Human Endeavor

Key Science Concepts and Skills

Activities Summaries	Earth Science Principles
Activity 1: Sources of Water in the World and in Your Community Students use water-filled milk jugs to understand how water is distributed in various reservoirs on Earth. They examine data on water use in their county to determine how much water their community uses and where the water comes from. Students then construct a diagram that illustrates how water moves within the hydrologic cycle.	• Distribution of salt and fresh water • Topographic maps • Water cycle
Activity 2: How Does Your Community Maintain Its Water Supply? Students set up a simple experiment to model the flow of water between surface reservoirs. They complete another experiment to model ground-water flow. Students then use their community's water-quality report to understand water-supply management in their community.	• Surface-water flow and supply systems • Ground-water flow and supply systems • Factors that affect water-supply systems
Activity 3: Using and Conserving Water Students design a method to determine how much water their school uses daily. They gather data on water use in their county and a neighboring county, and they compare water use in the two areas.	• Amount and purpose of water use • Consumptive and nonconsumptive water use • Water conservation
Activity 4: Water Supply and Demand: Water Budgets Students construct and interpret a water budget for their community. They make a climatograph for their community to understand how climate influences water use. Students then construct an irrigation water budget for the United States.	• Water budget • Climatograph
Activity 5: Water Pollution Students construct a ground-water model and infer how pollutants reach the ground-water table and move with ground-water flow. They investigate how extracting water from a ground-water well influences the movement of pollution in ground water. Students then test untreated and treated water samples from their community for nitrate. They also model how road salt pollutes water reservoirs.	• Ground-water pollution • Nitrate levels in water • Surface-water pollution • Kinds of pollutants
Activity 6: Water Treatment Students set up an experiment to model water treatment and then water filtration. They use their community's water-quality report to examine the water-treatment process in their community.	• Water treatment • Water filtration

Chapter 3

Equipment List for Chapter Three:

Materials needed for each group per activity.

Activity 1 Part A

- Five 1-gal. (clean) plastic milk jugs (or a 1-L beaker)
- Beaker for transferring water between containers
- Five small cups or other containers
- Measuring spoons or graduated cylinder
- Water
- Calculator

Activity 1 Part B

- Water-use data for your county*
- Local topographic map*

Activity 1 Part C

- Several blank sheets of paper
- Scissors
- Ruler or straightedge
- Poster board
- Markers
- Removable tape
- Colored pencils (blue, red, green)

Activity 1 Part D

- Calculator
- School blueprints (optional)
- Local annual precipitation data*

Activity 2 Part A

- 3-lb coffee can

- Clear, 750-mL (or 1-L) soda bottle with the top cut off
- Shallow container
- Coffee filter
- Sand
- Three flexible plastic tubes 1/4 to 1/2 in. diameter (each 25 cm long)
- Modeling clay
- C-clamp

Activity 2 Part B

- Plastic or rubber dish tub about 12 in. wide, 16 in. long, and 6 in. deep with a 1-cm hole cut into the bottom
- Duct tape
- Sharp pencil
- Sand
- Two pieces of rigid tubing (about 1/2 to 1 in. in diameter)
- Cheesecloth
- Chopstick with evenly spaced lines drawn on it (to serve as a ruler)
- Drinking straw
- Turkey baster

Activity 2 Part C

- Community water-quality report

Activity 3 Part A

- Calculator
- Copy of school water bill or water-use data

Activity 3 Part B

- Water-use data for your county and a neighboring county*
- Graph paper
- Colored pencils (eight different colors)
- Calculator

*The *EarthComm* web site provides suggestions for obtaining these resources.

Activity 4 Part A

- Internet access (or printouts of monthly temperature and precipitation data for the community)*
- Graph paper

Activity 4 Part B

- Several blank sheets of paper
- Scissors
- Tape

Activity 5 Part A

(Some of the materials from Activity 2 can be reused here.)

- Plastic or rubber dish tub about 12 in. wide, 16 in. long, and 6 in. deep with a 1 cm hole cut in the bottom
- Sandbox sand or concrete-making sand to fill the dish tub to a depth of 5 in.
- Two pieces of rigid plastic or metal tubing, 1/2 in. to 1 in. in diameter
- Cheesecloth
- Duct tape
- Measuring cup
- Drinking straw
- Turkey baster
- Food coloring
- Eye dropper
- Clear plastic drinking cups
- Stopwatch

Activity 5 Part B

- Water sample from a local stream (untreated)
- Tap water (treated)
- Nitrate test kit

Activity 5 Part C

- Calculator
- 1-gal. plastic milk jug
- Water
- Salt
- Teaspoon
- Plastic cups (one per student)

Activity 6 Part A

- Water samples from an untreated lake, swamp, or stream (2 L)
- Three 2-L plastic bottles with just enough of the top part removed to make a 5-cm wide opening, OR three 1.5-L beakers
- Silt or clay
- Alum (potassium aluminum sulfate—from grocery-store spice aisle or supply catalog)
- Other materials: balance, glass stirrers, scoop or spoon, paper towels, and white notebook paper

Activity 6 Part B

- 2-L plastic soda bottle with the bottom 5 cm cut off
- Two 500-mL beakers
- Nylon screen or a piece of nylon stocking
- Rubber band
- Clean aquarium gravel
- Clean sand
- Dropper filled with non-chlorine bleach mixed with water

Activity 6 Part C

- Copy of your community's water-quality report

*The *EarthComm* web site provides suggestions for obtaining these resources.

Water Resources
...and Your Community

Getting Started

A person requires less than one gallon of fresh water a day to survive. That's what you need to replace the fluids you lose daily. Yet the per capita use of fresh water in the United States is nearly 2000 gallons per day.

- Why is so much water used each day?
- How does your use of water change Earth systems?
- How do changes in Earth systems affect your use of water?

What do you think? Look at the diagram of the Earth systems at the front of this book. Think about where water is stored in the various spheres within the Earth system. Think about how water moves from one sphere to another. Use your reflections about water in the Earth system to answer the three questions above. Be prepared to discuss your answers with your small group and the class.

Scenario

Developers would like to build an industrial park, a mini-mall, and a planned residential area in your community. They estimate that the industrial park will use 100,000 gallons of water per day and the mini-mall will use 20,000 gallons per day. The planned residential area will include two city parks, one community hospital, and one golf course. As a member of a citizens' planning group, you have been assigned the task of providing the city or county planners with information related to these increased needs.

R144

Getting Started

Uncovering students' conceptions about Water Resources and the Earth System

Use **Getting Started** to elicit students' ideas about the main topic. The goal of **Getting Started** is not to seek closure (i.e., the right answer) but to provide you (the teacher) with information about the students' starting point and about the diversity of ideas and beliefs in the classroom. By the end of the chapter, students will have developed a more detailed and accurate understanding of how water as a resource relates to the Earth system.

Students most likely will not have a clear understanding of how much water they use in a single day. Encourage them to think about the various ways that they use water each day, starting with morning activities like bathing and brushing their teeth. They can then estimate how much water each of these activities uses, and sum them up. Have students compare their estimates, but remind them that you are not looking for the right answer.

Ask students to work independently or in pairs and to exchange their ideas with others. Avoid labeling answers as right or wrong. Accept all responses, and encourage clarity or expression and detail.

Students may struggle with the final two questions of **Getting Started** because they may not have thought about water from an Earth systems perspective. Ask students to look at the diagram on page Rxii of the textbook and determine which Earth system they will be discussing in this chapter. Have them think about the interactions of the hydrosphere with each of the other Earth systems. For the second question (How does your use of water change Earth systems?), they are likely to note that their use of water removes water from the hydrosphere, and that the water might be different (polluted) when it is returned to the hydrosphere. Encourage students to go deeper, thinking about how this removal affects all of the Earth systems. For the final question (How do changes in Earth systems affect your use of water?), they are likely to identify how changes in the atmosphere (e.g., drought, floods) affect the supply of water.

In time, *EarthComm* students will understand that because water is the primary component of the hydrosphere, use of water ultimately affects the atmosphere, biosphere, cryosphere, and geosphere. Changes in the other Earth systems can cause changes in the mode of operation of the hydrosphere; likewise, changes within the hydrosphere can affect the other Earth systems. Students will come to appreciate the importance of understanding these changes, because they can affect the availability of water—a precious and vital resource.

Chapter 3

Chapter Challenge

Your report will need to include enough information to answer the following questions:

- How much water does the residential development require?

- Does your town have enough extra water to supply the four new developments?

- If your town does not have enough extra water, what are ways that the town could increase its supply?

- Will the construction and operation of these developments pollute your town's drinking water? What factors would determine that?

- Will the addition of these developments become a strain on the water-treatment capabilities of your town? Will they add a lot of expense?

Assessment Criteria

Think about what you have been asked to do. Scan ahead through the chapter activities to see how they might help you to meet the challenge. Work with your classmates and your teachers to define the criteria for assessing your work. Devise a grading sheet for the assessment of the challenge. Your teacher may provide you with a sample rubric to help you get started. Record all this information. Make sure that you understand the criteria and the grading scheme as well as you can before you begin.

R145

Chapter Challenge and Assessment Criteria

Read (or have a student read) the **Chapter Challenge** aloud to the class. Allow students to discuss what they have been asked to do. Have students meet in teams to begin brainstorming what they would like to include in their **Chapter Challenge** reports. Request a brief summary in their own words of what they have been asked to do, and a description of attributes of a high-quality report.

Alternatively, lead a class discussion about the challenge and the expectations. Review the titles of the activities in the Table of Contents. To remind students that the content of the activities corresponds to the content expected for the Chapter Report, ask them to explain how the title of each activity relates to the expectations for the **Chapter Challenge**. Familiarize students with the structure of each activity. When you come to the section titled **Preparing for the Chapter Challenge**, point out that each activity contributes to the challenge in some way.

Guiding questions for discussion include:
- What do the activities have to do with the expectations of the challenge?
- What have you been asked to do?
- What should a good final report contain?

A sample rubric for assessing the **Chapter Challenge** is shown on the following page. You can copy and distribute the rubric as is, or you can use it as a baseline for developing scoring guidelines and expectations that suit your needs. For example, you might wish to ensure that core concepts and abilities derived from your local or state science frameworks also appear on the rubric. You might also wish to modify the format of the rubric to make it more consistent with your evaluation system. However you decide to evaluate the chapter report, keep in mind that all expectations should be communicated to students and the expectations should be outlined at the start of their work. Please review **Assessment Criteria** (pages xxiv to xxv of this Teacher's Edition) for a more detailed explanation of the assessment system developed for the *EarthComm* program.

Chapter 3

Assessment Rubric for Chapter Challenge on Water Resources

Meets the standard of excellence. **5**	_Significant_ information is presented about <u>all</u> of the following: • How much water is required by the residential development • Whether or not your town has enough extra water to supply the three new developments • How your town could increase its water supply (if there is not enough water) • Whether or not the construction and operation of these developments will pollute your town's drinking water, and how can this be determined • Whether or not the addition of these developments will strain the water-treatment capabilities of your town, and at what expense _<u>All</u>_ of the information is accurate and appropriate. The writing is clear and interesting.
Approaches the standard of excellence. **4**	_Significant_ information is presented about <u>most</u> of the following: • How much water is required by the residential development • Whether or not your town has enough extra water to supply the three new developments • How your town could increase its water supply (if there is not enough water) • Whether or not the construction and operation of these developments will pollute your town's drinking water, and how can this be determined • Whether or not the addition of these developments will strain the water-treatment capabilities of your town, and at what expense _<u>All</u>_ of the information is accurate and appropriate. The writing is clear and interesting.
Meets an acceptable standard. **3**	_Significant_ information is presented about <u>most</u> of the following: • How much water is required by the residential development • Whether or not your town has enough extra water to supply the three new developments • How your town could increase its water supply (if there is not enough water) • Whether or not the construction and operation of these developments will pollute your town's drinking water, and how can this be determined • Whether or not the addition of these developments will strain the water-treatment capabilities of your town, and at what expense _<u>Most</u>_ of the information is accurate and appropriate. The writing is clear and interesting.

Assessment Rubric for Chapter Challenge on Water Resources

Below acceptable standard and requires remedial help. **2**	<u>Limited</u> information is presented about the following: • How much water is required by the residential development • Whether or not your town has enough extra water to supply the three new developments • How your town could increase its water supply (if there is not enough water) • Whether or not the construction and operation of these developments will pollute your town's drinking water, and how can this be determined • Whether or not the addition of these developments will strain the water-treatment capabilities of your town, and at what expense <u>Most</u> of the information is accurate and appropriate. Generally, the writing does not hold the reader's attention.
asic level that requires remedial help or demonstrates a lack of effort. **1**	<u>Limited</u> information is presented about the following: • How much water is required by the residential development • Whether or not your town has enough extra water to supply the three new developments • How your town could increase its water supply (if there is not enough water) • Whether or not the construction and operation of these developments will pollute your town's drinking water, and how can this be determined • Whether or not the addition of these developments will strain the water-treatment capabilities of your town, and at what expense <u>Little</u> of the information is accurate and appropriate. The writing is difficult to follow.

Chapter 3

ACTIVITY I— SOURCES OF WATER IN THE WORLD AND IN YOUR COMMUNITY

Background Information

The Earth's Water

Water is the only common substance that exists at the Earth's surface as a solid, a liquid, and a gas. This fact is very important to the workings of the water cycle discussed below. Be sure to impress upon your students that water is an extremely unusual—almost unique—substance in the universe, whose special properties are largely a consequence of hydrogen bonding. It would be nice to present a concise and understandable account of the reasons why the H–O–H angle in the water molecule is not 180° (i.e., lined up in a straight line) but instead forms an angle of 108°. This angle is the basis of the electrically polar nature of the water molecule and therefore of the phenomenon of hydrogen bonding. However, your students will have to be content with a statement that this is so, without the benefit of a simple explanation for why it is so; this phenomenon can only be explained only with a lengthy and technically complex description on the basis of the quantum mechanics of covalent bonding.

In liquid water, a certain percentage of water molecules are hydrogen-bonded to other water molecules at any given time. These bonds are formed and then broken again on extremely short time scales. The local groups of hydrogen-bonded water molecules in the liquid are called fleeting clusters. Because these clusters form and dissipate so rapidly, the liquid water can flow like a fluid despite the existence of the hydrogen bonds.

The extremely high heat capacity of water is a direct consequence of the existence of hydrogen bonding in liquid water, in that the percentage of hydrogen bonds that exist at a given time decreases with increasing temperature. Therefore, heat energy has to be added to the water to increase its temperature, in addition to the heat energy needed just to increase the thermal agitation of the molecules. The extremely high heat of vaporization of liquid water is explained by the need for additional heat to break the remaining hydrogen bonds, in addition to the energy needed to expel water molecules from the liquid into the vapor.

There's more to the story than that, however: another important and extremely unusual property of water is the maximum in density of liquid water at about 4°C. This maximum is explained as the outcome of a balance between two competing and opposing effects: the decrease in the percentage of hydrogen bonds in the water, which tends to *increase* the density, and the thermal expansion of the liquid caused by the more vigorous thermal agitation of the water molecules, which tends to *decrease* the density. From 0°C to 4°C, the former effect outweighs the latter, and the density increases slightly; above 4°C, the opposite is the case, and the density decreases continuously up to the boiling point. This effect holds only for fresh water, however, not for seawater.

Yet another very unusual property of water is that the solid form, ice, is less dense than liquid water. That is why ice floats in water! Because of the arrangement that results from the positively charged hydrogen sides of the molecules being bonded to the negatively charged oxygen sides of the molecules, the geometry of the crystal structure of ice turns out to be somewhat more open than if all of the water molecules were allowed to pack themselves as closely as possible with no constraints imposed by the geometry of hydrogen bonding.

The Water Cycle

Water, in the form of liquid, solid, or vapor, is in a continuous state of change and movement. Water resides in many different kinds of places, and it takes many different kinds of pathways in its movement. The combination of all of these different movements is called the water cycle or the hydrologic cycle.

The water cycle is called a cycle because the Earth's surface water forms a closed system. In a closed system, material moves from place to place within the system but is not gained or lost from the system. The Earth's surface water is actually not exactly a closed system, because relatively small amounts are gained or lost from the system. Some water is buried with sediments and becomes locked away deep in the Earth for geologically long times. Volcanoes release water vapor contained in the molten rock that feeds them. Nonetheless, these gains and losses are very small compared to the volume of water in the Earth's surface-water cycle.

Evaporation (change of water from liquid to vapor) and precipitation (change of water from vapor to liquid or solid) are the major processes in the water cycle. The balance between evaporation and precipitation varies from place to place and from time to time. It's known, however, that there is more evaporation than precipitation over the surface of the Earth's oceans, and there is more precipitation than evaporation over the surface of the Earth's continents. That fact has a very important implication: there is net movement of water vapor from the oceans to the continents, and net movement of liquid (and solid) water from the continents to the oceans.

Elements of the Water Cycle

The oceans cover about three-quarters of the Earth. Ocean water is constantly evaporating into the atmosphere. If enough water vapor is present in the air, and if the air is cooled sufficiently, the water vapor condenses to form tiny droplets of liquid water. If these droplets are close to the ground, they form fog. If they form at higher altitudes, by rising air currents, they form clouds.

All of the solid or liquid water that falls to Earth from clouds is called precipitation. Snow, sleet, and hail are solid forms of precipitation. Rain and drizzle are liquid forms of precipitation.

When rain falls on the Earth's surface, or snow melts, several things can happen to the water. Some evaporates back into the atmosphere. Some flows downhill on the surface, under the pull of gravity, and collects in streams and rivers. This flowing water is called surface runoff. Most rivers empty their water into the oceans. Some rivers, however, end in closed basins on land. Death Valley and the Great Salt Lake are examples of such closed basins.

Some precipitation soaks into the ground rather than evaporating or running off. The water percolates slowly downward through the spaces of porous soil and rock material. Eventually the water reaches a zone where all of the pore spaces are filled with water. This water is called ground water. Some water, called soil moisture, remains behind in the surface layer of soil.

The roots of plants absorb some of the water that soaks into the soil. This water travels upward through the stem and branches of the plant into the leaves and is released into the atmosphere in a process called transpiration.

It's been estimated that each year about 36,000 km³ of water flows from the surface of the continents into the oceans, representing the excess of precipitation over evaporation on the continents. This surface water carries sediment particles and dissolved minerals that come to rest on the ocean bottom. When seawater evaporates, it leaves the dissolved materials behind. Over geologic time, this process has gradually made the oceans as salty as they are now.

Earth Systems Science and the Water Cycle
In Earth systems science, the water cycle is viewed as a flow of matter and energy. Each place that water is held is called a reservoir. The rate at which water flows from one reservoir to another in a given time is called a flux. Energy is required to make water flow from one reservoir to another. On average, the total amount of water in all reservoirs combined is nearly constant, but the various reservoirs involved in the water cycle do not hold a constant amount of water. For example, in many areas there may be more water in ground water during the spring (when precipitation is high, and water use and evaporation are low) than in the summer (when precipitation is low, and evaporation and water use are high).

More Information – on the Web
Visit the *EarthComm* web site www.agiweb.org/earthcomm to access a variety of links to web sites that will help you deepen your understanding of content and prepare you to teach this activity. Many of the sites also contain images that you can download.

Goals and Assessment

Clarify that the goals indicate what students should understand and be able to do as a result of the activity. Make sure students understand that Chapter Assessments are based upon these goals.

Goal	Location in Activity	Assessment Opportunity
Identify and analyze the various sources and distribution of salt water and fresh water on Earth.	**Investigate** **Part A** **Digging Deeper; Check Your Understanding** Question 1 **Understanding and Applying What You Have Learned** Question 3	Calculations for water distribution are reasonable; work is shown. Method for modeling fresh-water distribution is reasonable and valid. Responses to questions closely match those given in Teacher's Edition.
Interpret data and a topographic map to determine the water sources that your community uses for drinking water.	**Investigate** **Part B** **Understanding and Applying What You Have Learned** Question 4	Answers to questions reflect information given in county water-use data and on topographic maps.
Generate a graphical model of the transport of water between reservoirs within the water cycle.	**Investigate** **Part C** **Digging Deeper;** **Understanding and Applying What You Have Learned** Question 4	Model of the water cycle is accurate. Responses to questions closely match those given in Teacher's Edition.
Develop a method of determining the amount of fresh water that could be collected in one year from your school roof and on the entire area of your community.	**Investigate** **Part D**	Plan for estimating fresh-water runoff is reasonable and useful. Estimate is reasonable.

Chapter 3

Earth's Natural Resources Water Resources

Activity 1

Sources of Water in the World and in Your Community

Goals

In this activity you will:

- Identify and analyze the various sources and distribution of salt water and fresh water on Earth.

- Interpret data and a topographic map to determine the water sources that your community uses for drinking water.

- Generate a graphical model of the transport of water between reservoirs within the water cycle.

- Develop a method of determining the amount of fresh water that could be collected from your school roof and on the entire area of your community in one year.

Think about It

Imagine you are watching the news on television, and the meteorologist says, "We received one inch of rain yesterday."

- How much water was that?
- Where is all that water today?

What do you think? Record your ideas about these questions in your *EarthComm* notebook. Be prepared to discuss your responses with your small group and the class.

Activity Overview

Students model the distribution of water in the hydrosphere to gain an understanding of where most of the water on Earth is found and what proportion of water on Earth is available for human use. They examine water-use data for their county, along with a local topographic map, to determine where water used in the community comes from. Students then produce a poster to help them understand how water is cycled on the Earth's surface. Finally, they devise a plan to calculate how much fresh water could be collected from the roof of their school. **Digging Deeper** takes an in-depth look at the hydrologic cycle.

Preparation and Materials Needed

Part A

You will need five 1-gal. milk jugs for each student group. Note that the Student Edition text calls for five 4-liter milk jugs, but the 1-gallon jugs will work equally well. You should note, however, that there are only about 3.8 liters in a United States gallon (3785 ml). A few weeks before completing this activity in class, you may want to ask your students to begin collecting and bringing in the milk jugs. The jugs should be rinsed thoroughly.

A 1-L beaker could be used as an alternative to the milk jugs. Then, the volume of water represented by the oceans (97.54% of all water on Earth) becomes 975.4 mL.

Part B

In **Part B** of the investigation, students need water-use data for the county. The *EarthComm* web site offers a list of web sites that can help you in obtaining this data. You can either have your students use the Internet to find the data, or you can print it out and distribute copies.

Your should also obtain copies of a local topographic map before class. The *EarthComm* web site offers a list of resources to help you in acquiring the appropriate map. Consider laminating the maps to increase their longevity, because you will be using them in other activities in **Chapter 3**.

Part C

Aside from collecting the necessary materials, no advance preparation is required for this part of the activity. If you are pressed for time, however, you may wish to have students prepare (cut and color) the circles and boxes for the different reservoirs, processes, and definitions as a homework assignment, or you may wish to complete this step for them.

Part D

To complete **Part D** of the investigation, students will need to know the area of the school roof and the local annual precipitation. The dimensions of the roof can be determined by having students go outside and make estimates of roof height and width, or by using blueprints for the building, which should be available from a school official. The *EarthComm* web site offers links that will help you determine the average annual precipitation in your area.

Materials

Part A
- Five 1-gal (clean) plastic milk jugs (or a 1-L beaker)
- Beaker for transferring water between containers
- Five small cups or other containers
- Measuring spoons or graduated cylinder
- Water
- Calculator

Part B
- Water-use data for your county*
- Local topographic map*

Part C
- Several blank sheets of paper
- Scissors
- Ruler or straightedge
- Poster board
- Markers
- Removable tape
- Colored pencils (blue, red, green)

Part D
- Calculator
- School blueprints (optional)
- Local annual precipitation data*

*The *EarthComm* web site provides suggestions for obtaining these resources.

Think about It

Student Conceptions

If students seem to be having trouble visualizing what an inch of rain means, one idea is to show them rainfall data (in the form of graphs or maps) for cities in their state with which they are likely to be familiar. Links on the *EarthComm* web site can lead you to precipitation data for your area and other regions around the country and the globe. Another possibility is to measure out an inch of water and pour it into several containers of different sizes and ask students to compare them. Or, you could pour an inch of water into several containers of different sizes and ask students if the containers hold the same volume of water.

For the second **Think about It** question, they are likely to say that the water has soaked into the ground and/or has flowed into surface reservoirs, like lakes and rivers. They may also say that the water is in puddles. If so, you can ask them how long a puddle typically lasts, and what happens to the water eventually.

Answer for the Teacher Only

The main point of the first question is that although precipitation is measured in inches, we are most interested in the volume of water that falls to the ground. The second question should expose student understanding of the water cycle. The simple fact is that water does not just go away. Precipitation becomes part of a system (the water cycle or the hydrologic cycle) that has many parts. Water that falls to the ground can become surface runoff, flowing into rivers, streams, ponds, or ocean. It can infiltrate the ground to become ground water, or it can be evaporated and returned to the atmosphere.

Assessment Tool

Think about It Evaluation Sheet
Use this evaluation sheet to help students understand and internalize the basic expectations for the warm-up activity.

Chapter 3

Investigate

Part A: Water in the Hydrosphere

1. Fill five 4-L milk jugs with water. These five jugs of water represent all the water on the Earth.

   a) Calculate how many milliliters are in the 20 L (five milk jugs). Record this value. Note: the actual amount of water may not be exactly 20 L, but for the purpose of this model it will be satisfactory.

2. Ice (mostly in the form of glaciers) holds 1.81% of all the water on Earth (see the data table below).

   a) If 20 L represents all the water on Earth, calculate the number of milliliters that represents the water found in glaciers.

3. Remove this amount from the milk jugs and pour it into a separate container labeled "glaciers."

4. Repeat Steps 2 and 3 for the water found in ground water, saltwater and freshwater lakes and streams, and the atmosphere.

   a) Record your calculations.

   b) Calculate the number of milliliters of water in the oceans, but leave the water in the milk jugs.

   c) Find the sum of the six values. How do you account for the "missing" water?

 Clean up spills. Dispose of the water.

Distribution of Water in the Hydrosphere			
Reservoir	Percentage of total water	Percentage of fresh water (ice and liquid)	Percentage of fresh water (liquid only)
Oceans	97.54	—	—
Ice (mostly glaciers)	1.81	73.9	—
Ground water	0.63	25.7	98.4
Saltwater lakes and streams	0.007	—	—
Freshwater lakes and streams	0.009	0.36	1.4
Atmosphere	0.001	0.04	0.2

These figures account for 99.9% of all water. They do not add up to 100%, because some water is tied up in the biosphere and as soil moisture.

5. Develop your own method for making a model of the percentages of each category shown in the table for fresh water (ice and liquid) and liquid fresh water. Use something other than water as the physical material for your model.

   a) From your work with the models, write down several observations or discoveries that you found most surprising or striking. Explain your observations.

   ⚠ Have your teacher check the plan for your model.

R 147

Investigate

Assessment Tools

EarthComm **Notebook-Entry Checklist**
Use this checklist as a quick guide for student self-assessment and/or an opportunity to quickly score student work. Add further criteria specific to your classroom needs or to this particular investigation.

Investigate Notebook-Entry Evaluation Sheet
Point out the criteria listed on this evaluation sheet that are relevant to this particular investigation. Encourage students to internalize the criteria by making them part of your assessment conversations as you circulate around the classroom. For example, while students are working, ask them criteria-driven questions:
   • Is your work thorough and complete?
   • Are all of you participating in the activity?
   • Do you each have a role to play in solving this problem? And so on.

Part A: Water in the Hydrosphere

1. You may wish to fill the jugs and have them available for students at the start of class. Milk jugs actually hold about 3.8 L, but for the purposes of this experiment they can be considered to hold 4 L. If you wish to be more precise, use 19 L as the total for the 5 milk jugs.

   a) Remind students that there are 1000 mL in 1 L. Therefore, there are 20,000 mL in 20 L. (or 19,000 mL in 19 L if you are using 3.8 L/gal. in your calculations). Circulate around the room to make sure that students have done this calculation correctly, because it is essential to completing the rest of the investigation. If you have scaled down the activity (i.e., if you are using less than 20 L water total), remind your students to take this into account when making their calculations. If graduated cylinders are not available, then measuring spoons can be used instead.

Teaching Tip

There are 1280 tbsp. in five gallons of water, and there are three teaspoons per tablespoon. There are approximately 5 mL per teaspoon and 15mL per tablespoon. If you are using a tablespoon to measure the water, you may want students to calculate the number of tablespoons of water represented by each reservoir, rather than milliliters.

2. This is a relatively simple calculation:

$$20,000 \text{ mL } \times 0.0181 = 362 \text{ mL, or}$$

$$1280 \text{ tbsp. } \times 0.0181 = 23.2 \text{ tbsp.}$$

Chapter 3

Teaching Tip

If students seem to be having trouble doing the calculations, you may wish to do this first calculation together, as a class, on the blackboard. If you are using 19 L as the total volume, each of the mL calculations given here should be multiplied by 0.95.

3. Students can use a graduated cylinder or tablespoons to measure the water and transfer it to the glacier container (approximating for the tablespoon fraction).

4. a) Circulate around the room to make sure that your students are completing the exercise correctly and are not having any trouble with the calculations or measurements. Remind them to record all of their work and results in their notebooks. Once again, if using measuring spoons, approximations will have to be made for the tablespoon fractions. The calculations are as follows:

   Ground water: 20,000 mL x 0.0063 = 126 mL (or 8.1 tbsp.)
   Saltwater lakes and streams: 20,000 mL x 0.00007 = 1.4 mL (or 0.09 tbsp.)
   Freshwater lakes and streams: 20,000 mL x 0.00009 = 1.8 mL (0.11 tbsp.)
   Atmosphere: 20,000 mL x 0.00001 = 0.2 mL (or 0.01 tbsp.)

   b) **Oceans:** 20,000 mL x 0.9754 = 19,508 mL (or 1248.5 tbsp.)

   c) When students add up their calculations, they should equal approximately 20,000 mL (or 20 L):

   362 + 126 + 1.4 + 1.8 + 0.2 + 19508 = 19,999.4 (or 19.9994 L)

Teaching Tip

Discuss with students why their values do not add up to exactly 20 L. Where is the remaining 0.6 mL? The answer is given in the caption below the table.

5. Student models will vary. Encourage students to be creative here. For example, they could cut up a sheet of graph paper to represent the proportion of water in different reservoirs.

   a) Student responses will vary. They are often surprised by the small volume of fresh water, and the relatively high percentage of fresh water that exists as ground water.

Blackline Master Water Resources 1.1
The Water Cycle

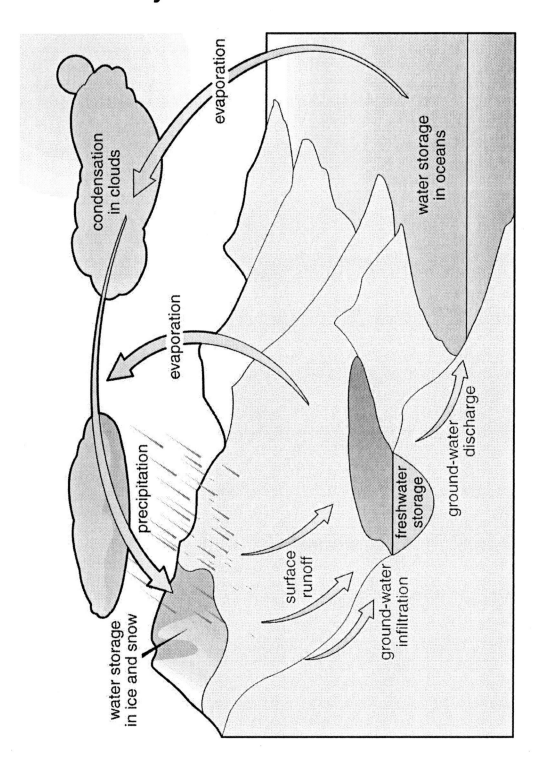

Earth's Natural Resources Water Resources

Part B: Local Water Use

1. Obtain a set of data on water use in your county. Use the data and a topographic map of your community to answer these questions:

   a) How much water does your city or county use each year?

   b) What and where are the sources of your community's drinking water?

   c) What are other water sources your community might use in the future? Record your ideas.

   d) Where are the water treatment facilities located?

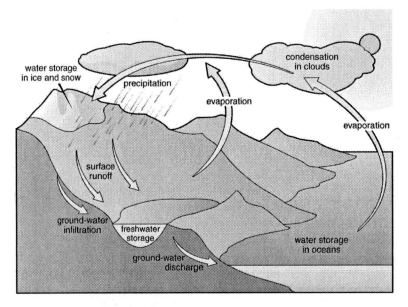

Part C: Modeling the Water Cycle

1. The total volume of water near the Earth's surface is almost constant, but the water is in constant motion. The water cycle describes how the Earth's surface water moves from place to place in an endless cycle. Study the diagram above that shows a simplified version of the water cycle.

2. On the following page is a more complete list of the components of the water cycle. There are also definitions of some terms with which you may not be familiar. The list is divided into two parts: *reservoirs* (places where water is stored) and *processes* (ways that water is moved from place to place).

   a) Using blank sheets of paper, draw a rectangular box for each reservoir item. Try to keep the dimensions of the boxes less than about 2.5 cm. Write the name of each reservoir in a box. You will have to write small.

   b) Draw a circle for each process item. Make the diameter of each circle less than 2.5 cm. Write the name of each process in a circle.

Part B: Local Water Use

1. Provide students with local water-use data and a local topographic map. If this is the first time that your students are using topographic maps, you may want to take a few moments to review how to read a map of this kind. For example, point out that water is shown in blue on topographic maps and that buildings (e.g., water-treatment facilities) are shown in black. The topographic maps will be used to locate water-treatment facilities, as well as surface-water reservoirs like rivers, lakes, or human-made water bodies that can serve as sources of drinking water. Answers to the questions will vary depending upon your community.

Part C: Modeling the Water Cycle

1. – 4. Use **Blackline Master Water Resources 1.1, The Water Cycle** to make an overhead of the diagram of the water cycle on page R148. Review the cycle with students briefly before they get started.

Circle around the room to answer any student questions that arise. Student posters will vary; a sample finished product is shown on the next page.

Chapter 3

Activity I   Sources of Water in the World and in Your Community

Reservoirs:
   oceans
   atmosphere
   clouds
   glaciers
   soil moisture
   ground water
   lakes
   rivers
   vegetation

Processes:
   evaporation from the ocean surface
   precipitation onto the ocean surface
   evaporation from the land surface
   precipitation onto the land surface
   precipitation onto glaciers
   condensation to form clouds
   melting of glaciers
   calving of glaciers
   surface runoff into rivers
   surface runoff into lakes
   infiltration of surface water
   ground-water flow
   river flow
   transpiration from plants
   uptake of water by plant roots

Definitions:
calving: Some glaciers end in the ocean. As the glacier ice moves forward into the ocean water, it breaks away from the glacier in huge masses, to float away as icebergs, which gradually melt.

ground water: Some of the liquid water at the Earth's surface moves downward through porous Earth materials until it reaches a zone where the material is saturated with water. This water flows slowly beneath the Earth's surface until it reaches rivers, lakes, or the ocean.

infiltration: Some of the rain that falls on the Earth's surface sinks directly into the soil.

soil moisture: Water, in the form of liquid, vapor and/or ice, resides in the Earth's soil layer. It is the water that remains in the soil after rainfall moves downward toward the ground water zone. Soil moisture is available for plants. What is not used by plants gradually moves back up to the soil surface, where it evaporates into the atmosphere.

surface runoff: Some of the rain that falls on the Earth's surface flows across the land surface, eventually reaching a stream, a river, a lake, or the ocean.

transpiration: Water taken up by the roots of plants is delivered to the leaves. Some of this water is used to make new plant tissue, and some is emitted from the leaves in the form of water vapor, by a process called transpiration.

3. Cut out all of the boxes and circles with a pair of scissors.

4. On a piece of poster board, draw a horizontal line lengthwise across the middle of the sheet. This represents the Earth's surface in a vertical cross-section view.

  a) On the left half of the poster board, draw some mountains to represent a continent.

  b) On the right half of the poster board, draw a small island or a sailboat to represent a large ocean.

  c) Using the simplified water-cycle diagram as a model, place the boxes and circles that you have created where you think they belong. Tape them to the poster board with small pieces of removable tape. Using removable tape allows you to adjust the positions of the boxes and circles as needed.

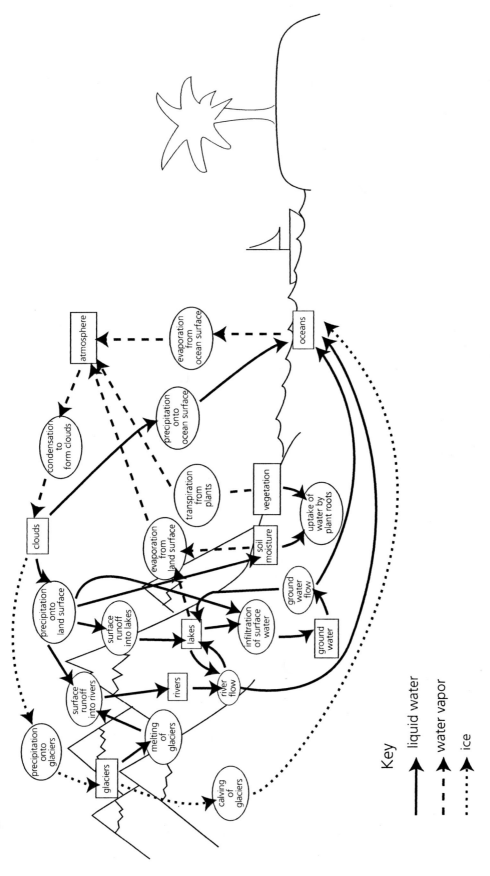

Key

liquid water

water vapor

ice

Chapter 3

Earth's Natural Resources Water Resources

d) With colored pencils, draw arrows between the various boxes and circles to show the movement or transport of water from place to place on or near the Earth's surface. Remember that a circle (representing a process) will be located in the middle of an arrow between two different boxes (representing two different storage places). Think about whether the movement or transport is in the form of liquid water, water vapor, or ice (or two or three of these at the same time). Use blue for liquid water, red for water vapor, and green for ice.

5. Once everyone in your small group has agreed upon the best version of the water cycle, compare your results with those of the other groups. Answer the following questions:

   a) Is there net movement of water vapor from the oceans to the continents, or from the continents to the oceans? Explain your answer.

   b) Is there net movement of liquid water from the oceans to the continents, or from the continents to the oceans? Explain your answer.

   c) How does the nature of the water cycle vary with the seasons?

Part D: Collecting Rainwater in Your Community

1. Suppose your community needed to consider ways of adding to their supply of fresh water. One alternative is to collect and store rain and melted snow that runs off from the roofs of buildings.

a) In your notebook, make a list of what you would need to know to calculate an estimate of the total amount of annual runoff from your school roof.

b) Develop a plan for making this estimate. Compare your plan with the plans of others. Refine it on the basis of your discussions.

c) Calculate the total runoff for one year's precipitation.

d) How does this amount compare to the amount of water that your city or county uses each year? What percentage is it?

e) What modifications to school grounds and buildings might need to be made to accommodate a runoff collection plan?

f) How could a plan for runoff collection benefit communities seeking to conserve available water?

g) How much money would this water save your school? Find out how much your community charges for each 1000 gallons of water. Assume that untreated roof water is worth half as much as treated city water. Calculate the value of the water that could be collected from the roof of your school for one year.

h) What could be some concerns or problems with using roof runoff as a source of water?

i) Calculate how much water falls as precipitation on the total area of your community in an average year, and compare that volume with your community's annual water use.

R 150

5. Once everyone has finished the exercise, pin the posters to the wall and have students compare them.

    a) Student answers may vary depending upon how they constructed their water cycle, but they should note that because there is a net excess of evaporation over precipitation over the oceans, there must be a net transport of water vapor from the oceans to the continents, rather than the other way around. Make sure that the students realize that water vapor nonetheless moves in both directions: from the oceans to the continents at some times and in some places, and from the continents to the oceans at other times and in other places.

    b) Students should note that there is net movement of liquid water from the continents to the oceans, as manifested by the flow of water from streams and ground water into the oceans. Water does not flow out of the oceans onto land.

    c) Student responses will vary. Some possible seasonal variations in the water cycle might include:
    - an increase in evaporation and transport of water vapor from the ocean to the continents during the summer months (relative to the spring, for example).
    - lower amounts of transpiration from plants during the winter relative to the spring and summer.
    - lower amounts of freshwater runoff from the continents to the oceans during the winter months (relative to the spring, for example) when water is stored on the continents in the form of snow and ice.
    - an increase in freshwater input to the oceans from the continents during the spring thaw.

Assessment Opportunities

Ask your students to prepare a sketch to illustrate the major elements of the water cycle. Have students exchange their sketches and provide constructive feedback.

Part D: Collecting Rainwater in Your Community

1. This question requires critical thinking and mathematics. You might want to review the calculation of the volume of a cube and area of a rectangle with your students. If the roof of your school has a complex architectural design, you may wish to provide some values to your students. You can model the calculation of the surface area and the volume of runoff for your students, or you can let them figure this out on their own.

    a) Students will need to determine the vertical projection area of the roof (i.e. looking straight down on the roof from above), in square feet or square meters, and local annual precipitation in inches per year. You can obtain the precipitation data online. If you have blueprints of the school, students should be able to use the dimensions from the prints to calculate the area of the roof. If you do not have blueprints, students will need to measure the

Chapter 3

outside dimensions of the building (length and width) to calculate the area. If your building is not a simple rectangle, students will need to divide the roof into two or more sub-rectangles and add their areas. The pitch of the roof is not relevant here; what's important is the area of the roof in vertical projection.

b) Student plans will vary. You may want to take students outside to look at the roof in order to plan appropriately. Plans should not involve climbing or scaling the building! One example would be to measure the length of one brick in the building, and then count the number of bricks to determine the width of the roof.

c) The total runoff from the roof will be equal to the vertical projection area of the roof times the inches of precipitation received locally in one year. Units of length might be a problem for your students. The area of the roof will probably be in square feet or square meters, and the precipitation data will probably be in inches. To compute the volume of water, the area and the vertical depth of precipitation have to be in the same units. You might need to help students convert the values to achieve this. The easiest would be to measure the roof area in square feet and convert precipitation from inches to feet by dividing by 12.

d) The volume of water collected from the roof will almost certainly be extremely small compared to the volume of water used per year by the community. To find the percentage, divide the volume of water collected on the roof by the community's water use and then multiply the result by 100.

e) Student answers will vary. One possibility is the adjustment of downspouts to collect water in containers.

f) Student answers will vary. Water collected from a roof could reduce the demands on the county water supply. One obvious way would be to use the collected runoff from the roof to water lawns and plants around the school grounds.

g) Students should divide the annual volume of roof water by 1000 gallons and multiply this by half of the cost per 1000 gallons. The actual cost savings will be low.

Teaching Tip

Discuss with students why they are asked to assume that the untreated roof water is worth half as much as treated city water. This will help them to begin thinking about the cost of water treatment.

h) Student responses will vary. Most likely, students will recognize that the water collected from the roof will need to be treated before it can be used for drinking. Another problem would involve how to add the roof runoff into the existing system for water distribution.

i) Answers will be dependent upon your community. A community with a large area, a small population, and abundant rainfall will probably have more precipitation than water use. A community with a small area, a large population, and little rainfall will almost certainly have less precipitation than water use.

Assessment Tools

EarthComm Notebook-Entry Checklist
Use this checklist as a quick guide for student self-assessment and/or an opportunity to quickly score student work. Add further criteria specific to your classroom needs or to this particular investigation.

Investigate Notebook-Entry Evaluation Sheet
Point out the criteria listed on this evaluation sheet that are relevant to this particular investigation. Encourage students to internalize the criteria by making them part of your assessment conversations as you circulate around the classroom.

Chapter 3

Reflecting on the Activity and the Challenge

You have learned that of all the water on Earth, less than 1% is available for drinking water. You also explored the nature of the water cycle, and how Earth system science can be used to characterize it. You have also learned where your community's drinking water comes from. This will help you determine whether or not your community has enough water for the additional development. It also provides you with some ideas about where extra water could be obtained.

Digging Deeper

THE WATER CYCLE

Water is the only common substance that exists at the Earth's surface as a solid, a liquid, and a gas. Water is present at or near the surface everywhere on Earth. In many places, the presence of water is obvious, in the form of lakes, rivers, glaciers, and the ocean. Even in the driest of deserts, however, it rains now and then, and although the humidity there is usually very low, there is at least some water vapor in the air.

Figure I Nearly two percent of Earth's water exists as ice.

Reflecting on the Activity and the Challenge

Discuss students' responses to the questions posed in the investigation. What did they find surprising or striking about the distribution of water on Earth? Point out that very little of the water on Earth (less than 1%) is readily available as fresh water and that most of our liquid fresh water is underground. Remind students that by examining the places where water is stored, they have begun to explore the water cycle. This would also be a good opportunity to find out what students think about where their drinking water comes from, and what they found most interesting in the local data.

Digging Deeper

As students read the **Digging Deeper** section, the relevance of the concepts investigated in **Activity 1** will become clearer to them. Assign the reading for homework, along with the questions in **Check Your Understanding** if desired.

Assessment Opportunity

Reword or restructure the questions in **Check Your Understanding** for a brief quiz. Use the quiz (or a class discussion of the questions in the textbook) to assess your students' understanding of the main ideas in the reading and the activity.

Assessment Tool

Check Your Understanding Notebook-Entry Evaluation Sheet
Use this sheet to evaluate the extent to which students understand the key concepts explored in **Activity 1** and explained in **Digging Deeper,** and to evaluate the students' clarity of expression.

Teaching Tip

Although only about 2% of water on the Earth's surface exists as ice, ice accounts for nearly 75% of all fresh water available. Most of the ice on the Earth's surface is found as glaciers. Antarctica holds 70% of Earth's fresh water, and 91% of Earth's ice.

Chapter 3

Earth's Natural Resources Water Resources

Geo Words

water cycle (or hydrologic cycle): the constant circulation of water from the sea, through the atmosphere, to the land, and its eventual return to the atmosphere by way of transpiration and evaporation from the land and evaporation from the sea.

closed system: a system in which material moves from place to place but is not gained or lost from the system.

evaporation: the change of state of matter from a liquid to a gas. Heat is absorbed.

precipitation: water that falls to the surface from the atmosphere as rain, snow, hail, or sleet.

Water, in the form of liquid, solid, or vapor, is in a continuous state of change and movement. Water resides in many different kinds of places, and it takes many different kinds of pathways in its movement. The combination of all of these different movements is called the **water cycle**, or the **hydrologic cycle**.

The water cycle is called a cycle because the Earth's surface water forms a **closed system**. In a closed system, material moves from place to place within the system but is not gained or lost from the system. The Earth's surface water is actually not exactly a closed system, because relatively small amounts are gained or lost from the system. Some water is buried with sediments and becomes locked away deep in the Earth for geologically long times. Volcanoes release water vapor contained in the molten rock that feeds the volcanoes. Nonetheless, these gains and losses are very small compared to the volume of water in the Earth's surface water cycle.

Evaporation and **precipitation** are the major processes in the water cycle. The balance between evaporation and precipitation varies from place to place and from time to time. It's known, however, that there is more evaporation than precipitation over the surface of the Earth's oceans, and there is more precipitation than evaporation over the surface of the Earth's continents. That fact has a very important implication. There is net movement of water vapor from the oceans to the continents, and net movement of liquid (and solid) water from the continents to the oceans.

The oceans cover about three-quarters of the Earth. Ocean water is constantly evaporating into the atmosphere. If enough water vapor is present in the air, and if the air is cooled sufficiently, the water vapor condenses to form tiny droplets of liquid water. If these droplets are close to the ground, they form fog. (See *Figure 2*.) If they form at higher altitudes, by rising air currents, they form clouds.

Figure 2 What part of the water cycle does the fog over the San Francisco Bay illustrate?

Teaching Tip

Have students discuss the question raised in the caption of *Figure 2* on page R152. The fog over the San Francisco Bay illustrates evaporation of water from the bay into the atmosphere. Because there is sufficient water vapor present in the air, and the air is sufficiently cooled, the water vapor condenses to form droplets of liquid.

Teaching Tip

Use **Blackline Master Water Resources 1.2, Ground-water Flow** to make an overhead of *Figure 4* on page R153. Incorporate this overhead into a discussion or lecture about the flow of ground water. Encourage students to think about how the elements of this diagram can be placed into the larger scope of the hydrologic cycle. Make sure that students are comfortable with the terms used on this diagram, because they will appear throughout the chapter.

Chapter 3

All of the solid or liquid water that falls to Earth from clouds is called precipitation. Snow, sleet, and hail are solid forms of precipitation. Rain and drizzle are liquid forms of precipitation.

When rain falls on the Earth's surface, or snow melts, several things can happen to the water. Some evaporates back into the atmosphere. Some water flows downhill on the surface, under the pull of gravity, and collects in streams and rivers. This flowing water is called **surface runoff**. Most rivers empty their water into the oceans. Some rivers, however, end in closed basins on land. Death Valley and the Great Salt Lake are examples of such closed basins.

Geo Words

surface runoff: the part of the water that travels over the ground surface without passing beneath the surface.

ground water: the part of the subsurface water that is in the zone of saturation, including underground streams.

Figure 3 Some of the water that falls to the Earth's surface collects in streams.

Some precipitation soaks into the ground rather than evaporating or running off. The water moves slowly downward, percolating through the open pore spaces of porous soil and rock material. Eventually the water reaches a zone where all of the pore spaces are filled with water. This water is called **ground water**. Some water, called soil moisture, remains behind in the surface layer of soil. (See *Figure 4*.)

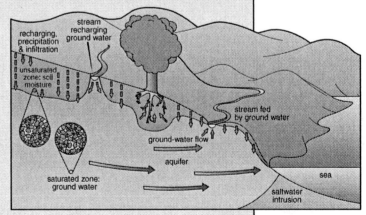

Figure 4 Schematic diagram of ground-water flow.

R 153

Blackline Master Water Resources 1.2
Ground-water Flow

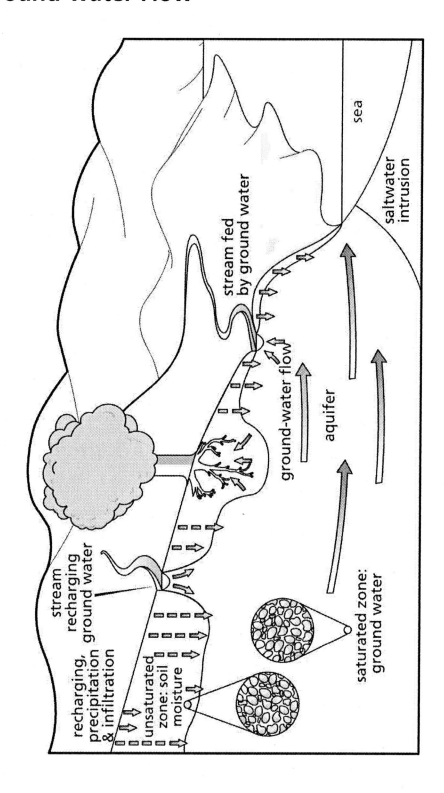

Earth's Natural Resources Water Resources

Geo Words

transpiration: the process by which water absorbed by plants, usually through the roots, is emitted into the atmosphere from the plant surface in the form of water vapor.

reservoir: a place in the Earth system that holds water.

flux: the movement of water from one reservoir to another.

The roots of plants absorb some of the water that soaks into the soil. This water travels upward through the stem and branches of the plant into the leaves and is released into the atmosphere as a vapor in a process called **transpiration**.

Figure 5 Plants like these broadleaf trees play an important part in the water cycle.

It has been estimated that each year about 36,000 km³ of water flows from the surface of the continents into the oceans. That represents the excess of precipitation over evaporation on the continents. This water carries sediment particles and dissolved minerals into the ocean. The sediment particles come to rest on the ocean bottom. When seawater evaporates, it leaves the dissolved materials behind. Over geologic time, this process has gradually made the oceans as salty as they are now.

In Earth systems science, the water cycle is viewed as a flow of matter and energy. Each place that holds water is called a **reservoir**. The rate at which water flows from one reservoir to another in a given time is called a **flux**. Energy is required to make water flow from one reservoir to another. On average, the total amount of water in all reservoirs combined is nearly constant. Although the data table in the investigation suggests that reservoirs have a constant amount of water in them, this is not the case. The amount of water stored in any one reservoir varies over time. For example, in many areas there may be more water in the form of ground water during the spring (when precipitation is high, and water use and evaporation is low) than in the summer (when precipitation is low, and evaporation and water use are high).

Check Your Understanding

1. In your own words, describe the water cycle.

2. Explain why the water cycle can be viewed as a closed system.

3. Describe three "paths" of the water cycle that precipitation can follow once it reaches the surface of the Earth.

Teaching Tip

Although transpiration occurs at all exposed parts of plants, the most transpiration occurs through tiny openings in leaves called stomata. For scientists looking at the water budget of a region, a crucial piece of missing data has been how much water is released to the atmosphere through plant transpiration. Previously, the processes of evaporation and transpiration (jointly referred to as evapotranspiration), could only be accurately measured using expensive instruments on uniform vegetation for areas no larger than a field. For larger areas calculations of evapotranspiration were made indirectly by simply quantifying the rest of the terms in the water budget and attributing the "missing water" to evapotranspiration processes.

More recently, however, a few scientists are honing a way to calculate this elusive water-budget parameter for areas as large as a regional watershed (also known as a drainage basin). These scientists hypothesize that evapotranspiration can be measured by looking at vegetative productivity, or "greenness". Greenness can be measured over large areas by satellite, and because the production of green matter involves using a specific amount of water, evapotranspiration can be calculated for large areas. The results for this work thus far show great promise for actually quantifying (to some extent) this previously elusive water-budget term. This is important because water-budget calculations are essential for watershed management, which is becoming increasingly important in today's society.

Check Your Understanding

1. The water cycle is the combination of all of the movements of water—in liquid, solid, or vapor form—among the various storage areas on and near the Earth's surface. Specifics of student responses will vary. Encourage them to use a diagram to answer the question. The main part of the water cycle involves evaporation over the oceans, movement of water as vapor to the continents, precipitation as rain over the continents, and flow in rivers back to the oceans—but there are many other pathways in the water cycle as well.

2. The water cycle can be viewed as a closed system because water moves from place to place in the system but is not gained or lost from the system (except for relatively small amounts that are buried for geologically long times along with sediments or released as water vapor from volcanoes).

3. When precipitation falls to the Earth, it can be evaporated back into the atmosphere, it can flow downhill and collect in rivers and lakes (as runoff), or it can soak into the ground to become ground water.

Chapter 3

Understanding and Applying What You Have Learned

1. Describe the different conditions on Earth under which water is a solid, a liquid, or a gas.

2. If 36,000 km^3 of water flow from the surface of the Earth into the oceans each year, how many cubic kilometers of water evaporate from the oceans each year?

3. The data table in the investigation defines the hydrosphere somewhat differently than the image shown in the front of the book. Explain any differences you note between the data table and the image.

4. Using the same techniques you used in Part C of the investigation, construct a model, with boxes and arrows, that shows sources, volumes, and transport paths and processes of water in your community. Use the water data that you collected in the investigation for your county or municipality. You will need to know the sources and volumes of water, and the uses.

5. How can the destruction of large areas of rain forest affect climate?

Preparing for the Chapter Challenge

Using what you have learned about water in your community and the water cycle, write a short essay on water resources in your community that can be included in your final **Chapter Challenge**. Include the following information in your essay:

• A complete description of the sources of drinking water for your community.

• An analysis of the stage or component of the water cycle that is most critical to your community.
• At least two new water sources that you have identified to provide for increased demands.
• A description of how this information can be applied to the questions raised by the **Chapter Challenge**.

Inquiring Further

1. **Volcanic eruptions and the water cycle**

   Volcanic eruptions release large amounts of water vapor. After you have done some research, construct a water cycle diagram that shows the pathways and flow of water and water vapor in a volcanic region.

2. **Drinking water from the ocean**

   How can the ocean be used as a source of drinking water? Visit the

EarthComm web site to do some research on the technology currently available.

3. **Dating water**

   Investigate how chlorofluorocarbons (CFCs) released from aerosols and tritium (hydrogen 3) released during global nuclear testing in the 1950s and early 1960s are used to determine the age of ground water.

R 155

Understanding and Applying What You Have Learned

1. Water exists as a solid in the form of solid ice in glaciers; as a solid in the form of ice crystals in the atmosphere at high altitudes; as a liquid in rivers, lakes, oceans, ground water, rain, and clouds (in the form of water droplets); and as vapor (a gas) in soil or rock and in the atmosphere.

2. Students may respond that the flow of water from the oceans (via evaporation) must exactly balance the amount of water entering the ocean each year as runoff (36,000 km³). That, however, would not be correct bookkeeping. Most of the water that evaporates from the ocean surface falls directly back to the oceans as precipitation rather than on the continents. Another complicating factor is that some of the surface runoff to the oceans from the continents is derived from evaporation on the continents and precipitation of that water back onto the continents. You might suggest to students that they imagine erecting an invisible screen from the shorelines of all the continents to the top of the atmosphere. Then, they can keep track of all of the water that passes in both directions across that screen. As an average over a long period of time, the mass of water passing the screen from the oceans to the continents is equal to the mass of water passing the screen from the continents to the oceans.

3. The main difference is that the data table lists water in the atmosphere and water in glaciers as being components of the hydrosphere, but the image on page Rxii of the student text places these components as belonging to separate systems (the atmosphere and cryosphere). This is a good opportunity for you to point out that the different Earth systems are all intricately linked.

4. Box models of the local water supply will vary. Models should label kinds of water reservoirs (e.g., ground water, surface water) and should label the fluxes of water to and from the various reservoirs. The reservoirs should include consumer end uses (think of them as sinks) but should not be limited to them, from the standpoint that water moves within the community in both natural ways and human-use ways.

5. Elimination of rain forests changes:
    -   transpiration
    -   evaporation from the surface
    -   surface runoff
    -   infiltration into the soil surface
    -   the balance between heating and cooling of the surface

Preparing for the Chapter Challenge

This section gives students an opportunity to apply what they have learned to the **Chapter Challenge**. They can work on this as a homework assignment or during class time within groups. This activity is designed as an introduction to water in the

Chapter 3

community and to the water cycle. Students should address all of the bulleted points, and the information they produce here can then be incorporated as an introduction to their Chapter Report.

In consideration of the rubric that you are using to assess this chapter challenge, you may want to assess how this **Chapter Challenge** relates to the overall goals and assessment of the Chapter Report. From the **Assessment Rubric for Chapter Report on Water Resources** on pages 504 and 505, the criterion of "How your town could increase its water supply (if there is not enough water)" is relevant to this **Chapter Challenge.**

Inquiring Further

1. Volcanic eruptions and the water cycle
This question requires students to focus on a different aspect of the water cycle and to think about how the water that is incorporated in rocks reenters the atmosphere. Some of the water vapor that is emitted from volcanoes is ground water in shallow sediment and rock adjacent to the volcano. Some is pore water in deeper rocks. Some is water that became dissolved in the magma itself when the magma was originally formed by melting of hydrous rock deep in the Earth.

2. Drinking water from the ocean
Desalination costs much more than the common methods of obtaining fresh water and treating it to the point where it is of drinking-water quality. In some areas of the world, where fresh water is very scarce, desalination is less expensive than obtaining needed supplies of fresh water from great distances away. Students can learn more about the processes involved in desalination and the countries that use desalination by visiting the *EarthComm* web site.

3. Dating water
By the age of water is meant the time elapsed since the water most recently condensed from vapor form to liquid form. By this definition, deep ground water is usually much older than water in rivers and lakes. In the last 50 years, human activities have released chemical substances and isotopes, like chlorofluorocarbons (CFCs) and tritium (a radioactive isotope of hydrogen, ^3H) into the atmosphere. These substances are incorporated into precipitation and take part in the water cycle. Their presence in ground water indicates that the ground water has been recharged within the last ~ 50 years. The age of ground water can be used to determine recharge rates in an aquifer, predict contamination potential, and estimate how long it will take to remove contaminants from the ground. CFCs can also be used to determine sources of contamination in ground-water systems.

NOTES

Chapter 3

ACTIVITY 2— HOW DOES YOUR COMMUNITY MAINTAIN ITS WATER SUPPLY?

Background Information

Ground Water as a Water Supply

Rainwater or snowmelt that infiltrate into the ground rather than flowing as surface runoff may continue percolating downward to become part of the body of ground water at some depth below the surface. Alternatively, it may remain in the soil, as what is called soil moisture.

Soil moisture may be taken up by plant roots, or it may gradually work its way back to the land surface. Some of the liquid soil water rises slowly back to the surface by capillary rise, a process similar to wicking. Some of the soil moisture that clings to soil particles evaporates, building up the humidity in the pore spaces between the particles, and then gradually diffuses upward toward the less humid air above the land surface.

Water that continues percolating downward eventually reaches the ground-water table. The ground-water table is an underground boundary surface between unsaturated soil and rock material above (called the vadose zone) and saturated soil and rock material below (called the phreatic zone). The ground-water table usually is located at the surface in places where there are rivers and lakes, and also in wetlands. In some areas, it can lie many tens or hundreds of meters below the surface.

Below the ground-water table, all of the pores and cracks in the soil material are filled with water. These pores and cracks are almost always very small, less than a millimeter, except in three kinds of materials:

- very coarse gravels
- certain kinds of volcanic rocks with large lava tubes left over from when the lava flowed and solidified
- various kinds of subterranean cavities and passageways caused by dissolution, including caverns, in regions underlain by limestones

In any given region, the ground-water table fluctuates up and down in its position, depending upon drainage sideways to rivers and lakes as well as replenishment from above by rainfall.

The ratio of open space to total volume in some region of the subsurface material is called its porosity. Porosity is a direct measure of the water-holding capacity of the material. Porosity ranges from almost 30% for very porous materials, like sand and gravel, to almost zero for tightly cemented or crystalline rock. Another very important characteristic of the porous material is its permeability, which is a measure of how easy it is to force water through the material by applying pressure. Materials with abundant large and connected pore spaces, like sand and gravel, have very high permeabilities, whereas materials like clay or solid rock have very low permeabilities. Fine sediments like clay have high porosity but low permeability, because the pore spaces, although abundant, are so small that it is difficult to force the water to move.

Bodies of subsurface material, whether sediment or rock, that have high porosity and permeability and can be tapped by wells for water supply are called aquifers. Materials that are not good suppliers of ground water because of either low porosity or low permeability, or both, are called aquicludes.

The concept of an aquifer is a relative and practical matter: there is no quantitative measure that tells you whether or not the subsurface material is a good aquifer. If you can extract adequate water from some underground body of material, you can call it an aquifer! The best aquifers consist of loose and porous sand and gravel, although fractured bedrock can also form good aquifers.

The ground-water table has topography that generally mimics the topography of the land surface, but its topography is more subdued than that of the land surface. Think about a hill that lies between two streams. If you drilled three wells down to the ground-water table, one at the crest of the hill and the others on each of the sides of the hill, you would ordinarily find that the elevation of the ground-water table is higher under the crest of the hill than on the hillsides. The depth to the ground-water table would be less, however, in the wells on the hillsides than in the well at the hilltop. The ground-water table actually comes to the surface at streams and rivers (streams flow not only because they receive surface runoff but also because they are being fed by the ground-water table).

Just as with surface water, you can think of ground water in terms of watersheds separated by ground-water divides. Usually, the positions of the ground-water divides are about the same as those of the surface-water divides. In some cases, however, they may be very different because of the particular arrangement of aquifers and aquicludes in the subsurface.

Because ground water must move through small pores, it flows very slowly. Ground-water speeds of a meter per day are considered high. Speeds as low as a meter per year are common. In general, the smaller the pore spaces between the grains, the slower the ground water flows.

Most aquifers, called unconfined aquifers, have a free connection upward to the surface. The water in these aquifers is replenished by downward percolation of surface water from directly above. Downward flow of surface water is called recharge. Some aquifers, however, are isolated from the surface by an aquiclude; layers of fine clay are especially effective aquicludes. These confined aquifers can't be recharged from directly above. The recharge area for a confined aquifer may be located far away, tens or even hundreds of kilometers.

Users of ground water would find life easier if the height of the ground-water table in the aquifer that supplied their well always stayed the same. However, variations in rainfall affect the height of the ground-water table. During dry spells, ground water keeps flowing toward lakes, rivers, and (in coastal areas) the ocean, but there is no recharge to keep the water table from falling. During rainy spells, on the other hand, recharge more than makes up for ground-water drainage, and the water table rises. In many areas, the difference in the height of the water table between dry spells and rainy spells can be many meters.

Before wells are drilled into a ground-water aquifer, discharge and recharge of the ground-water system is in long-term balance: the amount of water entering the system is balanced with the amount leaving, and the height of the ground-water table stays the same, on average. Pumping of ground water for water supply upsets this balance. As water is pumped from a well, the water table drops in the vicinity of the well. If water is being pumped from several wells in the same area, the ground-water table is depressed over the entire area. In fact, in many areas where ground water has been used for a long time, the land surface has subsided (sunk down) because so much water has been withdrawn.

Chapter 3

When ground water is pumped from a field of wells, one of two things can result:

- the ground-water table is lowered, but it becomes stabilized at some lower height as additional ground water flows into the area from surrounding areas
- inflow of ground water from surrounding areas is not great enough to stabilize the ground-water table, and the level keeps falling indefinitely.

In the latter case, the ground water can no longer be thought of as a renewable resource because it is being consumed faster than it can be replenished. In a real sense, the ground water is then being mined. The imbalance between recharge and discharge in some areas of the United States has lowered the ground-water table by as much as a 100 m over the last 50 years. Eventually, the expense of pumping from great depths becomes prohibitive (or, if the wells are not deep enough, they run dry!), and new sources must be developed.

Surface Water as a Water Supply

In most rivers, water flow is too variable from season to season to be a reliable, direct source of water supply. Most surface-water supplies are from large lakes or artificial reservoirs, which fluctuate less from season to season or from year to year.

Dams are beneficial in providing a water source and controlling floods, but they have disadvantages as well. Reservoirs behind dams displace wildlife and people, and they cover cropland. Dams can disrupt the natural migration of fish. The sediment carried by a river is deposited in the reservoir behind the dam. Over time, the reservoir slowly fills up with sediment, leaving less room for water.

Most cities and towns that use rivers for their water supply take out only a small fraction of the river discharge. In some places, however, the demand for river water is so great that the natural discharge of the river is greatly decreased. For example, so much of the discharge of the Colorado River in the Southwest is used for water supply that by the time the river reaches its mouth, in the Gulf of California, the flow is only a trickle!

Not all of the water in rivers comes from surface runoff. Rain also infiltrates the Earth to become ground water. Below a certain depth below the land surface, the rock and/or sediment is saturated with ground water. Flowing rivers, as well as lakes and springs, are places where the ground-water table comes out to the land surface. At times of high river flow, some of the river water feeds the ground water. At times of low river flow, however, the ground water feeds the river. This is why most large rivers flow even during long droughts. On average, ground water supplies as much as 40% of the water that flows in streams and rivers. During times of drought the figure is much greater, and during times of flood the figure is much smaller.

More Information – on the Web

Visit the *EarthComm* web site www.agiweb.org/earthcomm to access a variety of links to web sites that will help you deepen your understanding of content and prepare you to teach this activity. Many of the sites also contain images that you can download.

Goals and Assessment

Clarify that the goals indicate what students should understand and be able to do as a result of the activity. Make sure students understand that Chapter Assessments are based upon these goals.

Goal	Location in Activity	Assessment Opportunity
Create and manipulate physical models of surface-water and ground-water supply systems.	**Investigate** Part A and B	Models are set up correctly; observations are accurate.
Explain how a change in one part of the water supply system creates changes in other parts of the system.	**Investigate** Part A and B **Digging Deeper; Understanding and Applying What You Have Learned** Questions 1 – 3	Explanations are reasonable, and are based on observations made in experiments.
Understand the main ways that a community can increase its water supply.	**Investigate** Part C **Digging Deeper; Check Your Understanding** Question 1 **Understanding and Applying What You Have Learned** Questions 1 – 3, and 5	Answers to questions are reasonable, and reflect local water-quality data.
Compare and contrast surface-water systems and ground-water systems.	**Digging Deeper; Understanding and Applying What You Have Learned** Questions 1 – 2, and 4	Student responses are reasonable, and closely match those given in Teacher's Edition.
Analyze the water-supply system in your community.	**Investigate** Part C **Understanding and Applying What You Have Learned** Questions 3 and 5	Answers to questions are reasonable, and reflect local water-quality data.

Chapter 3

Earth's Natural Resources Water Resources

Activity 2

How Does Your Community Maintain Its Water Supply?

Goals

In this activity you will:

• Create and manipulate physical models of surface-water and ground-water supply systems.

• Explain how a change in one part of the water-supply system creates changes in other parts of the system.

• Understand the main ways that a community can increase its water supply.

• Compare and contrast surface-water systems and ground-water systems.

• Analyze the water-supply system in your community.

Think about It

You and others in your community expect to always have the water you need when you turn on the faucet.

• How can any community guarantee that there will be enough fresh water available to meet the needs for personal, recreational, business, industrial, and agricultural use?

• Suppose your region were experiencing a severe and prolonged drought. Would ground water or surface water be a more reliable water supply? Explain your response.

What do you think? Record your ideas in your *EarthComm* notebook. Be prepared to discuss your responses with your small group and the class.

Activity Overview

Students set up a simple experiment to model how surface-water reservoirs respond to increases and decreases in precipitation and outflow. Students then model the effect of increased supply and withdrawal on the flow of ground water. Finally, they look at their community's water-quality report to understand the community's water supply and associated costs. **Digging Deeper** reviews surface water and ground water as the two main sources of water supply. The reading also looks at other methods for supplying water for human use, including aqueducts, desalination, and conservation methods.

Preparation and Materials Needed

Part A

You may want to ask students to bring in 3-lb coffee cans and 750-mL (or 1-L) plastic soda bottles before class. To save time and increase safety for students, prepare the coffee cans and soda bottles in advance by cutting holes in both, as shown in the diagram on page R157 in the student text. Insert the tubing into the holes and seal with modeling clay to prevent leakage.

Part B

To save time and in the interest of student safety, it is recommended that you cut the 1-cm hole in the bottom of the dish tub. Cover the hole from the inside with a couple of pieces of duct tape and use a sharp pencil to punch a small hole in the tape. Fill the tub with sand.

Part C

For **Part C** of the investigation, you will need to obtain and make copies of your community's water-quality report. The *EarthComm* web site contains links that will help you to acquire this information.

> ## Teaching Tip
>
> You may wish to set up one or two stations each for the surface-water and ground-water models. Students can then cycle through the stations, thereby reducing the amount of materials needed. Also, setting up the materials before class will save time. Students can work on **Part C** of the investigation in between stations, so that there is no lag time.

Materials

Part A
- 3-lb coffee can
- Clear, 750-mL (or 1-L) soda bottle with the top cut off
- Shallow container
- Coffee filter
- Sand
- Three flexible plastic tubes 1/4 – 1/2 in. diameter (each 25 cm long)
- Modeling clay
- C-clamp

Part B
- Dish tub with a 1-cm hole cut into the bottom
- Duct tape
- Sharp pencil
- Sand
- Two pieces of rigid tubing (1/2 – 1 in. diameter)
- Cheesecloth
- Chopstick with evenly spaced lines drawn on it (to serve as a ruler)
- Drinking straw
- Turkey baster

Part C
- Community water-quality report*

Think about It

Student Conceptions

After completing **Activity 1**, students should have a good idea of how much water their community uses in a year. They will need to think about how this water is distributed.

Student responses to the second **Think about It** question will vary. There is no clear answer, because it depends upon the extent of the drought. You may wish to set some parameters. For example, what if the dry spell lasted a week, a month, a year, or a decade? Which supply would be most reliable? Students are likely to say that the ground-water supply would be affected least, because it is not exposed directly to the ground surface. Ask them to consider how they would modify the answer to the first **Think about It** question in light of the hypothetical drought.

*The *EarthComm* web site provides suggestions for obtaining these resources.

Answer for the Teacher Only

Again, there are no clear answers to these questions: the answers will depend upon your individual community. Ground-water systems must be replenished by water that infiltrates from above. If a drought covers an extensive region, ground-water levels will eventually fall. Surface-water reservoirs may be stable in times of drought if they are replenished from sources far removed from the drought region. If a region experiences a prolonged drought, communities would have to ration water by:

• eliminating luxury uses like washing cars
• regulating the watering of lawns
• discouraging wasteful uses (allowing water to run from the cold-water tap until it's cold or from the hot-water tap until it's hot, running partial loads of clothes in washing machine, etc.)

Assessment Tool

EarthComm **Think about It Evaluation Sheet**
Use this evaluation sheet to help students understand and internalize the basic expectations for the warm-up activity.

Investigate

Part A: Modeling a Surface-Water Supply System

1. Set up a coffee can, a clear, 750 mL soda bottle (top cut off), and a shallow container as shown in the diagram.

2. Line the inside of the bottle with coffee filter paper. (This keeps sand from entering the plastic tubing.)

3. Pour sand into the filter until the bottle is one-third full of sand.

4. One kind of surface-water reservoir is one in which a river is dammed to make a lake that a community can use for its water supply. Assume that the ideal situation is a balance between a flow of water into the reservoir and the flow out of the reservoir. Water flows out of the reservoir not only for community use but also to return to the river downstream of the community. Water flow for both of these uses can be controlled by a system of valves.

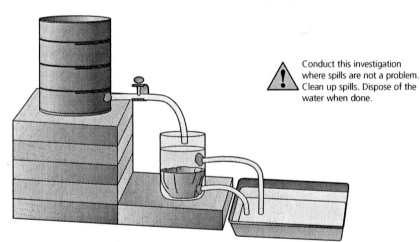

 Conduct this investigation where spills are not a problem. Clean up spills. Dispose of the water when done.

Identify the following parts of the water supply system on the model that you set up:

- precipitation, river flow, reservoir inflow;

- surface-water reservoir;

- ground-water reservoir;

- ground-water withdrawal/outflow (for the community);

- surface-water withdrawal/outflow (for the community);

- total community consumption.

a) Draw a box model (systems diagram) of the physical model, with reservoirs and flows. Label the reservoirs and flows.

Investigate

Part A: Modeling a Surface-Water Supply System

1. - 3. If you have prepared the materials before class, students can begin with **Step 4**. If you are having the students set up the experiment on their own, circulate around the room to make sure that the setup is being done correctly. Students can use the illustration on page R157 in the text as a guide.

4. Precipitation is represented by the addition of water to the large upper supply area (the coffee can). Both river flow and reservoir inflow are represented by the flow in the upper tube leading from the upper supply vessel. The surface-water reservoir is represented by the water in the upper part of the lower vessel (the cut-off soda bottle). The ground-water reservoir is represented by the water in the pore spaces of the sand in the lower part of the lower vessel. Ground-water withdrawal for the community is represented by the flow through the lower tube leading from the lower vessel. Surface-water withdrawal for the community is represented by the flow through the upper tube leading from the lower vessel.

Note: The description of the inflow and outflow from a surface reservoir is more detailed than the setup of the model shown in the diagram, in the following respects:

- The flow through the upper tube leading from the lower vessel (the cut-off soda bottle) represents both withdrawal for community use and flow from the reservoir back into the river downstream of the reservoir.
- The flow through the lower tube leading from the lower vessel represents both ground-water withdrawal for community use and ground-water flow to areas downslope from the reservoir.

a) Make sure that the students understand the principle behind the technique of constructing and depicting a box model.

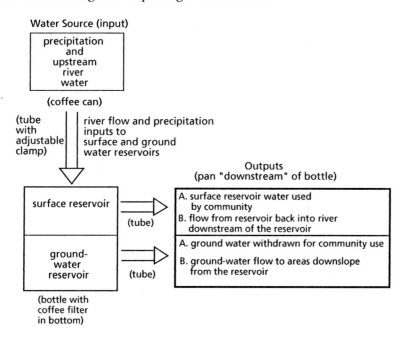

Chapter 3

Earth's Natural Resources Water Resources

5. Release the clamp slightly to allow water to flow into the bottle. Observe what happens with the flow of water out of the bottle. Adjust the flow until you have a steady-state system (in which reservoir inflow equals reservoir outflow).

a) Record what you see.

b) How did the two water-supply reservoirs respond to the increase in precipitation?

6. Release the clamp completely to allow a full flow of water from the can. Observe what this does to the flow of water out of the bottle.

a) Record what you see.

b) How did the water-supply reservoirs respond to a decrease in precipitation?

7. Tighten the clamp so that the flow comes to a trickle. Again, allow the system to reach a steady state. Observe what this does to the flow of water out of the bottle.

a) Record what you see.

b) Which reservoir lasted longer during the "drought"?

8. Finally, tighten the clamp to stop the flow of water.

a) Observe and record the results.

9. In your model, you did not control community consumption (water flowed out of the reservoirs despite a decrease in recharge into the reservoirs).

a) Where would you place additional clamps on the model to keep water in both reservoirs during a drought?

b) What would these clamps represent in the real world?

10. What would you add to the model to show the return of water from the community to the system shown in your model and to other parts of the system not shown (for example, communities downstream)?

a) Make a diagram to show your ideas.

11. How could you use a graduated cylinder and a stopwatch so that your model allows you to measure the rate of flow of water from one system reservoir to another (flux)?

a) Record your ideas in your *EarthComm* notebook.

b) Add flux arrows to the box model you drew.

Part B: Modeling Ground-Water Supply

1. Your teacher will cut a hole about 1 cm across in a dish tub, on the bottom and very close to where the surface curves upward to one of the end walls of the tub.

2. Cover the hole from the inside with two or three pieces of duct tape.

With a sharp pencil, punch a circular hole in the duct tape, from the inside of the tub, to be about 3 mm in diameter.

3. Pour sand evenly into the dish tub, making a small depression (a low spot) near the center of the sand surface. Make the depression about 5 cm deep and about 10 cm across, with gently sloping sides. The depression represents a shallow lake

 Wear safety goggles. Clean up spills immediately. Dispose of water when you are done.

R 158

5. a) Students should record their observations in their notebooks. They should be able to follow the flow of the water from precipitation into the surface reservoir. From here, water flows both out of the surface reservoir (for consumption) and into the ground-water reservoir. Water from the ground-water reservoir then flows out (for consumption). Encourage students to use sketches to illustrate what they observe.

   b) The increase in precipitation will cause the water levels in both of the water-supply reservoirs to rise (although this may be clearly visible only for the surface-water supply in this model.

6. a) Students should observe that the water level in the surface water-supply reservoir rises. Both the flow of surface water out of the lower vessel and the flow of ground water out of the lower vessel will increase.

   b) A decrease in precipitation would cause the water level in the water-supply reservoirs to fall.

7. a) Once the system has reached a steady state, students should observe that the surface water-supply reservoir quickly becomes depleted until level drops to that of the upper outflow tube, which is the outflow for the surface water-supply reservoir. After that, if the flow from the ground water is sufficient, the level of surface water will continue to drop until it is completely depleted. If the ground water outflow is less than the input, the system will reach a steady state with the water-supply reservoirs at a new equilibrium level and the total outflow equal to that of the total input.

   b) The ground-water reservoir will last longer during the drought.

8. a) Students should observe that the surface-water reservoir will empty more quickly than the ground-water reservoir.

9. a) Clamps would be placed over the tubes leaving the surface-water and ground-water reservoirs to keep water levels in both reservoirs up during a drought.

   b) In the real world, these clamps would represent dams, or control over well-water discharge.

10. Students would have to rebuild the model so that both of the tubes leading from the reservoir vessel branch into two tubes, with one set of branches leading to community use and the other to the natural river and ground-water systems. Water used in the community and then returned to the natural environment could then be represented by additional siphon tubes leading from the lower pan and back into downstream rivers and downslope ground water.

11. Catch a small volume of water from one of the tubes for a time interval measured with the stopwatch.

Part B: Modeling Ground-Water Supply

1-2. If you have prepared the materials before class, students can begin with **Step 3.** If you are having the students set up the experiment on their own, circulate around the room to make sure that the setup is being done correctly. Students can use the illustration on page R159 in the text as a guide.

NOTES

Chapter 3

or wetland. The diagram illustrates the arrangement in the tub.

4. Place the tub where water can drain harmlessly from the hole in the bottom.

5. Cover one end of a piece of the rigid tubing with a layer of cheesecloth and tape it to the sides of the tubing with duct tape.

   Bury the tubing to stand vertically in the sand, about 10 cm away from the center of the depression. The lower end of the tubing should be about 5 cm from the bottom of the tub. This represents a town water supply well.

6. Cover one end of another piece of rigid tubing with a layer of cheesecloth and tape the cheesecloth to the sides of the tubing.

   Bury the tubing to stand vertically in the sand, about 10 cm away from the center of the depression. The lower end of the tubing should be only about 0.5 cm from the bottom of the tub. This represents a monitoring well.

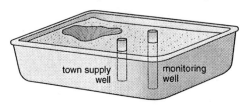

7. Each member of your group needs to assume one of the following roles: a water supplier, a consumer, a well monitor, and a data recorder.

8. Establish an equilibrium ground-water system with a water table high enough for there to be water in the lake. Follow these steps:

   • The water supplier starts supplying water at a rate of 1000 mL/min. The water supplier pours a steady supply of water onto the sand at the "upstream" end of the tub (the end opposite the hole in the bottom of the tub). The best way to do this is to measure out a constant volume of water in a measuring cup and pour that water into the tub every 15 or 30 seconds.

   • The data recorder can then record the rate of supply. This represents ground-water recharge from infiltration of precipitation into the ground surface.

   • The well monitor will monitor the test well by dipping a chopstick ruler gently into the well until it rests on the bottom and then reading the position of the water surface in the well.

9. Continue supplying water and monitoring the well until the level of the water table stops changing.

   If there is no surface water in the depression, increase the rate of water supply and monitor the test well until the height of the ground-water table stops changing.

   Repeat the process until there is water in the depression.

3. - 6. Students should make the depression in the sand and place the wells as shown on page R159 in the text.

7. Make sure that each member of the group knows what their respective role is and what they are expected to do. Each member of the group needs to participate.

8. - 9. Students should record how much time and water is required to maintain a constant ground-water table (meaning that the level of water in the monitor well remains constant). They should also note how much time and water is required to fill the depression with water.

Chapter 3

Earth's Natural Resources Water Resources

10. Now it's time to start extracting water from the supply well.

    To do this, tape a drinking straw to the end of the turkey baster and seal the joint with a wrap of duct tape.

11. The consumer should then squeeze the bulb of the baster, insert the straw into the bottom of the well, and **slowly** release the bulb.

12. Keep track of the volume of water extracted per minute. Try to keep the rate of extraction constant, minute after minute.

    The data recorder should record the rate of withdrawal.

13. Meanwhile, the well monitor should monitor the test well and record the height of the water table every minute.

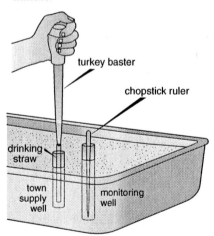

turkey baster

chopstick ruler

drinking straw

town supply well

monitoring well

14. Depending on the rate of extraction relative to the rate of recharge, three outcomes of this part of the investigation are possible:

 Mark the baster and chopstick ruler as laboratory equipment. Dispose of the drinking straw.

- A new equilibrium height of the water table in the test monitoring well is established, and some water remains in the depression.

- A new equilibrium height of the water table in the test monitoring well is established, but the depression has lost its water.

- The supply well runs dry; that is, the rate of withdrawal cannot be maintained.

  Try to reproduce each of these outcomes, by adjusting the rate of withdrawal from relatively low to relatively high.

15. Use the results of this part of the investigation to answer the following questions:

   a) What is the difference between a renewable ground-water supply and a nonrenewable ground-water supply? What do you think it means to "mine" ground water?

   b) Under what conditions in this part of the investigation were you using a renewable supply of ground water? Under what conditions might you have been using a nonrenewable supply of ground water?

   c) Do you think it would have been less likely to lower the water table if you had obtained a given rate of supply by pumping water from two nearby supply wells instead of only one? Explain your answer.

 R 160

10-13. If students seem confused by the instructions, they can use the illustration on page R160 in the student text to help them understand how to extract water from the supply well. Remind them to record the rate of water withdrawal.

14. Students should attempt to produce each of the listed outcomes and should record the rate of withdrawal required to attain each. They will find that the first scenario (where a new equilibrium of the ground-water table is established and water remains in the depression) requires the lowest rate of withdrawal, followed by the second scenario. Finally, the highest rate of withdrawal will be required to get the supply well to run dry. Remind students to record the height of the ground-water table for each instance.

15. a) If the rate of withdrawal is less than the rate at which ground water flows toward the well to replenish the water removed, then the water level in the well reaches a steady level, and the ground-water supply can be considered to be renewable. If, on the other hand, the rate of withdrawal exceeds the inflow, the water level in the well continues to fall. Then the ground-water supply can be considered to be nonrenewable. If the water is extracted faster than it can be replenished, causing the water level to fall, the ground water can be considered to be mined.

    b) If the water level in the wells reached a steady state in which the water level was well above the bottom of the tub, the supply could be considered to be renewable. If the water level in the wells dropped to the bottom of the tub, the supply could be considered to be nonrenewable.

    c) If the two wells had been very close together, the water level in the wells would have been about the same as if the rate of pumping had been from only one well. If the wells had been sufficiently far apart, so that the water was drawn from a much larger area, the water level in the two wells would have been higher than if the rate of pumping had been from only one well.

Assessment Tool

EarthComm Notebook-Entry Checklist
Use this checklist as a quick guide for student self-assessment and/or an opportunity to quickly score student work. Add further criteria specific to your classroom needs or to this particular investigation.

Investigate Notebook-Entry Evaluation Sheet
Point out the criteria listed on this evaluation sheet that are relevant to this particular investigation. Encourage students to internalize the criteria by making them part of your assessment conversations as you circulate around the classroom.

Teaching Tip

After students have completed the investigation, have them set the tubs of sand aside for **Activity 5**.

Chapter 3

Part C: Water Supply in Your Community

1. You will be able to find the answers to some of these questions in your water supplier's consumer confidence report, also called a water quality report. For some of the other questions, appoint one student in your group or class to call your community's water supplier to ask. The *EarthComm* web site also provides links.

   a) What kind of municipal water system does your community use? What factors do you think affected their selection of the system?

   b) How does your community evaluate potential development projects in light of water supply?

   c) What contingency plans does your community have in order to deal with potential droughts?

   d) Every freshwater system requires maintenance and upkeep. What does this mean for your community, and what does it cost?

   e) Determine the average cost per gallon of water for your community and three others in or near your county. List the factors that account for the differences in the cost per gallon. Does the cost vary? Why?

   f) Determine the gallons-per-person use for your own community and two neighboring communities. List the factors that could account for the differences.

Reflecting on the Activity and the Challenge

You learned that many factors affect the operation of a community's water supply system. Some of these factors are drought, increased demand, and the needs of other communities. You also learned that a system consists of many parts and that a change in one part of the system affects other parts of the system. You also took a closer look at the impact of consumption on ground-water supplies. You then analyzed the water-supply system in your community. Understanding these factors will help you evaluate whether your community's water-supply system is capable of handling the extra demands of the three new developments proposed in the **Chapter Challenge**.

Part C: Water Supply in Your Community

1. Answers to these questions will be dependent upon your community.

Reflecting on the Activity and the Challenge

Have students read this brief passage and share their thoughts about the main point of **Activity 2** in their own words. Hold a class discussion about how this investigation relates to what they are being asked to accomplish in the **Chapter Challenge.** Students should be gaining a greater understanding of where water used by humans comes from and how the activities of humans can affect these supplies.

Chapter 3

Earth's Natural Resources Water Resources

Digging Deeper

WATER SUPPLIES

Sources of Water Supplies

In colonial times in the United States, most people took water from rivers or dug their own wells. Today, cities and towns need reliable and safe supplies of water for their citizens. Water must be collected, stored, and treated. There must be enough water to see a community through times of drought and times of increased water use. The two main water sources are surface waters, from rivers and lakes, and ground water.

Figure 1 Surface water from freshwater lakes is a valuable source for human use.

Figure 2 Large pumps are used to supply water to a community.

Digging Deeper

Assign the reading for homework, along with the questions in **Check Your Understanding** if desired.

Assessment Opportunity

Reword or restructure the questions in **Check Your Understanding** for a brief quiz. Use the quiz (or a class discussion of the questions in the textbook) to assess your students' understanding of the main ideas in the reading and the activity. A few sample questions are provided below:

Question: Name three ways that a community can increase its water supply.
Answer: Answers will vary, but should include three of the following:
- withdrawing water from ground-water aquifers
- withdrawing water directly from nearby rivers or lakes
- building dams to create reservoirs to store runoff
- transporting water from a distant area by means of aqueducts
- converting salt water to fresh water
- improving the efficiency of water use through water conservation

Question: Name and define the two main zones of the ground-water system. Use an illustration to show where each is located.
Answer: The saturated zone is a subsurface zone in which all of the pore spaces are filled with water. It is located below the water table. The unsaturated zone is the zone above the saturated zone (but below the ground surface), where not all of the pore spaces are filled with water. Student sketches should resemble the illustration in *Figure 4* on page R164.

Question: What is the difference between porosity and permeability?
Answer: Porosity is a measure of the percentage of pores in a material; permeability is a measure of how easy it is to force water to flow through a porous material.

Teaching Tip

Figure 1 (the body of water on page R162) shows the Quabbin Reservoir, which was built in the 1930s to supply water to eastern Massachusetts and the Boston metropolitan area. It took the Swift River seven years to fill this 412-billion-gallon reservoir, which is the largest in the world devoted solely to water supply. At an elevation of 530 feet, the Quabbin Reservoir is high enough that its water can be delivered to Boston (which is near sea level) under the force of gravity. This, however, is not always the case. *Figure 2* shows a modern well field constructed to supply water for the city of Spokane, Washington. In this case, energy must be expended to pump ground water out of the Rathdrum Prairie Aquifer.

Chapter 3

There are six ways to increase the supply of water to a community:

- withdrawing water from ground-water aquifers;
- withdrawing water directly from nearby rivers or lakes;
- building dams to create reservoirs to store runoff;
- improving the efficiency of water use through water conservation;
- transporting water from a distant area by means of aqueducts;
- converting salt water to fresh water.

Many of these choices affect other communities. For example, if one town takes water from a river, it decreases the amount of water available to towns downstream.

Surface Water

In most rivers, water flow is too variable from season to season to be a reliable direct source of water supply. Most surface-water supplies are from large lakes of artificial reservoirs, which fluctuate less from season to season or from year to year.

Dams, such as the one shown in *Figure 3*, are beneficial in providing a water source and controlling floods, but they have disadvantages as well. Reservoirs behind dams displace wildlife and people, and they cover cropland. Dams can disrupt the natural migration of fish. The sediment carried by a river is deposited in the reservoir behind the dam. Over time, the reservoir slowly fills up with sediment, leaving less room for water.

Ground Water

Ground water for water supply is pumped from porous material below the surface. Three concepts are important in understanding ground-water supplies. **Porosity** is a measure of the percentage of pores (open spaces) in a material. **Permeability** is a measure of how easy it is to force water to flow through a porous material. An **aquifer** is any body of sediment or rock that has sufficient size and sufficiently high porosity and permeability to provide an adequate supply of water from wells. The best aquifers consist of loose and porous sand and gravel, although fractured bedrock can also form good aquifers.

Geo Words

porosity: a measure of the percentage of pores (open spaces) in a material.

permeability: a measure of how easy it is to force water to flow through a porous material.

aquifer: any body of sediment or rock that has sufficient size and sufficiently high porosity and permeability to provide an adequate supply of water from wells.

Figure 3 A large reservoir was created from the Colorado River at the Glen Canyon Dam, Arizona.

Teaching Tip

Glen Canyon Dam, shown in *Figure 3* on page R163, was constructed to provide water storage in the Upper Colorado River Basin. The dam created Lake Powell, the second largest man-made reservoir in the United States It took 17 years for the lake to fill to a level of 3700 feet above sea level. Of the water in Lake Powell, 85% is used for agricultural production. Before construction of the dam, water temperatures in the river ranged from 80°F to near freezing. Since the dam was constructed, water temperatures in the river average 46°F year-round. The construction of the dam has also resulted in a change in plant and animal species in the area.

The allocation of water resources from the Colorado River is a subject of great importance and concern for many people in the western United States. Many states, including California, Nevada, Arizona, Colorado, New Mexico, Utah, and Wyoming, depend on water from the Colorado River Basin. As the population in the arid southwestern United States continues to grow, the need for managing of water resources will become increasingly important.

Teaching Tip

Use **Blackline Master Water Resources 2.1, The Main Two Zones of the Ground-water System** to make an overhead of *Figure 4* on page R164. Incorporate this overhead into a lecture about the different zones of the ground-water system.

Teaching Tip

Use **Blackline Master Water Resources 2.2, The Water Table** to produce an overhead of *Figure 5* on page R164. Incorporate this overhead into a lecture or discussion about the flow of ground water. Students should understand that gravity pulls water downhill. They should realize that where the water table meets the ground surface, a surface-water reservoir (river or lake) is found. In *Figure 5*, ground water is recharging surface water. Ask students to consider how the illustration could be altered so that surface water is recharging the ground-water system.

Chapter 3

Earth's Natural Resources Water Resources

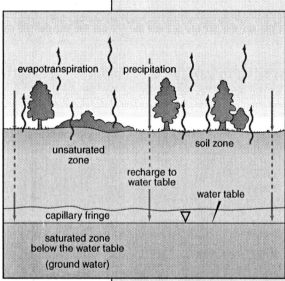

Figure 4 The main two zones of the ground-water system—saturated and unsaturated. The water table marks the upper surface of the saturated zone.

Down to a certain depth below the surface, the pores in the sediment and rock are mostly filled with air, except when water is percolating downward after a heavy rain. This is called the unsaturated zone. Eventually the downward-moving water reaches a zone called the **saturated zone**, where all of the pores are filled with water. The top of the saturated zone is called the **water table**. These zones are illustrated in *Figure 4*. The water table can be located at the surface in places next to rivers and lakes, and also in wetlands. In some areas it can lie many tens of meters below the surface.

Because ground water must move through small pores it flows very slowly. Ground water speeds of a meter per day are considered high. Speeds as low as a meter per year are common. In general, the smaller the pore spaces between the grains, the slower the ground water flows. Ground water moves from areas where the water table is relatively high to areas where it is relative low. *Figure 5* shows the flow of ground water in a typical landscape.

Geo Words

saturated zone: the zone, beneath the water table where all of the pores are filled with water.

water table: the surface between the saturated zone and the unsaturated zone (zone of aeration).

evapotranspiration: loss of water from a land area through transpiration of plants and evaporation from the soil and surface water.

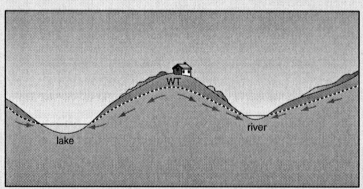

Figure 5 The water table (WT) is shown as a dashed line. The blue arrows show the direction of ground-water flow.

Blackline Master Water Resources 2.1

The Main Two Zones of the Ground-water System

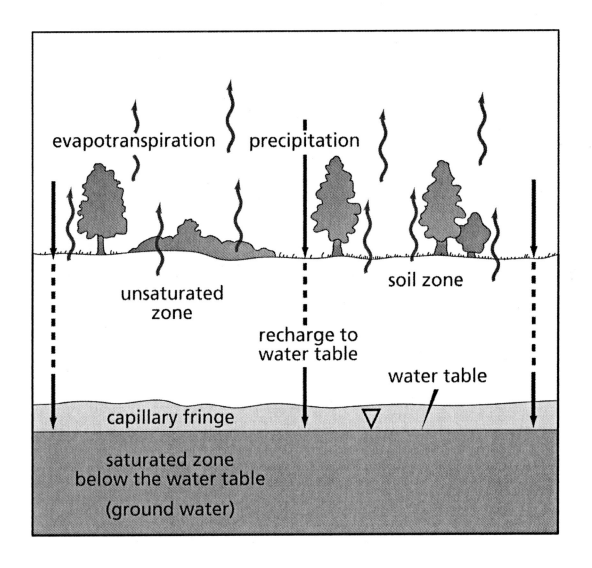

In rural areas, most homes have their own wells to tap ground water. Many towns, and even some small cities, obtain part or all of their water supply from several large wells that pump ground water from large aquifers.

Most aquifers, called **unconfined aquifers**, have a free connection upward to the surface. The water in these aquifers is replenished by downward percolation of surface water from directly above. The sand in your model was an unconfined aquifer. Addition of new water to an aquifer by downward flow of surface water is called **recharge**.

Some aquifers are isolated from the surface by an overlying layer of very impermeable material, called an **aquiclude**. Layers of fine clay are especially effective aquicludes. Confined aquifers cannot be recharged from directly above. The recharge area for a confined aquifer may be located far away, tens or even hundreds of kilometers.

In many areas where ground water has been used for a long time, the land surface has subsided (sunk down) because so much water has been withdrawn. For example, at Edwards Air Force Base in California, the land has subsided more than two meters, damaging some of the runways once used by the Space Shuttle.

Figure 6 The signs show the approximate position of the land surface in 1925, 1955, and 1977 in the San Joaquin Valley, California. Switching to surface water slowed the rate of land subsidence, but new ground-water pumping during a drought from 1987 to 1992 caused further subsidence.

Geo Words

unconfined aquifer: an aquifer that has a free connection upward to the surface.

recharge: addition of new water to an aquifer by downward flow of surface water.

aquiclude: a body of rock that will absorb water slowly, but will not transmit it fast enough to supply a well.

Blackline Master Water Resources 2.2
The Water Table

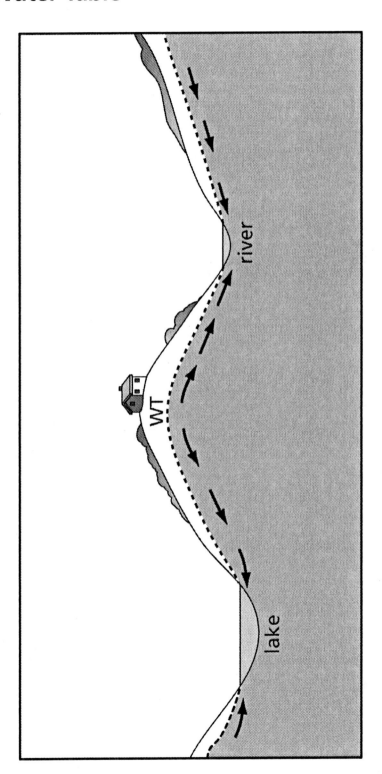

Earth's Natural Resources Water Resources

Geo Words

aqueduct: a system of large surface pipes and channels used to transport water.

desalination: the process of removing dissolved salts from sea water in order to make it potable.

Check Your Understanding

1. Describe six possible ways to increase the supply of water to a community.

2. What are the advantages of building a dam to provide a source of surface water? What are the disadvantages?

3. Explain how porosity and permeability of Earth materials are important when considering ground water as a water source.

4. Why is desalination of ocean water not a practical source of water supply?

Aqueducts

In areas where water use is greater than local supplies, as in southern California, water must be brought in from distant areas where water is abundant. About two-thirds of California's precipitation falls in the north, but about two-thirds of the population lives in the south. A system of reservoirs in northern California supplies southern California with water that is transported through a system of large surface pipes and channels, called **aqueducts**, like the one shown in *Figure 7*. Over long distances the water in the aqueducts flows downhill under gravity, but in some places enormous pumping plants must raise the water up over hills and mountains.

Figure 7 This aqueduct in southern California carries much needed water over long distances.

Desalination

Converting sea water to fresh water, a process called **desalination**, is still too expensive to be widely used. It is used in some arid countries that are located near the ocean. Israel and Saudi Arabia obtain much of their water from desalination. As new and less expensive techniques for desalination become developed, the process will be a more and more important source of fresh water in many countries.

Water Conservation

Conservation is a great way for a community to stretch its water supplies further without having to develop new supplies. Although the water supply stays the same, if the community uses less, then there will be more water for new development. You will learn more about conservation in later activities.

Teaching Tip

The photograph in *Figure 6* on page R165 is a striking example of how withdrawing too much water from the ground can result in land subsidence.

Teaching Tip

The photo in *Figure 7* on page R166 shows one of the many aqueducts in California that transports water over large distances to the densely populated areas of southern California. The demands on the water resources of California are so great that only one major river system in the state (the Smith River in northern California) still flows freely and naturally, without a single dam, for its entire length.

Check Your Understanding

1. Six ways to increase the supply of water in a community include:
    - withdrawing water from ground-water aquifers
    - withdrawing water directly from nearby rivers or lakes
    - building dams to create reservoirs to store runoff
    - transporting water from a distant area by means of aqueducts
    - converting salt water to fresh water
    - improving the efficiency of water use through water conservation

2. Dams are beneficial in providing a water source, controlling floods, generating electricity, and providing areas for recreation. However, dams can also displace people and wildlife and disrupt the natural migration of fish. Infilling behind the dam can eventually result in less room for water storage.

3. Porosity is a measure of the percentage of pores in a material; permeability is a measure of how easy it is to force water to flow through a porous material. For an aquifer to be a viable source of water, it must have a sufficiently high porosity (i.e., be able to hold large amounts of water) and high permeability (i.e., water should be easily removed and recharged).

4. Desalination is rather expensive to be widely used. New and less expensive techniques for desalination are under development.

Assessment Tool

Check Your Understanding Notebook-Entry Evaluation Sheet
Use this sheet to evaluate the extent to which students understand the key concepts explored in **Activity 2** and explained in **Digging Deeper**, and to evaluate the students' clarity of expression.

Understanding and Applying What You Have Learned

1. Look at the diagram of the surface-water reservoir.

   a) Write down the major parts of the water-supply system. Also, think about the parts that may not be shown.

   b) Which part of the system do people have the least control over? Why?

   c) Which part of the system do people have the greatest control over? Why?

   d) How might the volume of water entering the reservoir from the river vary from season to season?

   e) How might the amount of fresh water that a community needs vary from season to season?

   f) Assume there is a severe drought. How might the system respond in order to guarantee the amount of fresh water needed by the community?

   g) What other factors can be manipulated in times of drought to make the system operate as efficiently as possible?

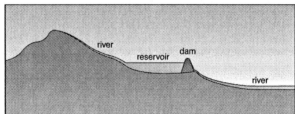

A surface-water reservoir.

2. Look at the diagram of the confined aquifer.

   a) What factors would affect the level of fresh water in this aquifer?

   b) What will happen if the volume of water entering at the recharge area decreases and the demand for fresh water from the wells remains the same? When would this situation be likely to occur? Which wells would be affected first?

   c) Assume that the ground water enters the aquifer at a constant rate. What would the community need to know before agreeing to a development project that would result in a significant increase in water use?

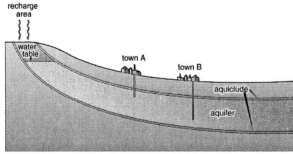

A confined aquifer.

Understanding and Applying What You Have Learned

1. **a)** The major parts of the water-supply system include:
    - river or rivers supplying the water
    - reservoir
    - dam that creates the reservoir
    - river downstream of the reservoir

   Not shown is the pipeline or aqueduct that carries the water from the reservoir to the pumping station, or distribution facility, that treats the water and distributes it to the community.

   b) People have the least control over the discharge (volume rate of flow) in the river that feeds the reservoir, because that is dependent upon rainfall in the watershed of the river.

   c) People have the greatest control over the rate of withdrawal of water from the reservoir, because that is entirely under human control, up to some maximum beyond which the water level in the reservoir would decrease to nothing.

   d) Volume of water entering the reservoir varies from season to season in three ways:
    - precipitation in the watershed
    - human water use
    - evaporation from the water surface of the reservoir

   e) Demand would probably be greatest in the warm season, for watering of lawns and plants and filling of swimming pools.

   f) As the water level in the reservoir is drawn down, there would be increased inflow of water from the surrounding still-high ground-water table into the bed of the reservoir.

   g) Leaks in piping can be repaired; demand can be reduced by public information campaigns or increased water prices.

2. **a)** The level is determined by the relative magnitude of withdrawal from the lower part of the aquifer and the recharge at the upper end of the aquifer.

   b) The water table would fall. This is likely to occur during prolonged drought. The well for Town A would be affected before the well for Town B.

   c) The community would need to know that the increased rate of withdrawal, leading to a lowered water table, would not be so great as to draw down the water level in the well to the point where withdrawal at the required rate would not be possible.

Earth's Natural Resources Water Resources

3. a) Suppose that the capacity of the ground-water supply is sufficient to meet the increased needs of a new development in your community. The development is approved. How will this affect other communities using the same ground water in the future?

   b) Should this have any effect on your community's decision to approve the development? Explain why or why not.

4. Compare and contrast the advantages and disadvantages of using ground water versus surface water.

5. Suppose your community's water supplier proposed building a dam on a nearby river to increase water supply. Make a list of the pros and cons of damming a river in your community.

Preparing for the Chapter Challenge

Using what you have learned in this activity and your community's water quality report, determine how much water the proposed developments will use per year. As a class, decide on the number of houses that will be included in the area.

Does your community have an adequate water supply to meet this demand now and over the next 25 years?

Inquiring Further

1. **First American reservoirs**

   Visit the *EarthComm* web site to do some research on the first reservoirs constructed in America. When and where were they built? How did they work?

2. **First American ground-water systems**

   Visit the *EarthComm* web site to conduct some research on the first ground-water systems developed in America. When and where were they built? How did they work?

3. **Water supplies in desert cities**

   Pick a large desert city, such as Las Vegas or Phoenix. Conduct some research to find out how these cities maintain a water supply.

4. **Water needs of a golf course**

   How much water does it take to run a golf course? You have already established an estimate for your community's water use. Pick several communities in parts of the United States with a very different climate from yours. Find out how much water an average golf course uses in those communities. What accounts for the differences in water use?

3. a) The water level in wells in the other communities would fall.

   b) If there were danger of unacceptable lowering of the water table in the wells of surrounding community, the given community would need to worry about litigation to prevent the planned increase in pumping.

4. Surface water is a good choice as a water source because it is easily accessible. However, the water flow in some natural surface-water reservoirs, like rivers, is too variable from season to season to serve as a reliable, direct source of water. Ground-water supplies are more consistent than natural surface-water supplies. However, if withdrawal from ground-water supplies is too great, they can be overdrawn to the point of being dry. Ground-water supplies also take a longer time to replenish than surface-water supplies.

5. Pros:
    - greater water supply and potential source of power generation

   Cons:
    - cost of new construction
    - flooding of land
    - displacement of preexisting buildings, cemeteries, etc.
    - cost of additional water treatment

Preparing for the Chapter Challenge

Encourage students to draw upon what they have learned, and to make accurate estimates and gather the necessary data for their final reports. You will need to decide, with your students, how large the residential development will be. Student responses to the questions posed in this section are directly tied to the **Chapter Challenge**.

Review the rubric that you are using to assess the **Chapter Challenge** with your students. From the **Assessment Rubric for Chapter Report on Water Resources** on pages 504 and 505, the following criteria would be relevant for assessment:
   - how much water the residential development requires
   - whether or not your town has enough extra water to supply the three new developments

Chapter 3

Inquiring Further

1. First American reservoirs

Students' findings will vary. The establishment of man-made surface reservoirs in America generally didn't become common until the 19th century, even though dam construction existed well before that time. Early storage dams were used to supply water for locks, but by the latter part of the 19th century, the development of urban areas and the need for irrigation increased the need for established water supplies. This made storage dams and surface reservoirs economically desirable. The advent of hydroelectric power near the end of the 19th century greatly stimulated development of storage dams and surface reservoirs in the 20th century. Students can visit the *EarthComm* web site to get them started on their research.

2. First American ground-water systems

The use of ground water has existed for thousands of years. Its earliest use took advantage of natural springs, but the advent of wells greatly expanded the utilization of this resource. In America, ground-water systems were used to supply water to urban areas and for use in irrigation beginning in the 19th century. The development of ground-water resources advanced dramatically in 1909 with the introduction of the deep-well turbine pump. Its adaptability and operating characteristics placed efficient water wells within the economic reach of a great many people. Students can visit the *EarthComm* web site to get them started on their research.

3. Water supplies in desert cities

Student findings will vary. Las Vegas is an excellent example of the needs and problems associated with the development of adequate water supplies for metropolitan areas located in desert regions. Students can visit the *EarthComm* web site to help them get started with their research.

4. Water needs of a golf course

Answers to questions will vary depending upon the area in which you live. Direct students to the *EarthComm* web site to view information about different golf courses around the United States.

NOTES

ACTIVITY 3— USING AND CONSERVING WATER

Background Information

Water Resources

Human society needs water for many purposes: drinking, cleaning, industrial and commercial processes, irrigation, cooling for electrical power generation, and recreation. Historically, people have clustered in localities and developed communities where water resources were readily available. Many communities, large and small, use water from a lake or a river. Some communities build reservoirs (which are just artificial lakes) for storing water before it is distributed to homes, businesses, or farms. In rural areas, and also in small- to medium-sized communities, ground water pumped from wells is the main source of water.

In most situations, a community's water source is derived from the local watershed area. Whether the water is taken from a natural or artificial lake, a river, spring, or well, the water has collected from precipitation over the watershed area. Some large cities, like New York, Boston, and Los Angeles, have a very large population and are not near abundant sources of fresh water. They need so much water that water must be conveyed. But in all of these situations, water being used by humans comes from a particular watershed area.

In most regions of the United States, rainfall varies considerably from season to season. For example, in some locations, riverbeds are dry during much of the year and flow only during storms. In other places, rivers flow all year

long, even during months with little rain. One advantage of using ground water, the other major source of water supply in the United States, is that it is not as vulnerable to short-term changes in precipitation as surface water.

In many areas of the world, communities are located in arid regions but are not greatly distant from the ocean. In some such areas, the cost of water is so great that it is economical to develop desalination plants along the shoreline. The salt is removed from the seawater by various processes to make it usable as fresh water. There have even been proposals to tow gigantic icebergs from the high-latitude oceans to be parked near coastal cities for water supply!

Water Supplies

In colonial times in the United States, most people took water from rivers or dug their own wells. With the great increase in population and industry since those early days, cities and towns now require far more sophisticated water-supply systems to ensure reliable and safe water for their citizens. Water must be collected, stored, and treated. There must be enough water to see a community through times of drought and times of increased water use. The two main water sources are surface waters from rivers and lakes, and ground water.

There are several ways to increase the supply of water to a community:
- withdraw water from ground-water aquifers
- withdraw water directly from nearby rivers or lakes
- build dams to create reservoirs to store runoff
- transport water from a distant area by means of aqueducts
- convert salt water to fresh water

Note that many of these choices affect other communities. For example, if one town takes water from a river, it decreases the amount of water available to towns downstream. There is also one very important way to make more water available to a community without actually increasing the supply: improving the efficiency of water use through water conservation.

Water Use

Uses of fresh water are highly varied:
- domestic use
- public use for street cleaning, maintenance of parkland, fire-fighting, public buildings
- irrigation for agriculture
- industrial and commercial use
- cooling for electrical power generation processes

Irrigation is the largest single use of fresh water in the United States More than a hundred billion gallons are used to irrigate crops every day, on average. Public and domestic use of water accounts for only about 10% of fresh-water use.

Consumptive water is water that is returned to the atmosphere as water vapor. The water that evaporates before soaking into the ground when you sprinkle your lawn and garden, for example, is consumptive water. Nonconsumptive water is water that is returned, in liquid form, to the natural environment after use. Most of the total water use is said to be nonconsumptive. Some nonconsumptive water goes into rivers, lakes, or the ocean, and some soaks into the ground. In your home, for example, almost all the water you use (except for what you drink and then return to the atmosphere by breathing and perspiring) is nonconsumptive. Think about what happens to water used in bathing, clothes washing, and toilet flushing: it either goes into either a municipal sewer system or into a home septic system.

More Information – on the Web
Visit the *EarthComm* web site www.agiweb.org/earthcomm to access a variety of links to web sites that will help you deepen your understanding of content and prepare you to teach this activity. Many of the sites also contain images that you can download.

Chapter 3

Goals and Assessment

Clarify that the goals indicate what students should understand and be able to do as a result of the activity. Make sure students understand that Chapter Assessments are based upon these goals.

Goal	Location in Activity	Assessment Opportunity
Design a method for determining how much fresh water your school uses every day.	**Investigate** Part A	Plan is reasonable, and is carried out effectively.
Analyze statistics on water use for your county and an adjacent county, and explain any differences in per-person water use.	**Investigate** Part B	Data table is filled in completely and correctly. Calculations of per capita water use for both counties are done correctly.
Differentiate between uses of fresh water in the United States and identify these uses as consumptive and nonconsumptive.	**Digging Deeper; Check Your Understanding** Question 4 **Understanding and Applying What You Have Learned** Question 1	Responses are reasonable, and closely match those given in Teacher's Edition.
Explore methods of water conservation and make suggestions about which methods would most benefit your school and your community.	**Investigate** Parts A and B **Digging Deeper; Check Your Understanding** Questions 2 – 3	Answers to questions are reasonable. Proposed methods of water conservation for the school and community are logical.

NOTES

Chapter 3

Activity 3   Using and Conserving Water

Goals

In this activity you will:

- Design a method for determining how much fresh water your school uses every day.

- Analyze statistics on water use for your county and an adjacent county and explain any differences in per-person water use between your county and an adjacent county.

- Differentiate between uses of fresh water in the United States and identify these uses as consumptive or nonconsumptive.

- Explore methods of water conservation and make suggestions about which methods would benefit your school and home.

Think about It

You use water for everyday activities, probably without considering how much water you are actually using.

- How much water does it take to brush your teeth?
- Devise a plan to arrive at an accurate estimate of how much water you typically use to do this.

What do you think? Record your ideas in your *EarthComm* notebook. Be prepared to discuss your responses with your small group and the class.

R 169

EarthComm

Activity Overview

Students list all the ways that water is used in their school. They devise and carry out a plan to determine how much water each of these uses requires in a day. Using their findings, students make suggestions for ways the school could conserve water. They examine data on water use for their county and a neighboring county and construct a bar graph to illustrate the data. Students then consider methods of water conservation for both areas. **Digging Deeper** reviews the ways that fresh water is used in the United States and explains the difference between consumptive and nonconsumptive water use. The reading also introduces common methods of water conservation.

Preparation and Materials Needed

Part A

You will need to obtain data on the amount of water your school uses each day. School water bills should be available from your school office or school system headquarters.

Part B

You will most likely want to gather the needed data on the water use in your county and a neighboring county for your students before class. The *EarthComm* web site contains links that can help you find the needed information. If the necessary data is not available for your county, select a nearby county that has a similar lifestyle (i.e., rural or urban). Your students will also need to know the populations of each county.

Materials
Part A
  • Calculator
  • Copy of school water bill or water-use data

Part B
  • Water-use data for your county and a neighboring county*
  • Graph paper
  • Colored pencils (eight different colors)
  • Calculator

*The *EarthComm* web site has suggestions for obtaining these resources.

Think about It

Student Conceptions

Students are not likely to have thought much about how much water is needed for brushing their teeth. Some of them may turn the water off while brushing, but others may leave the water running. Have students share their plans for estimating how much water is needed for brushing their teeth. An example of a possible method might include collecting the water in a container while brushing and then measuring the amount using a measuring cup or beaker. Encourage students to be creative.

Answer for the Teacher Only

Answers will vary, but on average, water flows from the tap at a rate of 1.5 gallons/minute.

Assessment Tool

Think about It Evaluation Sheet
Use this evaluation sheet to help students understand and internalize the basic expectations for the warm-up activity.

Blackline Master Water Resources 3.1
Data Table for Water Use

Water Use in Two Local Counties	Ground water (units)	Surface water (units)	Ground water (units)	Surface water (units)
County name:				
Total population of county:				
Total use				
Domestic (residential)				
Commercial				
Industrial: mining				
Industrial: power plants				
Agricultural: farming				
Agricultural: livestock				
Agricultural: irrigation				

Chapter 3

Earth's Natural Resources Water Resources

Investigate

Part A: Water Use in Your School

1. Working in small groups, brainstorm and develop a list of the various ways that water is used at your school each day.

    a) Record your list in your notebook.

2. Share your list with the class and arrive at a combined list.

3. As a class, rank all the items on the board from the highest estimated water use to the lowest.

    a) Record this list and ranking.

4. Each group should take one category of water use and devise a plan for determining the volume of water used daily for this activity.

    a) Design your plan and write it out, detailing specifically how you will calculate a fairly accurate estimate.

5. Share your plan with the class. Be prepared to make changes on the basis of comments from the class.

6. With the approval of your teacher, carry out your plan.

    a) Write down all of the steps that you took to determine the average daily water use for your category.

    b) Write down your group's estimate of water use.

    c) Write down ways that your school could reduce this amount.

 Any water collected should be disposed of. Wipe up spills immediately.

d) Estimate how much daily water use could be reduced if these water conservation methods were adopted in your school.

7. Combine the estimates from each group to come up with one estimate for the average daily water use at your school.

8. Find out how much water the school really does use every day. This is something you might be able to find out from the office, or the school system headquarters, or the school's water supplier. You might need to average a few monthly water bills and then divide it by the number of days in a month to come up with the daily water use.

    a) How close was your group's estimate to the actual number?

    b) How close was your class's estimate to the actual number?

    c) What might have caused the difference?

9. Working in small groups, brainstorm and develop a list of the various ways that water use could be conserved at your school.

    a) Record your list in your notebook.

10. As a class, discuss how much water could be saved at your school each day by using the methods each group came up with for saving water.

    a) How much water would these measures save for a whole school year?

    b) Calculate the cost savings that your plan would have.

Investigate

Part A: Water Use in Your School

1. Encourage students to think of at least 10 different uses of water in their school.

   a) Student lists will vary. Possible uses include:
      - drinking
      - flushing toilets
      - washing hands
      - watering the school lawn
      - cleaning floors
      - washing dishes
      - in laboratories
      - cooking

2. After a class discussion, students should write a composite list on the blackboard. If some of the items seem similar, encourage students to combine them into a single category. Each student group will be investigating one of these uses in detail, and you will want to cover as much ground as possible.

3. Make sure that the class is able to reach a consensus on ranking the items from highest to lowest in terms of water use.

4. Have each group choose a category of water use to investigate. If you have more categories of water use than you have student groups, assign the categories that ranked highest first.

   a) Circulate around the room and answer any questions that arise as students devise their plans.

5. After the student groups have presented their plans to the class, have them modify their plans according to the suggestions for improvement that they received from their classmates.

6. If the student plans require them to leave the classroom, you may want have them carry out their plans as homework assignments. Students will most likely have to make some assumptions. For example, if they are attempting to determine how much water is used daily in school bathrooms, they will need to determine the amount of water used by one student, and then determine how many students are in the school and estimate how many times each student uses the bathroom in a single day.

   a) Remind students that it is important for them to document how they made their estimates. This is particularly important if the work is done outside the classroom, without your direct supervision.

   b) Answers will vary.

c) Answers will vary. Encourage students to come up with reasonable methods of water conservation, like alternating days to water the lawn or having the school install low-flush toilets.

d) Answers will vary. Remind students to record any steps they took to come up with their estimate.

7. Once students have made their estimates, they can sum them to get an estimate of water use for the entire school.

8. Provide students with data on school water use.

a) Answers will vary.

b) Answers will vary.

c) Answers will vary. Differences between the two may arise because student calculations are not exact; rather, these calculations are estimates, often based on assumptions rather than precise measurements. Also, there are likely some water uses which students did not include on their list, and which were therefore not included in the final tally.

9. a) Answers will vary. Encourage students to be creative.

10. a) Answers will vary.

b) Answers will vary.

Assessment Opportunity

Your school may already be using methods of water conservation. Have students research this. They can estimate how much more water the school would use without these methods. If your school does not use any water conservation methods, your students may want to propose their ideas to the school board. To help determine whether their water conservation plans would be feasible, students can also research how expensive it would be to carry out their plans. They can then share their findings as a short presentation.

NOTES

Chapter 3

Part B: Water Use in Your Community

1. Gather data on the water use in your county and one additional nearby county. If this information is not available for your county, use the two counties closest to your community.

2. Construct a copy of the data table shown below for each county.

   a) Enter the data value for each category. You may wish to change or add categories that are characteristic of the counties you are studying.

3. Construct a bar graph showing the volume of use for surface water and ground water for each of the following categories: domestic, commercial, industrial, and agricultural.

   a) Use different-colored bars for each category. Have separate bars for each category's sub-use area. The vertical axis of the bar graph should be used to show the volume of water used per year.

Water Use in Two Local Counties				
County name:				
Total population of county:				
	Ground water (units)	**Surface water (units)**	**Ground water (units)**	**Surface water (units)**
Total use				
Domestic (residential)				
Commercial				
Industrial: mining				
Industrial: power plants				
Agricultural: farming				
Agricultural: livestock				
Agricultural: irrigation				

4. Calculate the per capita water use of ground water and surface water for both your county and your neighboring county. Use the equation below.

   $$\frac{\text{Volume used per year}}{\text{Population of county}} = \begin{array}{l}\text{per capita}\\\text{water use}\\\text{per year}\end{array}$$

   a) Show your calculations and the result in your notebook.

5. Use the results of this part of the investigation to answer the following questions:

   a) Which county has the higher per capita water use?

   b) Consider the uses of water in both counties. Why is the per capita water use of each so high? What could account for a difference in water use by communities in the same area?

Part B: Water Use in Your Community

1. You can collect this data before class and provide students with copies. Try to select an additional county that all the students are likely to be familiar with.

2. Students may need to adapt the data table shown on page R171 to fit their area, particularly if they live in an urban area that makes little use of water for agricultural purposes. A copy of the data table is available as **Blackline Master Water Resources 3.1, Data Table for Water Use.**

3. Bar graphs will vary depending upon the area in which you live. Students should plot data for both counties, according to use (domestic, commercial, etc.) and source (surface water or ground water). Sub-areas (farming, livestock, irrigation) should also be plotted. A sample, hypothetical graph is shown below.

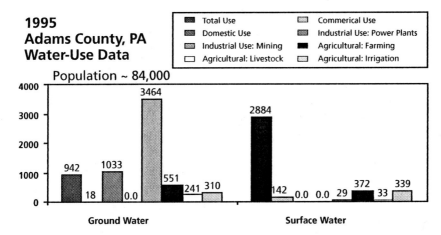

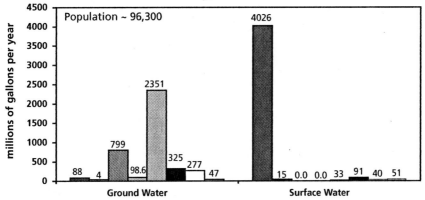

Chapter 3

County Name	Adams, PA		Northumberland Co., PA	
Population (thousands)	84.0		96.3	
1995 Usage Data (millions of gallons/year)	Ground Water	Surface Water	Ground Water	Surface Water
Total Use	942	2884	88	4026
Commercial Use	18	142	3.7	15
Domestic Use	1033	0.0	799	0.0
Industrial Use: Power Plants	0.0	0.0	98.6	0.0
Industrial Use: Mining	3464	29	2351	33
Agricultural: Farming	551	372	325	91
Agricultural: Livestock	241	33	277	40
Agricultural: Irrigation	310	339	47	51

4. a) Students can use the given equation to make the calculations. Remind them that they will need to calculate the per capita water use of surface water (total) and the per capi[ta] use of ground water (total) for each county. In other words, they will do four calculations.

5. Answers to the questions are dependent upon the area in which you live. Because there a[re] many different uses for water, many possibilities exist. For example, one county might be[ ] largely agricultural, with little population but a large need for water. This would cause th[e] per capita water usage to be quite high. Another nearby county might be largely residenti[al] with little agricultural or industrial demands for water. In such a case, one might expect t[he] per capita water use to be low.

Assessment Tool

EarthComm Notebook-Entry Checklist
Use this checklist as a quick guide for student self-assessment and/or an opportunity to quickly score student work. Add further criteria specific to your classroom needs or to this particular investigation.

Assessment Tool

Investigate Notebook-Entry Evaluation Sheet
Point out the criteria listed on this evaluation sheet that are relevant to this particular investigation. Encourage students to internalize the criteria by making them part of your assessment conversations as you circulate around the classroom.

NOTES

Chapter 3

Earth's Natural Resources Water Resources

c) Consider the uses of water in your county and the other county. What water conservation methods would you suggest for them?

d) Share your ideas on water conservation with the class. Add one or two methods to your list besides those your group has already recorded.

e) If your community conserved 20 gallons per person per year, how many more people could live there and not exceed your current yearly water use?

Reflecting on the Activity and the Challenge

You learned approximately how much water your school uses every day and came up with ways to reduce this amount. Reducing water use day to day is a good way of conserving water for future use. You also analyzed your county's water consumption and compared it to a neighboring county.

This provides you with knowledge of areas that you may consider when developing a conservation plan. Together, this information could help your community accommodate the new developments proposed in the **Chapter Challenge**.

Reflecting on the Activity and the Challenge

Students should now have a good idea of how much water a facility like a school uses in a single day, and they should also realize how much water their community needs. They should be able to use this information to determine whether the water use in the community will need to be reduced to accommodate the proposed construction. If water conservation will be required, then students will have begun to think about ways that this can be done. Discuss with your students how the results of the investigation are tied to what they are being asked to do in the **Chapter Challenge**.

Digging Deeper

FRESH-WATER USE

Types of Water Use

Figure 1 shows the source, use, and disposition of fresh water in the United States in 1995. The amounts shown are in millions of gallons per day (Mgal/day). Irrigation is the largest single use of fresh water in the United States. An estimated 134 billion gallons were used to irrigate crops every day in 1995 (that's equal to ten stacks of one-gallon milk jugs that reach to the moon!). The domestic use of water (26,100 Mgal/day) accounts for only about 7.7% of daily fresh water use. Irrigation and thermoelectric use make up 78% of daily demand for fresh water. That is why it is important to develop methods of irrigation that minimize consumption (reduce water loss).

- **Public supply** implies water used for street cleaning, fire fighting, public facilities like swimming pools or fountains, and supply to public buildings like a city hall or municipal museum.
- **Domestic** implies water used in homes.
- **Commercial** implies water used in businesses, large and small, but exclusive of manufacturing processes.
- **Industrial** implies water used in manufacturing processes.
- **Thermoelectric** implies generation of electricity by oil-fired, coal-fired, gas-fired, or nuclear-powered power plants. Almost all of this water is used for cooling the steam that is used to produce the rotational motion that drives the electrical generators. Most of the water is said to be self-supplied; that is, it is taken from special reservoirs or wells connected to the power plant rather than from municipal supplies. Also, most of the water is returned to rivers, streams, or the ocean, warmer than before it was used for cooling.

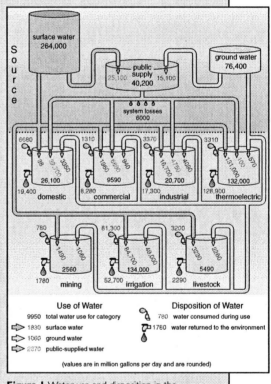

Figure 1 Water use and disposition in the United States in 1995.

Digging Deeper

Assign the reading for homework, along with the questions in **Check Your Understanding** if desired.

Assessment Opportunity

Reword or restructure the questions in **Check Your Understanding** for a brief quiz. Use the quiz (or a class discussion of the questions in the textbook) to assess your students' understanding of the main ideas in the reading and the activity.

Teaching Tip

Use **Blackline Master Water Resources 3.2, Water Use and Disposition** to make overhead of *Figure 1* on page R173. Use this overhead to discuss the major uses of water in the United States Review the terms introduced in the illustration and make sure that students understand the flow of the diagram.

Teaching Tips

Water used by the cooling towers of a nuclear power plant, like the one shown in *Figure 2* on page 174, falls into the category of thermoelectric use. Ask your students to look at the diagram in *Figure 1* on page R173 and consider how much water is used for thermoelectric purposes relative to other uses.

Loss of water in the fountain shown in *Figure 3* on page R173 is nearly entirely through evaporation. As such, this use of public supply waters would be considered to be mostly consumptive.

Watering crops, as shown in *Figure 4* on page R174, is both consumptive and nonconsumptive. Some of the water can be evaporated back into the atmosphere during irrigation, and some soaks into the soil.

Chapter 3

Earth's Natural Resources Water Resources

Figure 2 Cooling tower of a nuclear power plant.

Figure 3 About 12% of the demand for fresh water is for public supply.

Consumptive Use versus Nonconsumptive Use

Most of the total water use shown in *Figure 1* is said to be nonconsumptive. **Consumptive water** is water that is returned to the atmosphere as water vapor. **Nonconsumptive water** is water that is returned, in liquid form, to the natural environment after use. Some nonconsumptive water goes into rivers, lakes, or the ocean, and some soaks into the ground. In your home, for example, almost all the water you use, except for what you drink and then return to the atmosphere by breathing and perspiring, is nonconsumptive. The portion of water used in lawn and garden sprinkling that evaporates before soaking into the ground is also consumptive water. Think about what happens to water used in bathing, clothes washing, and toilet flushing: it goes either into a municipal sewer system or into a home septic system.

Figure 4 Is this an example of consumptive or nonconsumptive use?

Geo Words

consumptive water: water that is returned to the atmosphere as water vapor.

nonconsumptive water: water that is returned, in liquid form, to the natural environment after use.

R 174

Blackline Master Water Resources 3.2
Water Use and Disposition

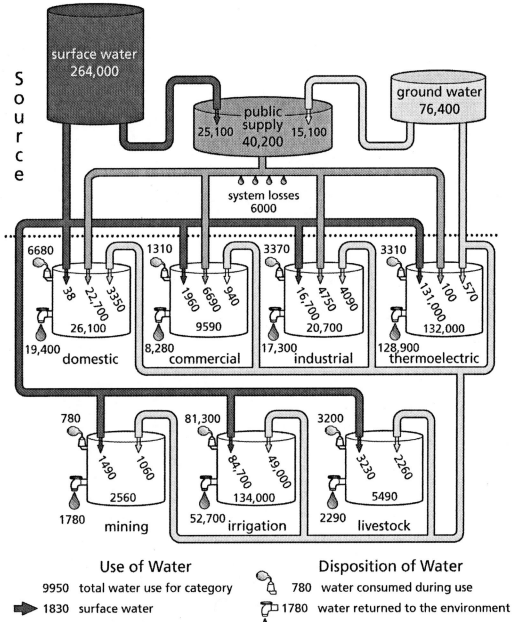

source, use and disposition of freshwater
in the United States in 1995

surface water 264,000

ground water 76,400

public supply 40,200 — 25,100 / 15,100

system losses 6000

	6680		1310		3370		3310

domestic: 38 / 22,700 / 3350 — 26,100 — 19,400

commercial: 1960 / 6690 / 940 — 9590 — 8,280

industrial: 16,700 / 4750 / 4090 — 20,700 — 17,300

thermoelectric: 131,000 / 100 / 570 — 132,000 — 128,900

mining: 780 — 1490 / 1060 — 2560 — 1780

irrigation: 81,300 — 84,700 / 49,000 — 134,000 — 52,700

livestock: 3200 — 3230 / 2260 — 5490 — 2290

Use of Water

9950   total water use for category

1830   surface water

1060   ground water

2370   public-supplied water

Disposition of Water

780   water consumed during use

1780   water returned to the environment

(values are in million gallons per day and are rounded)

Chapter 3

Water Conservation

As you learned in the last activity, common ways of increasing water supply are construction of dams and reservoirs, transportation of water from one place to another through pipelines or aqueducts, and development of ground-water sources. In many parts of the country, however, population growth is outstripping the ability to meet the growing demand for water. In communities where new water sources are limited, water conservation programs are widely encouraged. Such programs will become more and more important as the years go by, even in the United States, which is blessed with a water supply that is more abundant than in most countries of the world.

Some examples of ways citizens can save water at home are low-flush toilets, low-flow showerheads, and faucet aerators. In public bathrooms, faucets with motion sensors ensure that the water is turned off when no one is using it.

In agriculture, farmers can conserve water by using **drip irrigation** for crops, as shown in *Figure 5*. Traditional methods of irrigation are very inefficient: much of the water evaporates in the air or soaks into the soil between the crop rows rather than being

Figure 5 In a drip irrigation system water flows into main tubing and then into smaller tubes that branch out along rows of crops.

delivered to plant roots. In drip irrigation, water is supplied slowly to the soil through special piping and tubing that is punctured with many small holes. The water drips or oozes out around the plant roots. Xeriscaping is an excellent way of saving water in cities and towns in arid regions. **Xeriscaping** means planning residential or commercial landscaping to use as little water as possible. It involves using plants that are naturally adapted to the climate of the area rather than plants like grasses, which need frequent watering during hot, dry weather. *Figure 6* shows an example of xeriscaping. Other ways of conserving water in landscaping is to use mulch around the bases of plants, as well as drip irrigation.

Figure 6 Landscaping with plants that require little water.

Geo Words

drip irrigation: a form of irrigation in which water is supplied slowly to the soil through special piping and tubing that is punctured with many small holes that allow the water to drip or ooze out.

xeriscaping: residential or commercial landscaping planned to use as little water as possible.

Check Your Understanding

1. What categories account for the major water use in the United States?

2. How can water conservation effectively increase the water supply of a community?

3. Why does drip irrigation save water?

4. What is the difference between consumptive and nonconsumptive water use?

R 175

Teaching Tips

Figure 5 on page R175 is a close-up of the tubing used to supply water to grapes by using a drip irrigation system.

Figure 6 on page R175 shows an example of xeriscaping in Florida. The plants in the photograph require little water for survival, thereby conserving water while maintaining an attractive landscape.

Check Your Understanding

1. Irrigation and thermoelectric use account for 80% of daily fresh-water use.

2. In some communities, new water sources are limited. Water conservation programs that encourage using as little water as possible for certain purposes, like irrigation, can increase the amount of water available for the community.

3. In drip irrigation, water is supplied slowly to the soil through special piping and tubing that is punctured with small holes. The water oozes out around the plant roots. This is much more efficient than traditional methods of irrigation because the water is delivered directly to the plant roots and does not have an opportunity to evaporate or soak into the soil between crop rows.

4. Consumptive water is water that is returned to the atmosphere as water vapor, whereas nonconsumptive water is water that is returned, in liquid form, to the natural environment after use.

Assessment Tool

Check Your Understanding Notebook-Entry Evaluation Sheet
Use this sheet to evaluate the extent to which students understand the key concepts explored in **Activity 3** and explained in **Digging Deeper**, and to evaluate the students' clarity of expression.

Earth's Natural Resources Water Resources

Understanding and Applying What You Have Learned

1. Examine *Figure 1* on page R173. Each of the seven cylinders has two small faucets. The top faucet shows the amount of water that is "used up," or consumed. The bottom faucet shows the amount of water returned back to the environment (nonconsumptive use). Use the diagram to answer the following questions:

   a) For each of the seven major uses of water, calculate the percentage of water consumed during use and returned to the environment.

   b) Which use of water has the highest percentage of nonconsumptive use? (highest percentage returned to the environment).

2. Think about the water cycle. Some of the water used by plants is returned to the atmosphere by transpiration. Describe two other ways that irrigation water reenters the water cycle.

3. Suppose that irrigation water is withdrawn from a confined aquifer. Why might this cause a problem in the water supply?

Preparing for the Chapter Challenge

1. Would you expect the proposed developments in your community to increase demand, decrease supply, or a combination of both? Explain your answer. You will use this material in your **Chapter Challenge**.

2. Using what you've learned about water conservation and water use in your community, write a short essay about three water conservation methods that you would suggest for the proposed developments. Defend your choices. Explain why your methods might be useful if water demand increases or if water supply decreases.

3. If your town does not have enough extra water for the new developments, what are some ways the town could get more water? Include water conservation ideas as well as ideas for building more capacity. You can incorporate these ideas into your final report.

Inquiring Further

1. **Water use in your home**

   Design a plan for conducting a water audit in your home. Calculate a reasonable estimate for the total amount of water used in your home daily, monthly, and over a period of one year.

Be sure to obtain parental permission before carrying out your plan.

2. **Water use in a different community**

   Find the gallons-per-day-per-person water use for a town in a different part of the United States. Is it very different from your town? What could account for the difference?

R 176

Understanding and Applying What You Have Learned

1. a) For each cylinder, divide the volume associated with each faucet by the total volume that leaves the cylinder, and multiply by 100 to obtain the percentage. Note that, because of rounding, the sum of the stated input volumes is not quite the same as the sum of the stated output volumes.

    b) Thermoelectric.

2. Water used in irrigation can be evaporated directly into the atmosphere. The water can also soak into the ground, to join ground water and eventually flow into a surface reservoir like a stream, a lake, or the ocean.

3. Irrigation water drawn from confined aquifers is recharged only in the source or recharge area. The recharge area is limited to the region where the aquifer reaches the surface. Impermeable barriers (aquicludes) or barriers with relatively low permeability (aquitards) prevent water from recharging the aquifer from above or below. This situation prevents any of the irrigation water from soaking into the ground and returning to the aquifer. This maximizes any depression of the water table in this situation.

Preparing for the Chapter Challenge

Activity 3 not only provides students with insight into the amount of water they use and how to measure the amounts used: it also helps them recognize ways to conserve and help balance supply and demand. They will also be able to identify how water is used for commercial and agricultural needs. They will be able to identify the reasons for variations due to the seasons, weather conditions, and population size, and they will understand how topography can influence the amount of water available. This helps them formulate their report to the development group in an effort to project future use and estimate the quantities of water which each different segment in the proposed development will need.

Criteria for assessment, as presented in the sample rubric in this Teacher's Edition (see **Assessment Rubric for Chapter Report on Water Resources** on pages 504 and 505), include discussions of:

- whether or not your town has enough extra water to supply the three new developments
- how your town could increase its water supply (if the town does not have enough water)

Understanding and Applying What You Have Learned

1. The proposed developments would certainly increase demand. Development might decrease supply somewhat, if it affected surface runoff in the watershed area or decreased infiltration to feed ground-water supplies.

Chapter 3

2. Possible methods for water conservation include:
   - encouraging planting of drought-resistant grasses, flowers, shrubs, and trees
   - restricting outdoor watering to certain days or hours
   - adjusting water rates by arranging a sliding scale whereby unit price increases with total water use
   - encouraging home conservation practices like using low-flush toilets and low-flow shower heads

3. The town could get more water by:
   - drilling new wells
   - tapping additional surface-water supplies, perhaps from some distance away in different watersheds
   - increasing conservation efforts

Inquiring Further

1. Water use in your home

Students can use methods similar to those used in **Part A** of **Investigate** to determine the amount of water used in their homes. It is important that students have the approval of their parents to share any data on personal water use from their homes with you or the class. If possible, collect student data and provide it to the class without identifying the households audited.

Students can conduct their home water audit in one of two ways:

- They can time and measure household water use. To do this, they can measure how much water comes out of a given faucet or shower head in one minute (for example), and then multiply this by the average time that faucet would run in a day. Students can also look at the water-use information for the water-use devices (dishwasher, washing machine, toilet) in their house. This information is commonly available for modern appliances and fixtures.

- If they live in a town or city, they can read the water meter on a given day at, say, 5 p.m., and then read it again 24 hours later.

Have students calculate the estimated use for one week. Then have them take readings for a month and compare the actual value to their estimates. Have them explain variations between households or variations between given days of the week. Once they have determined their monthly use, have them predict yearly use. If possible, they could go to the city water department and find out what they used for a year and compare this to their calculations. Once again, have them explain differences between actual use and estimated use.

2. Water use in a different community

You may wish to give students a list of cities from which to chose. They should investigate a city that has a climate different from their own. For example, students living in Florida might want to investigate water use in a town in Arizona. The

EarthComm web site contains suggestions to help you find the necessary data. To determine per capita use, divide the population by the gallons per day used by the community.

ACTIVITY 4— WATER SUPPLY AND DEMAND: WATER BUDGETS

Background Information

Water Supply and Demand

The factors that determine the water needs of a community or region include:

- geography
- topography
- climate
- population size
- land use
- economic activities
- human activities

The quantity of water available for a region is directly related to climate and geography. Population centers tend to be located where large quantities of water are available, although some modern cities (like Las Vegas, Phoenix, or Los Angeles) have grown up in arid or semiarid regions located far from water sources.

Although a large supply of water is available in most parts of the United States, water resources require careful management. Weather extremes, pollution, changes in water use, and changes in weather patterns can cause the quantity of available water to vary. By plotting inches of rainfall for each month on a graph, students can see how water use and supply match up: do they, or do they not?

The experience that students have gained in **Activities 1, 2,** and **3** in analyzing water use should help them suggest ways of bringing the supply and demand curves into balance. Looking over the amount of water used by commercial, industrial, agricultural, and domestic users offers a truer picture of how much water is used than looking only at domestic usage (which most people are more familiar with). This information helps students to develop suggestions for conservation practices by each group of users and to identify the amounts that can be conserved by each group.

Water Budgets

The term budget is used somewhat differently in the context of science than in the more common financial context. A water budget, for example, is an accounting of the sources of water supply and water demand, and of how the supply is divided among the various uses that make up the demand. Studying a water budget is useful because it organizes thinking about water use and can help to identify aspects of water use that need the attention of natural scientists, engineers, planners, or government officials.

Water Conservation

Common ways of increasing water supply include:

- construction of dams and reservoirs
- transportation of water through pipelines or aqueducts
- development of ground-water sources

In many parts of the country, however, population growth is outstripping the ability to meet the growing demand for water. In communities where new water sources are limited, water conservation programs are widely encouraged. Such programs will become more and more important as the years go by, even in the United States, which is blessed with a more abundant water supply than most countries of the world.

Some ways citizens can save water at home include:
• low-flush toilets
• low-flow showerheads
• faucet aerators

In public bathrooms, faucets with motion sensors ensure that the water is turned off when no one is using it.

Traditional methods of irrigating agricultural crops are very inefficient: much of the water evaporates in the air or soaks into the soil between the crop rows rather than being delivered to plant roots. Farmers can conserve water by using drip irrigation for crops. In drip irrigation, water is supplied slowly to the soil through special piping and tubing that is punctured with many small holes. The water then drips or oozes out around the plant roots.

Xeriscaping, an excellent way of saving water in cities and towns in arid regions, involves planning residential or commercial landscaping to use as little water as possible. Xeriscaping uses plants that are naturally adapted to the climate of the area rather than plants like grasses that need frequent watering during hot, dry weather. Other ways of conserving water in landscaping are to use mulch around the bases of plants, as well as drip irrigation.

More Information – on the Web
Visit the *EarthComm* web site
www.agiweb.org/earthcomm to access a variety of links to web sites that will help you deepen your understanding of content and prepare you to teach this activity. Many of the sites also contain images that you can download.

Chapter 3

Goals and Assessment

Clarify that the goals indicate what students should understand and be able to do as a result of the activity. Make sure students understand that Chapter Assessments are based upon these goals.

Goal	Location in Activity	Assessment Opportunity
Construct a water budget for your community from data sets.	**Investigate** Part A	Climatograph is correctly drawn, axes are labeled; graph has a title. Answers to questions are correct and are based on data.
Explain the influence of local climate on the water budget in your community.	**Investigate** Part A **Understanding and Applying What You Have Learned** Questions 1 and 3	Answers to questions are correct, and are based on data.
Identify times of year when both supply and demand are greatest and lowest.	**Investigate** Part A **Understanding and Applying What You Have Learned** Question 1	Answers to questions are correct, and are based on data.
Describe key controls on the quantity and availability of surface water and ground water in your community.	**Investigate** Part A **Check Your Understanding** Questions 1 – 3	Responses to questions are reasonable, and are based on available data.
Construct and analyze a box model of an irrigation water budget.	**Investigate** Part B	Box model is drawn correctly; percentages add up correctly.
Explain how ground-water development affects the ground-water system.	**Digging Deeper; Check Your Understanding** Question 3	Responses to questions closely match those given in Teacher's Edition.

NOTES

Chapter 3

Activity 4

Water Supply and Demand: Water Budgets

Goals

In this activity you will:

- Construct a water budget of your community from data sets.

- Explain the influence of local climate on the water budget.

- Identify times of year when supply and demand of water are greatest and lowest.

- Describe key controls on the quantity and availability of surface and ground water in your community.

- Construct and analyze a box model of an irrigation water budget.

- Explain how ground-water development affects the ground-water system.

Think about It

Take a moment to consider the source and amount of money you have during a month's time and how you spend it.

- Are there times of the year when you have more money and times of the year when you have less money?
- What are the advantages of planning or budgeting for lean times? How would you go about doing this?

What do you think? Record your ideas in your *EarthComm* notebook. Be prepared to discuss your responses with your small group and the class.

Activity Overview

Students collect data on the average temperature and precipitation in their community and use the data to construct a climatograph. They answer a series of questions using the climatograph that help them to determine the water budget of their community throughout the year. Students then construct a box model to illustrate the irrigation water budget for the United States. **Digging Deeper** reviews the concept of a water budget and examines fluctuations in water resources, including those that occur in rivers and ground water.

Preparation and Materials Needed

Part A

Students will need to use the Internet to collect the data (temperature and precipitation for their community over a 12-month period) needed to construct their climatographs. If your students do not have Internet access, you will need to obtain the data for them ahead of time and distribute copies in class. The *EarthComm* web site contains links that will help you or your students find the necessary data.

Part B

No advance preparation is required for this portion of the activity.

Materials

Part A:
- Internet access (or printouts of monthly temperature and precipitation data for the community)*
- Graph paper

Part B:
- Several blank sheets of paper
- Scissors
- Tape

Think about It

Student Conceptions

The **Think about It** questions prompt students to consider the advantages of budgeting and planning for times when resources are scarce. Students at this level may have begun to be concerned with their finances. It is likely that they will say they have less money around holidays (like Christmas), around birthdays of family

*The *EarthComm* web site has suggestions for obtaining these resources.

members and friends, and perhaps during times when they go on vacation. Students are likely to see the value of budgeting to ensure that money is available even during lean times. It is not likely that they have a great deal of familiarity with balancing budgets, but they may have some sense of what steps are needed to plan efficiently. You may wish to construct a sample yearly budget with your students to give them a sense of what you are looking for. After completing **Activity 4**, students will see the relevance of these questions.

Assessment Tool

Think about It Evaluation Sheet
Use this evaluation sheet to help students understand and internalize the basic expectations for the warm-up activity.

NOTES

Earth's Natural Resources Water Resources

Investigate

Part A: Your Local Water Budget

1. Work with members of your group to construct and interpret your local water budget. Find data for the average temperature and amount of precipitation in or near your community on a monthly basis. Visit the *EarthComm* web site or your school library to help you to find this information.

2. Make a climatograph to show the monthly average temperature and precipitation for your community. Use the sample shown to help you construct your graph.

   a) Keep a copy of your graph in your *EarthComm* notebook. Be sure that the axes are correctly labeled and that the graph has a title.

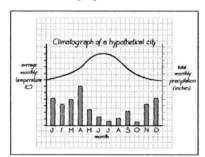

3. Using your graph as a resource, answer the following questions:

   a) During which month or months does your community receive the least amount of precipitation?

   b) During which month or months does your community receive most of its precipitation, or is precipitation spread evenly through the year?

   c) Does your community receive a great deal of snowfall during the winter? If yes, in which month, or months, is snowfall heaviest? In which month or months is snowmelt likely to be greatest?

   d) From your work in previous activities, does your community rely mostly on surface water or on ground water?

   e) In what months would you expect the surface water to be the lowest?

   f) Suggest ways the community would have to adjust if these levels became extremely low.

   g) In what months would you expect ground water to be the lowest?

   h) Suggest ways the community would have to adjust if these levels became extremely low.

   i) Predict the month or months for which there is the greatest demand for water. Explain why. Do these times of increased demand coincide with high water supply or low water supply?

Part B: Constructing the Irrigation Water Budget for the United States

1. To irrigate crops, farmers divert water from existing supplies. Crops use some of the water, some is lost in transit, some evaporates, and some returns to the sources of supply in liquid form. All of the water can be accounted for, in one way or another, in terms of a budget. As with any budget, all of the amounts have to add up in a consistent way (that is, "be accounted for").

R 178

Investigate

Part A: Your Local Water Budget

1. If possible, give students some time to find the needed data. They will want to collect monthly temperature and precipitation data for their community for one year. If time or access to resources is limited, provide this data for the students.

2. a) Students can use the sample climatograph on page R178 in the text to help them set up their graphs. Remind them to label the axes and give their graph a title.

3. Answers to these questions will be dependent upon your community. Students will need to recall information from previous activities and tie it into the data and graph they have constructed here. Remind them to think about earlier activities in which they examined how ground-water and surface-water reservoirs respond to decreased inflow and/or increased consumption. Students may believe that ground-water aquifers react instantaneously to changes in recharge. Remind them of the slow speed at which water flows through soil and into aquifers, which may cause a delayed response in the water levels in wells (i.e., ground-water levels may in fact be lower for several months after a rainy period). Have them compare rainfall with average monthly temperatures. In order to plan a water budget, temperature and rainfall must be in balance with use by the end of the year.

Teaching Tip

Some cities have annual precipitation and temperature records from the past 100 years. Students can plot this data and compare any trends they see—like wet and dry years, and cold and warm years—and look for connections between the two.

Assessment Opportunity

Visit the *EarthComm* web site to download a climatograph from a city besides your own. Provide students with copies of the climatograph and ask them to use the graph to prepare a water budget for this city. Students should consider during what months demand for water would be highest and lowest, and think about how the community can work around this variability. Encourage students to keep their responses short. This will help you further assess whether or not students clearly understand how climate can affect a community's water budget.

Chapter 3

Part B: Constructing the Irrigation Water Budget for the United States

1. The goal of **Part B** is to help students see how a large-scale water budget works. This is a bit of a brain teaser and will challenge some students more than others.

    a-c) A completed sample box model is shown below. Student interpretations may differ. Remind students that the percentages must add up accordingly.

Box Model of U.S. Irrigation

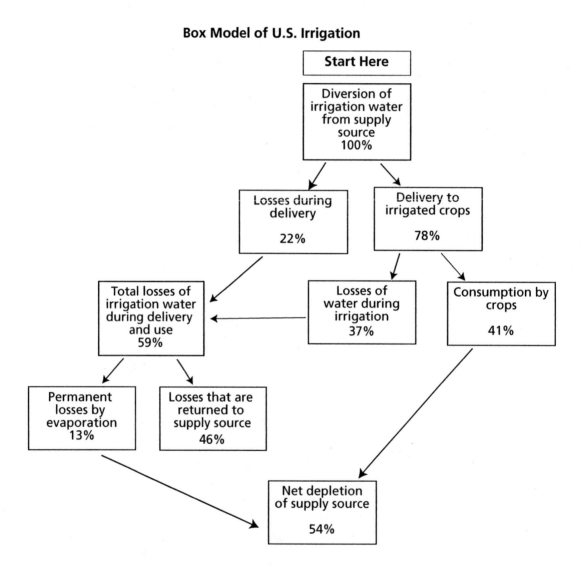

NOTES

Chapter 3

Below, in alphabetical order, are several components of the water budget for irrigation. Together, these components account for all of the water involved in irrigation. The significance of the percentages is described following the list.

- consumption by crops: 41%
- delivery to irrigated crops: 78%
- diversions of irrigation water from supply sources: 100%
- losses during delivery: 22%
- losses of water during irrigation use: 37%
- losses that are returned to supply sources: 46%
- net depletion of supply sources: 54%
- permanent losses by evaporation: 13%
- total losses of irrigation water during delivery and use: 59%

a) On blank sheets of paper, draw a box for each of the items above and write in the name of the item and the percentage.

b) Cut out the boxes and arrange them on a large sheet of paper in a way you think shows the natural "flow" of the irrigation water.

c) Draw arrows from one box to another box to show the pathways followed by the irrigation water. The starting box is the one titled

"diversion of irrigation water from supply sources."

To understand the meaning of the percentages, read the two notes below:

- If two arrows come into a given box, the percentages in the boxes where the arrows come from have to add up to the percentage in the given box.
- If two arrows leave from a given box, the percentages in the boxes where the arrows go have to add up to the percentage in the given box.

2. Within your small group, discuss the nature of the water budget and try to agree upon a "flow diagram."

a) Compare your result with those of the other student groups.

3. Use the class results to answer the following questions:

a) What do you think are the important kinds of sources for irrigation water?

b) What processes do you think are involved in the various water losses listed above?

c) In what ways can farmers minimize these water losses?

d) How is the ground-water system affected by irrigation?

Reflecting on the Activity and the Challenge

You learned how much precipitation falls in your community during the different months of the year. You also speculated about which months your community uses more water than usual. You also examined an irrigation water budget, which should make you think about the water budget in your own community. This will help you anticipate the possible impact of the new developments on your community's water supply.

2. **a)** Post the completed box models on the wall for comparison. Discuss with
    students any discrepancies between the models. Have them try to come up with
    an agreement on a flow diagram as a class.

3. **a)** Ground-water wells; rivers; lakes; artificial reservoirs.

    b) Possible responses might include spillage during delivery, and evaporation.

    c) Keep canals and supply ditches in good repair; keep irrigation water close to
    crop roots to help cut down on evaporation loss; avoid overhead sprinkling
    where possible.

    d) Much irrigation water is derived from ground-water supplies; some of the
    irrigation water infiltrates the ground surface to rejoin the ground-water supply
    rather than being evaporated or taken up by plant roots.

Assessment Tools

EarthComm Notebook-Entry Checklist
Use this checklist as a quick guide for student self-assessment and/or an
opportunity to quickly score student work. Add further criteria specific to your
classroom needs or to this particular investigation.

Investigate Notebook-Entry Evaluation Sheet
Point out the criteria listed on this evaluation sheet that are relevant to this
particular investigation. Encourage students to internalize the criteria by making
them part of your assessment conversations as you circulate around the classroom.

Reflecting on the Activity and the Challenge

Students should now have a clear sense of what a water budget is and why it is
important. Review these concepts and discuss why construction of a water budget
will be an important part of their final reports. Students should recognize the need
to plan water use in conjunction with the development of the new properties,
especially if the community water supply will be significantly affected by their
construction. They should see that planning will help the community get by with
adequate water even during lean times when available water is less plentiful. Have
students revisit the **Think about It** questions at the start of **Activity 4**.

Chapter 3

Earth's Natural Resources Water Resources

Geo Words

water budget: an accounting of the sources of water supply and water demand, and of how the supply is divided among the various uses that make up the demand.

discharge: the volume of water that flows past a point on the river per unit of time.

Digging Deeper

NATURAL FLUCTUATIONS IN WATER RESOURCES

Water Budgets

In the financial world, a budget is a plan for the finances of a government or an organization or a person for a period of time, like a year, giving expected expenses and the income to cover those expenses. The term is used somewhat differently in science. A **water budget**, for example, is an accounting of the sources of water supply and water demand, and of how the supply is divided among the various uses that make up the demand. In this activity you dealt with your community's water budget, and also the irrigation water budget for the entire United States. Studying a water budget is useful because it organizes your thinking about water use and can help to identify aspects of water use that need the attention of natural scientists, engineers, planners, or government officials.

In most regions of the United States, rainfall varies considerably from season to season. For example, in some locations, riverbeds are dry during much of the year and flow only during storms. In other places, rivers flow all year long, even during months with little rain. Can you explain these differences? One advantage of using ground water, the other major source of water supply in the United States, is that it is not as vulnerable to short-term changes in precipitation as surface water. As you will see below, however, the ground water table fluctuates up and down as well.

Rivers

The volume of water that flows past a point on the river per unit of time is called the **discharge** of the river. It is measured in cubic feet per second or cubic meters per second. Large rivers, like the Missouri–Mississippi river system can have discharges as great as 16,800 m^3/s. In most rivers, discharge varies greatly between times of drought and times of floods.

Figure 1 A river with a large discharge.

Figure 2 River bed during a drought.

Digging Deeper

Assign the reading for homework, along with the questions in **Check Your Understanding** if desired.

Assessment Opportunity

Use a quiz to assess student understanding of the concepts presented in this activity. Three sample questions are listed below:

Question: Use a sketch to show how the ground-water table changes when ground water is pumped from a well.
Answer: Student drawings should resemble *Figure 5A/5B* on page R182 of the student text.

Question: Use a sketch to show how a river can still have water in it during a drought.
Answer: Student drawings should resemble *Figure 4B* on page R181 of the student text.

Question: What is a water budget and why is it important to a community?
Answer: A water budget is an accounting of the sources of water supply and water demand, and of how the supply is divided among the various uses that make up the demand. A water budget is important because communities need water throughout the year, but the amount of water available can fluctuate from season to season.

Teaching Tip

Use **Blackline Master Water Resources 4.1, The Water Table and Rivers** to make overheads of *Figures 4A/B* on page R181. Incorporate these overheads into a discussion on what the water table represents and how it fluctuates during times of high and low river flow.

Chapter 3

Most cities and towns that use rivers for water supply take out only a small fraction of the river discharge. In some places, however, the demand for river water is so great that the natural discharge of the river is greatly decreased. For example, so much of the discharge of the Colorado River in the Southwest is used for water supply that by the time the river reaches its mouth in the Gulf of California, the flow is only a trickle!

Not all of the water in rivers comes from surface runoff. Rain also infiltrates into the Earth to become ground water. Remember from previous activities that after a certain depth below the land surface, the rock and/or sediment is saturated with ground water. Flowing rivers, as well as lakes and springs, are places where the water table comes out to the land surface (*Figure 3*). At times of high river flow, some of the river water feeds the ground water as shown in *Figure 4A*. At times of low river flow, however, the ground water feeds the river as in *Figure 4B*. This is why most large rivers flow even during long droughts. On average, ground water supplies as much as 40% of the water that flows in streams and rivers. During times of drought the percentage is much greater, and during times of flood the percentage is much smaller.

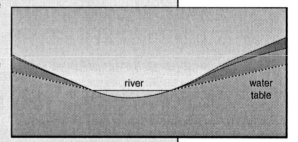

Figure 3 The water table comes to the land surface at flowing rivers.

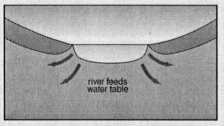

Figure 4A High river flow.

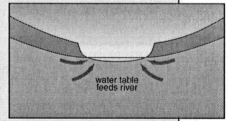

Figure 4B Low river flow.

Ground Water

At some depth below the Earth's surface, the porous rock and sediment are saturated with ground water. The upper limit of saturation is called the water table. Above the water table, in the unsaturated zone, most of the pore spaces are occupied by air. Below the water table, in the saturated zone, all of the pore spaces are filled with water, which flows slowly toward areas where the water table is lower.

Blackline Master Water Resources 4.1
The Water Table and Rivers

A. high-river flow

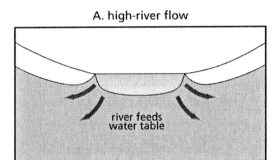

river feeds
water table

B. low-river flow

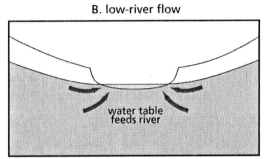

water table
feeds river

Blackline Master Water Resources 4.2
The Water Table and Wells

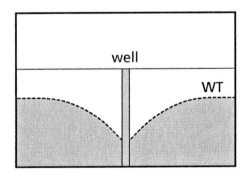

well

WT

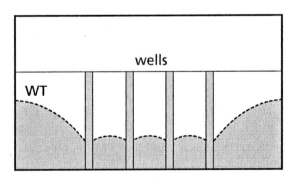

wells

WT

Chapter 3

Earth's Natural Resources Water Resources

Geo Words

cone of depression: the conical shape of the water table near a well, caused by the drawing down of the water table as water is pumped from the well.

Users of ground water would find life easier if the height of the water table in the aquifer that supplied their well always stayed the same. Variations in rainfall affect the height of the water table. During dry spells, ground water keeps flowing toward lakes, rivers, and (in coastal areas) the ocean, but there is no recharge to keep the water table from falling. During rainy spells, on the other hand, recharge more than makes up for ground water drainage, and the water table rises. In many areas the difference in the height of the water table between dry spells and rainy spells can be many meters.

Before wells are drilled into an aquifer, discharge and recharge of the ground-water system is in a condition of long-term balance: the amount of water entering the system is balanced with the amount leaving, and the height of the ground-water table stays the same, on average. Pumping of ground water for water supply upsets this balance. As water is pumped from a well, the water table drops in the vicinity of the well and takes on a conical shape called a **cone of depression**, as shown in *Figure 5A*. If water is being pumped from several wells in the same area, the cones of depression merge with one another, and the water table is depressed over the entire area, as shown in *Figure 5B*.

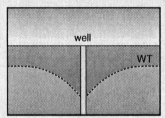

Figure 5A   **Figure 5B**

Check Your Understanding

1. How can a river still have water in it during a drought?

2. What is the difference between the time for surface water and ground water to respond to precipitation?

3. What happens to the water table when ground water is pumped from a well?

As you may have realized from the investigation in Activity 2, when ground water is pumped from a field of wells, one of two things can result:

- The water table is lowered, but it becomes stabilized at some lower height as additional ground water flows into the area from surrounding areas.
- Inflow of ground water from surrounding areas is not great enough to stabilize the water table, and the level keeps falling indefinitely.
- In the latter case, the ground water can no longer be thought of as a renewable resource, because it is being consumed faster than it can be replenished. In a real sense, the ground water is then being mined. In some areas of the United States the imbalance between recharge and discharge has lowered the water table by as much as a 100 m over the last 50 years. Eventually the expense of pumping from great depths becomes prohibitive (or, if the wells are not deep enough, they run dry!), and new sources must be developed.

Teaching Tip

Use **Blackline Master Water Resources 4.2, The Water Table and Wells** to make an overhead of *Figure 5A/B* on page R182. Incorporate this overhead into a discussion on how the water table fluctuates in response to withdrawal of water from wells.

Check Your Understanding

1. A river can still have water in it during times of drought if it is being fed by ground water.

2. Surface water responds very quickly to precipitation, whereas ground water takes much longer to respond to either an increase or a decrease in precipitation.

3. When ground water is pumped from a well, the ground-water table falls if the rate of discharge exceeds the rate of recharge.

Assessment Tool

Check Your Understanding Notebook-Entry Evaluation Sheet
Use this sheet to evaluate the extent to which students understand the key concepts explored in **Activity 4** and explained in **Digging Deeper**, and to evaluate the students' clarity of expression.

Chapter 3

Understanding and Applying What You Have Learned

1. Visit the *EarthComm* web site or the school library to find out the number of acres in your county. One inch of rain falling on one acre is equal to 27,154 gallons (102.9 cubic meters) of water. This is called one acre-foot of water.

   a) Calculate the number of gallons of water that reach the Earth in your county in a year's time.

   b) Is that more or less than the amount of water that your county uses in a year?

   c) Assume for a moment that your county's demand for water exceeded the amount of precipitation that fell. How is it possible that streams are still flowing and that water is still in the ground?

2. How would the "income" and "expenditures" in the water budgets of developing countries that lack the advances in technology that have occurred in the Unites States differ from our country?

3. How much does the distribution of annual rainfall affect your community's water supply?

Preparing for the Chapter Challenge

In your community, what times of the year do high use and low availability limit your water resources? How would the addition of the three proposed developments affect this situation? Write answers to these questions in a short essay to include with your **Chapter Challenge**.

Inquiring Further

1. **Water use in other countries**

   Calculate and compare the total volume of rainfall, per unit of land area (for example, per square kilometer) for two other countries: one that has an arid climate, and one that has a rainy climate. Determine the types of water resources these countries have available. Compare their use and rainfall patterns with those of your community. Explain any differences.

2. **High Plains Aquifer**

   The High Plains Aquifer (formerly the Ogallala Aquifer), an enormous confined aquifer that stretches from Kansas to Texas, is one of the largest sources of ground water for irrigation in the United States. Go to the *EarthComm* web site to research the magnitude of lowering of the water table in the aquifer from 1950 to 2000. What do you think might be potential alternative sources of irrigation water, if and when it becomes too costly to pump water from the High Plains Aquifer to the surface?

Understanding and Applying What You Have Learned

1. Answers to these questions will vary. One inch of rain falling on one acre is equal to 27,154 gallons of water.

   a) The number of gallons of water that reach the Earth in your county in a year's time is equal to the number of inches of precipitation, times 27,514 gallons for each acre, times the number of acres in the county.

   b) Answers will vary depending on the data for each individual county.

   c) It is possible for a county to use more water than falls in it as precipitation, without completely draining the supplies of surface water or ground water. This is because any given county may utilize water collected over a greater area than the county borders, and water may also be stored in surface reservoirs. Recharge zones for aquifers can be far away from the regions where the ground water is being withdrawn, and drainage basins that feed surface streams and rivers can be quite extensive.

2. Countries lacking certain technologies, like drip irrigation, may have higher water consumption than what is necessary for similar industrial or agricultural applications in areas of the United States that possess the technology. Additionally, technological advances in agriculture that have improved the harvest yield per acre of land allow for a reduction in the area needed to be farmed, saving additional water. Conversely, some of the modern conveniences commonly found in United States homes (indoor plumbing, dishwashers, well-watered lawns, etc.) can lead to higher domestic water use than might be expected in a developing country.

3. If water comes from ground-water wells or naturally occurring lakes, local annual rainfall is important to the community because it is needed to recharge aquifers. If water comes from reservoirs fed by river flow, local rainfall may be less important than regional rainfall because water can travel great distances along a river to supply a reservoir.

Preparing for the Chapter Challenge

Analysis and comparison of income and expenditures for their local water budget will provide each group with the ability to prepare a report for the **Chapter Challenge**. The report to the development committee should reflect whether the current water budget is balanced and has a reserve amount of water available for extremes and emergencies. It should also indicate whether the water budget can support the proposed developments' projected increases in use, yet still provide adequate quantities of water for emergencies.

Several aspects of this **Chapter Challenge** are relevant to the **Assessment Rubric for the Chapter Report** (See **Assessment Rubric for Chapter Report on Water Resources** on pages 504 and 505 of this teachers edition). These relevant considerations include:

- How much water is required by the residential development
- Whether or not your town has enough extra water to supply the three new developments

Inquiring Further

1. Water use in other countries

You may wish to assign students to different countries for this question. They can compare their findings for the two countries with the water use in their own community. Encourage students to pick countries that are located in extreme conditions: e.g., a country in an arid region, like Israel or Saudi Arabia, and a country in a humid region, like Costa Rica.

2. High Plains Aquifer

At the rate of drawdown in much of the aquifer, within a few decades it will become infeasible to obtain the needed supplies of ground water from the aquifer. One potential source is diversion of some of the discharge of the Mississippi–Missouri river system, but this water would have to be diverted over very long distances.

NOTES

Chapter 3

ACTIVITY 5— WATER POLLUTION
Background Information

Water Pollution

The quality of the water is affected by the land area over which the water flows and the subsurface material through which the water flows. Water dissolves or picks up material from rocks and sediments. It is very important to know the location of a community's watershed area so that present land-use patterns can be studied to determine whether the water is being polluted. It is equally important for community leaders to be informed about where the watershed area is located so that future land use will be compatible with protection of the water supply. Water pollution may or may not be visible or have an odor. Long-term consumption of water with undetected pollutants can have devastating effects on the health of humans, both before birth and throughout life.

Pollution in Surface Water and Ground Water

Both surface water and ground water can become polluted. Surface water generally contains a greater variety of pollutants, from a greater variety of sources. That's because there are so many ways a pollutant can get into a river or lake. Polluted ground water may contain fewer kinds of pollutants, but there may be a larger amount of each pollutant than in a typical surface-water supply. It's usually much more difficult to solve problems of ground-water pollution than problems of surface-water pollution. That's because ground water moves far more slowly than surface water. It might take many human lifetimes for pollution to clear from a large aquifer by natural ground-water flow.

Kinds of Pollutants

The most common pollutants generally fall into one of two categories:

Organic:
- sewage
- livestock wastes
- pathogenic (disease-causing) microorganisms

Non-organic:
- nutrients in fertilizers (mainly nitrates and phosphates)
- industrial and commercial chemicals
- road salts
- agricultural pesticides
- acidic mining wastes
- waste heat
- radioactive waste

Domestic Sewage

About three-quarters of homes in the United States are served by municipal sewage systems. The remaining one-quarter discharge their sewage into home septic systems (or even directly into the ground!). The human wastes in sewage are not themselves especially harmful. However, the disease-causing microorganisms commonly found in sewage are hazardous to health. Many illnesses, like cholera and typhoid, are caused by contact with sewage. Such illnesses are much less common in the United States than in areas of the world where people are more exposed to untreated sewage. Coliform bacteria, a common type of bacteria that live in the intestinal tracts of all warm-blooded animals (including humans), are generally not harmful. However, their presence in water supplies is commonly used as a signal of sewage contamination.

Nutrients from Fertilizers

Plants need many inorganic chemical nutrients for vigorous growth. The most important of these nutrients are two elements nitrogen and phosphorus. Nitrogen is the most abundant gas in the Earth's atmosphere, but it cannot be used directly by plants. Certain soil bacteria convert nitrogen gas into soluble forms of nitrogen, mainly nitrate ions, which can be used by plants. Lightning bolts do the same thing, and of course human-made chemical fertilizers contain soluble nitrogen. The problem with nitrogen-containing fertilizers is that they can be too much of a good thing: excess nitrogen promotes the growth of algae in natural surface waters, and when the algae die and decay they rob the waters of the oxygen needed by aquatic animals. Runoff from croplands is not the only culprit; fertilizer on lawns, golf courses, and home gardens often carries even more excess nitrogen than croplands.

Human wastes contain water-soluble nitrogen compounds. In areas where untreated or inadequately treated sewage finds its way into ground-water supplies, nitrate levels in drinking water can be excessive. Human infants can be harmed by drinking water that has high nitrate levels because nitrate reduces the amount of oxygen carried by red blood cells. Too much nitrate can cause blue-baby syndrome, which can lead to suffocation.

Phosphorus is used by plants in the form of phosphate ions. The most common form of phosphorus in fertilizers is ground-up phosphate rock, a type of sedimentary rock. Deposits of a material called guano—the excrement of bats and seabirds—are also mined for phosphorus. The problem with soluble phosphorus is similar to that with nitrogen: it causes algal blooms in natural waters. Another problem with phosphorus is

that the phosphates contained in household soaps and detergents can pollute ground-water supplies. In many heavily populated communities that rely on ground water, restrictions have been placed on the phosphate content of detergents.

Toxic Chemicals

In earlier times, the number of artificial chemicals and materials used by people in their daily lives was small. Today, however, such substances number in the tens of thousands, and the list grows rapidly. A large percentage of these are potential, or definite, health hazards. Many of these chemicals were found to be toxic only after long use. Some of the major chemical pollutants are:
- gasoline
- chemical solvents
- agricultural pesticides and herbicides
- compounds of heavy-metal elements like copper, zinc, lead, mercury, and cadmium used in industrial processes

Unfortunately, the rate of development of new substances is far greater than the ability of government and private testing laboratories to investigate their toxicity. Inevitably, many of these toxic substances find their way into surface-water supplies and ground-water supplies. Partly, this is from everyday use; in many places, however, toxic chemicals have been dumped, illegally, on the land surface or stored in decaying drums at dumpsites. Cleaning up such sites, which are found all around the United States, is a multi-billion-dollar task.

Waste Heat

Most of the electricity in the United States is generated in power plants where burning of fossil fuels produces steam to drive turbines, which in turn drive electrical generators. Enormous quantities of water are used to

Chapter 3

cool and condense the steam back into liquid water, to be recycled in the power plant. The cooling water is discharged into rivers, lakes, or the ocean. It is typically 5 to 10°C warmer than the water bodies it enters. This causes three major interrelated problems:

- Because warm water cannot hold as much dissolved oxygen as cool water, less oxygen is available for aquatic animals.
- The higher temperatures tend to lead to the disappearance of some species of organisms and replacement with warmer-water organisms.
- Sudden temperature changes can kill organisms outright, by what is called thermal shock.

Water Hardness

As water moves through soil, sediment, or bedrock, it dissolves some of the minerals. Water hardness reflects the concentration of dissolved solids—mainly calcium and magnesium—in the water. Ground water tends to be harder than surface water, because it has had more time to be in contact with the solid materials. Hard water is also more common in areas where the bedrock is limestone; limestone consists of the calcium carbonate mineral calcite, which is slightly soluble in water.

Some degree of hardness in drinking water is considered healthy, because calcium and magnesium are essential nutrients. Hardness also lowers the concentration of lead dissolved from the lead-based solder that is often used in copper plumbing systems. A disadvantage of hard water is that it reduces soapsuds. It can also leave hard, white deposits in teapots, on shower walls, and in water heaters and boilers. Water softeners reduce hardness by replacing the calcium and magnesium ions with sodium ions. However, the high sodium levels in softened water render it undesirable as drinking water for persons with high blood pressure.

More Information – on the Web

Visit the *EarthComm* web site www.agiweb.org/earthcomm to access a variety of links to web sites that will help you deepen your understanding of content and prepare you to teach this activity. Many of the sites also contain images that you can download.

Goals and Assessment

Clarify that the goals indicate what students should understand and be able to do as a result of the activity. Make sure students understand that Chapter Assessments are based upon these goals.

Goal	Location in Activity	Assessment Opportunity
Construct and analyze a physical model of the movement of pollutants in ground water. Determine how the pumping of water from wells influences this movement.	**Investigate** Part A **Understanding and Applying What You Have Learned** Question 4	Experiment is completed correctly; data are recorded in notebook. Responses are reasonable and closely match those given in Teacher's Edition.
Measure the level of nitrates in a stream in your community.	**Investigate** Part B	Tests are completed correctly; answers to questions are reasonable relative to findings.
Conduct a mathematical analysis (case study) of pollution of surface waters by road salt based on a map, quantitative data, and initial assumptions.	**Investigate** Part C	Questions are answered correctly; all work is shown. Experiment is completed correctly; observations are recorded.
Identify and describe ways that human activity affects surface-water and ground-water quality in your community.	**Investigate** Parts A, B, and C **Digging Deeper; Check Your Understanding** Questions 1, 3 – 4 **Understanding and Applying What You Have Learned** Questions 1 – 4	Responses to questions are reasonable and closely match those given in Teacher's Edition.

Chapter 3

Earth's Natural Resources Water Resources

Activity 5   Water Pollution

Goals

In this activity you will:

- Construct and analyze a physical model of the movement of pollutants in ground water and determine how the pumping of water from wells influences this movement.

- Measure the level of nitrates in a stream within your community.

- Conduct a mathematical analysis (case study) of pollution of surface waters by road salt.

- Identify and describe ways that human activity affects surface water and ground water quality.

Think about It

We rely on clean sources of water to sustain ourselves.

- How do pollutants get into surface water?
- How do pollutants get into ground water?

What do you think? Record your ideas in your *EarthComm* notebook. Be prepared to discuss your responses with your small group and the class.

Activity Overview

Students begin by modeling the movement of pollution (represented by food coloring) in ground water, using the same setup that they used in **Activity 2**. They also model the effect of withdrawal of well water on the movement of pollutants through the ground. Students then test samples of treated (tap) water and untreated (stream) water for nitrate concentrations to compare the quality of the two. Finally, they develop mathematical and physical models of the effect of road salt on a water reservoir. **Digging Deeper** begins with a review of the importance of computational models in representing natural processes. The reading then focuses on the types and effects of various pollutants, including sewage, fertilizers, toxic chemicals, and waste heat.

Preparation and Materials Needed

Part A

If you have reserved the tubs from **Activity 2**, no advance preparation is required for this part of the activity beyond gathering the necessary materials. If you need to prepare the tubs, see the **Preparation** section for **Activity 2, Part B**.

Part B

Students need nitrate test kits for this segment. Please take note of safety precautions provided with the kits. You may want to outline the procedures for the chemical test on the blackboard and review them with the class. Clarify any questions students may have regarding these procedures.

Safety precautions for teams collecting water samples:
- Inform parents of where you are going and your estimated return time.
- Go to the site in teams of three.
- Collect samples from the shore only—do not wade into streams.
- Label the samples.
- Secure lids tightly and rinse the outside of the containers with clean water.
- Wash and dry hands after collecting samples.
- Keep samples refrigerated and use them as soon as possible.

Note: If it is not possible for students to collect samples, you can prepare dilute nitrate-contaminated water. Add 1 gram of potassium nitrate to 1 liter of water. Alternatively, you can collect samples from a local stream for the students to use, but keep them refrigerated and use them as soon as possible. You must comply with local and federal safety standards when working with chemicals. Follow Environmental Protection Agency guidelines for disposing of them. Review the instructions with students before having them test the water samples. You must collect the waste materials from all tests in a plastic 1-gal. jug. Contact your state department of natural resources about proper disposal; do not discard them down the drain.

Part C

If you are low on supplies, you may wish to ask your students to bring in 1-gal. milk jugs for this part of the activity. **Part C** includes a math problem that may be a bit daunting for students at first. It is recommended that you complete the exercise before class, so that you can be aware of how the problem should work and can anticipate any difficulties that may arise.

Materials

Part A
(Some of the materials from **Activity 2** can be reused here.)
- Plastic or rubber dish tub about 12 in. wide, 16 in. long, and 6 in. deep
- Sandbox sand or concrete-making sand to fill the dish tub to a depth of 10 cm
- Two pieces of rigid plastic or metal tubing, 1/2 in. to 1 in. in diameter
- Cheesecloth
- Duct tape
- Measuring cup
- Drinking straw
- Turkey baster
- Food coloring
- Eye dropper
- Clear plastic drinking cups
- Stopwatch

Part B
- Water sample from a local stream (untreated)
- Tap water (treated)
- Nitrate test kit

Part C
- Calculator
- 1-gal. plastic milk jug
- Water
- Salt
- Teaspoon
- Plastic cups (one per student)

Think about It

Student Conceptions

Now that they have completed **Activities 1** to **4**, students should have a better sense of how surface water and ground water move. Imagining how pollutants get into surface water will most likely be within the grasp of most students.

Answer for the Teacher Only

The movement of pollutants in ground and surface waters is discussed in the **Background Information** section of **Activity 5**.

Assessment Tool

Think about It Evaluation Sheet
Use this evaluation sheet to help students understand and internalize the basic expectations for the warm-up activity.

Investigate

Part A: Ground-Water Pollution

1. Prepare the dish tub the same way you did in the investigation in Activity 2. Fill the tub with about 10 cm of sand. You are constructing a model that will enable you to infer how pollutants reach the water table and then move with the ground-water flow.

2. Wrap a small piece of cheesecloth around the end of two pieces of rigid tubing, and tape it in place with duct tape. Bury the pieces of tubing vertically in the sand. The lower end of the tubing should be about 5 cm from the bottom of the tub. Locate the wells as shown in the diagram below.

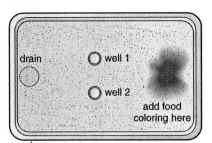

3. As you did in the investigation in Activity 2, attach a drinking straw to the end of the turkey baster.

4. With a measuring cup, pour water onto the "upstream" end of the dish tub (the end opposite the drain hole) at a constant rate of about 1000 mL/min. Wait until the outflow from the drain hole is constant.

5. With an eyedropper, squirt food coloring onto the surface of the sand at the place where you add water.

Use as much food coloring as the eyedropper can hold. Start the stopwatch as soon as the food coloring begins to be washed into the sand by the water supply.

6. With the baster, extract small volumes of water, about 1 mL each time, from Well 1 every five seconds and squirt the water into a clean plastic cup.

   a) When you first detect the food coloring in the water samples, note the time elapsed. Keep taking samples until there is no noticeable food coloring in them, and note the time elapsed.

7. With the eyedropper, squirt several more drops of food coloring into Well 1. Begin taking small water samples from Well 2 with the baster. Continue to extract water from the well until the water draining from the tub shows some food coloring.

   a) Did you detect any of the food coloring in Well 2?

8. With the eyedropper, squirt several more drops of food coloring into Well 1. With the baster, extract about 100 mL/min from Well 2 for a time long enough that the water draining from the tub shows some food coloring.

   a) Again, determine whether you can detect any food coloring in Well 2.

9. Use the results of your investigation to answer the following:

   a) What factors do you think determine how long it took for the food coloring to be detected in Well 1 in Step 6?

 Wear goggles throughout. Clean up spills. Mark baster as lab equipment.

R 185

EarthComm

Investigate

Part A: Ground-Water Pollution

1. If you have retained the tubs from **Activity 2** or have prepared them in advance, students can proceed to **Step 2**. If students need to set up the dish tub, they can refer to **Activity 2, Part B**, for directions.

2. Students should prepare two pieces of tubing as described. They can use the diagram on page R185 in the text to help them to place the wells.

4. It is important to provide a consistent supply of water to the model. It may be useful for students to try pouring water into the drain before conducting the experiment in order to get a feel for approximately what a 1000 mL/min. pouring rate is like.

5. The food coloring represents a pollutant. Students should add the food coloring at the location indicated on the diagram on page R185. After adding the food coloring, students should continue to add water at a constant rate (1000 mL/min.) and should start their stopwatch with the first addition of food coloring. Have them add the food coloring over as short a time period as possible, because the addition of food coloring is supposed to represent a one-time slug of pollutant.

6. a) Students should record the time when they first notice food coloring as well as the time when they notice no more food coloring in the extracted samples. The food coloring is likely to appear fairly abruptly but later disappear only slowly. Students are likely to have trouble deciding when there is no more food coloring in the water. If that troubles them, convey to them that that's the way things work, and that they might start to think about how to explain their concern in terms of our struggles to deal with water pollution.

7. a) Make sure that fresh water is added continuously at the upstream end of the tub for this step. Students should not detect colored water from Well 2, because the rate of withdrawal of water from Well 2 was not sufficient to redirect the flow past Well 1 toward Well 2.

8. a) Make sure that fresh water is added continuously at the upstream end of the tub for this step. Students should now detect some food coloring in Well 2. The reason is that the flow direction in the vicinity of the wells is mainly from Well 1 to Well 2, because of the temporarily much lower water level in Well 2 owing to the rapid withdrawal. If students still do not detect any food coloring in Well 2, have them increase the rate of withdrawal from Well 2 even more and/or decrease the rate of addition of fresh water at the end of the tub. Students should record the time that they add the food coloring to Well 1 and the time that they first detect coloring at Well 2.

9. **a)** As the water table builds up beneath the point of addition of water, the water table takes on a slope from right to left in the drawing on page R185. That causes the ground water to flow from the point of addition toward the point of drainage. The speed of flow increases with the slope of the water table, but it always takes a nonzero time for the water and the pollutant contained in it to make the trip from the point of addition to the point of sampling.

b) The time of detection is much longer for a combination of two reasons:
- Longitudinal dispersion: on the scale of individual sand grains, the various threads of flow through the pore spaces among the grains travel at somewhat different speeds as they branch and rejoin, so the slug of pollutant gradually lengthens in the streamwise direction as it moves.
- Diffusion: the contact between the polluted water and the unpolluted water slowly becomes blurred by the random motions of all the molecules in the liquid, both water molecules and food-coloring molecules. In this brief experiment, the first effect is much greater than the second.

c) The rate of withdrawal from Well 2 was not great enough to divert the path of ground-water flow so that the pollutant plume from Well 1 impinged upon Well 2.

d) The rate of withdrawal from Well 2 was great enough to divert the path of ground-water flow enough that the pollutant plume from Well 1 impinged upon Well 2. It would be a good idea to have students sketch their conception of the path of the pollutant in **Questions 9(c)** and **9(d)**. See the figures below:

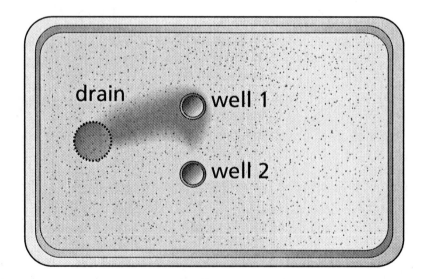

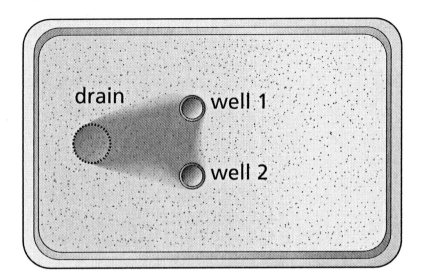

Earth's Natural Resources Water Resources

b) Why do you think that the time during which some food coloring could be detected in Well 1 in Step 6 was much longer than the time it took to squirt the food coloring on the sand surface originally?

c) Why do you think that no food coloring could be detected in Well 2 in Step 7?

d) Why do you think that some food coloring could be detected in Well 2 in Step 8? Dispose of water and straw after use.

Part B: Testing Water for Nitrate
How does the quality of water from local streams, ponds, or untreated wells compare to water that has been processed by a local water treatment plant? To find out, you will test for nitrate chemical levels in two types of water. One source will be an untreated water sample from one of your community's streams. The second will be tap water that was treated by your community's water treatment facility.

1. If you have a stream near your school, collect a water sample and label its location. Your teacher will provide you with a safety protocol for collecting the sample. If such water sources are not available nearby, your teacher will provide you with a sample.

2. Read over the instructions for the nitrate test kits. Be sure you are familiar with the testing procedures as outlined.

3. Test your untreated water sample for nitrate.

   a) Record your results.

   Follow the test-kit instructions carefully. Dispose of chemicals immediately and in the manner described in the directions.

4. Test the treated tap water for nitrate.

   a) Record your results.

5. Use the results to answer the following question:

   a) How does the quality of water from local streams, ponds, or untreated wells compare to water that has been processed by a local water treatment plant?

Part C: Modeling Water Pollution from Road Salt
1. The sketch map shows a reservoir that is located next to an interstate highway. It's winter. The highway crew salt the highway three times in one day. Two days later a storm dumps 2.5 cm of rain on the area.

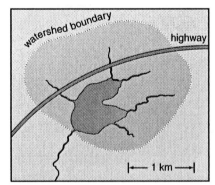

You have the following information:

• The trucks spread 70 kg of salt per kilometer on each traffic lane of the highway.

• The highway has three traffic lanes in each direction.

• The length of the highway that runs through the watershed area of the reservoir is 5 km.

Part B: Testing Water for Nitrate

Teaching Tip

Students may not be familiar with nitrate as a pollutant. You may wish to review briefly the source of nitrates (mainly from agricultural chemicals and fertilizers). Have them record their predictions about which sample will have the higher nitrate concentration. Students living in urban areas may not think that fertilizers will have a strong impact on their water supply, but you can point out that very large quantities of nitrate-containing fertilizer is used on private lawns, public parks, and golf courses.

1. Students should collect, or obtain from you, the untreated water sample.

2. Review safety procedures for using the nitrate testing kits.

3. Students will need about 10 minutes to label and run the nitrate test on the untreated sample.

4. Students should follow the same procedures to test their tap (treated) water sample.

5. a) The degree of variation between the two samples will vary depending upon where you live, but students should find that the treated sample has lower nitrate concentrations than the untreated water sample.

Teaching Tip

After students have tested both samples, have a class discussion about the possible sources for the pollution (nitrate). Students can refer to the information they collected in **Activity 1**. If the untreated water samples have come from a variety of sources, have students plot the locations on their maps and attempt to identify any correlation between location and nitrate levels.

Part C: Modeling Water Pollution from Road Salt

1. Students may be a bit intimidated by the math for this part of the investigation. You may want to review the given information and assumptions with the class before they begin.

Chapter 3

- The watershed area of the reservoir is 10 km².

- All of the rain that falls on the watershed area of the reservoir runs off into the reservoir. (The ground is frozen solid, so the part of the rain that infiltrates into the ground to become ground water is negligible.)

- The average depth of the reservoir is 5 m.

- The surface area of the reservoir is one square kilometer.

Possible assumptions:

- The reservoir water that is displaced by the new runoff from the rainstorm contains none of the salt.

- The mixing time of the reservoir is shorter than the time elapsed before the next major rainstorm.

Additional helpful facts:

- There are 1000 L in a cubic meter.

- There are 1000 mL in a liter.

2. Using this information, answer the following questions:

a) What is the average concentration of salt in the runoff that enters the reservoir from its watershed? Express your answers in grams of salt per liter of water.

b) What is the concentration of salt in the reservoir after the rainstorm and before the next rainstorm? Express your answers in milligrams of salt per liter of water.

c) The United States government has issued what is called a health advisory about sodium in public water supplies. A health advisory provides guidance about levels of pollutants, without specifying mandatory maximum concentrations. According to the health advisory, levels of up to 20 mg of sodium per liter of water should not result in adverse health effects over short time periods for most people. How does your

2. a) To find the concentrations of salt in the runoff, students will need to know the total amount of salt deposited on the highway, in grams, and the total volume of the runoff, in liters.

It is given that the trucks spread 70 kg of salt per kilometer on each lane, and that there are 5 km of highway in the area. Therefore, 70 kg x 5 km = 350 kg of salt are deposited on each lane. There are three lanes going in two directions, so six lanes total. Therefore, in one application, the trucks deposit 350 kg x 6 = 2100 kg of salt. Note, however, that the trucks dump salt three separate times, meaning that the total amount of salt applied in the area is 2100 x 3 = 6300 kg of salt. There are 1000 g in 1 kg, so this is (6300 kg x 1000 g/1 kg) = **6,300,000 g** of salt.

It is given that 2.5 cm of rain fell in over the watershed, which has an area of 10 km^2 (given). Students will need to assume that any precipitation on the ground before the rainstorm is negligible, so the total volume of the runoff is equal to the area of the watershed (10 km^2) times the depth of rain (2.5 cm). Students will need to make some conversions to get the proper units. They will need to know that there are 1,000,000 m^2 in 1 km^2 and that there are 100 cm in 1 m. Therefore, the area of the watershed is (10 km^2 x 1,000,000 m^2/1 km^2) = 10,000,000 m^2 and the depth of water (rainfall) is (2.5 cm x 1 m/100 cm) = 0.025 m. The total volume of the runoff is therefore (10,000,000 m^2 x 0.025 m) = 250,000 m^3. There are 1000 L in 1 m^3, so the total volume of the runoff is (250, 000 m^3 x 1000 L/1 m^3) = **250,000,000 L** of water.

Therefore, the salt concentration of the runoff is (6,300,000 g/250,000,000 L), or **0.0252 g/L**.

b) To answer this question, students will need to know the volume of water in the reservoir and the amount of salt that enters the reservoir.

From **Question 2**, it is known that 6300 kg of salt was deposited in the area. There are 1000 g in 1 kg and 1000 mg in 1 g, so this 6300 kg of salt is (6300 kg x 1000g/1 kg x 1000 mg/1 g) = 6,300,000,000 mg of salt.

The volume of water in the reservoir is equal to its depth (given at 5 m) times its area (given at 10 km^2). Students will need to do some conversions to get a volume of (5 m x 1,000,000 m^2) = 5,000,000 m^3. There are 1000 L in 1 m^3, so 5,000,000 m^3 is equal to 5,000,000,000 L of water in the reservoir.

The salt concentration of the reservoir is therefore (6,300,000,000 mg/5,000,000,000 L) = **1.26 mg/L**.

c) Students will need to calculate the percent of sodium found in their concentration (1.26 mg/L x .40). They will find that it is well below levels constituting a health advisory.

Earth's Natural Resources Water Resources

calculated concentration of salt in the reservoir compare with this figure for sodium? Keep in mind that the composition of the salt is sodium chloride (NaCl) and that the sodium accounts for about 40% of the weight of salt.

d) Can you think of any problems with the assumptions you were given above? If so, how do you think they would affect the results from your model? In other words, how realistic is your model?

3. Fill a one-gallon plastic milk jug with water. The jug holds about 3.8 L, but assume that it holds a full 4 L. A teaspoon of salt weighs about 0.5 g. Put a teaspoon of salt in the jug, shake it well, and pour small amounts into plastic drinking cups,

one for each member of your group. Taste the water, and decide whether you can taste any of the salt. If not, add another teaspoon of salt and try again. Repeat the process until everyone can taste the salt. Compare the salt concentrations (in grams per liter) at which each member of your group was first able to taste the salt.

a) What is the minimum concentration of salt in water that causes the water to taste salty to members of your group?

b) How do these values compare with the salt concentration you calculated for the reservoir?

c) How do they compare with the United States government health advisory on sodium given earlier?

Reflecting on the Activity and the Challenge

You modeled the pollution of a ground water aquifer, modeled pollution of a reservoir by road salt, and tested the quality of water in your community. Understanding ways that water can

become polluted will help you understand how the three developments proposed in the **Chapter Challenge** might affect the quality of your community's water supply.

d) Students had to make several assumptions here. They had to assume that there was no snow or ice on the ground that could contribute to the runoff (i.e., the volume of the runoff was equal to the volume of rainfall). They had to assume that all of the runoff entered the reservoir and that none of the salt was lost (i.e., all of the salt entered the reservoir). They had to assume that when the runoff entered the reservoir, an equal amount of water was displaced and did not dilute the salt concentration of the runoff.

3. a) Answers will probably vary quite a lot from student to student, depending upon their sensitivity to a salty taste. The developer of this exercise could not detect a salty taste with only one teaspoon of salt but sensed a slight saltiness with two teaspoons of salt.

   b) On the assumption that one teaspoon of salt weighs about 0.5 g, one teaspoon of salt per 4 L of water is equivalent to about 0.1 g/L, or 100 mg/L, which is greater by a factor of about a hundred than the concentration calculated for the reservoir.

   c) A salt concentration of 100 mg/L is five times the health advisory level of 20 mg/L.

Assessment Tools

EarthComm **Notebook-Entry Checklist**
Use this checklist as a quick guide for student self-assessment and/or an opportunity to quickly score student work. Add further criteria specific to your classroom needs or to this particular investigation.

Investigate Notebook-Entry Evaluation Sheet
Point out the criteria listed on this evaluation sheet that are relevant to this particular investigation. Encourage students to internalize the criteria by making them part of your assessment conversations as you circulate around the classroom.

Reflecting on the Activity and the Challenge

Review with students what they learned in the investigation. Discuss the implications of removing water from a well with respect to water pollution and how this ties into the **Chapter Challenge**. Have them consider the kinds of locations that the water samples they tested in **Part B** came from. Ask students to consider how nitrates might get into ground-water or surface-water systems in cities and towns. This can be tied into **Part A** of the investigation. Discuss with students possible places where the pollutants end up. Do they stay within the local community or affect those more remote? How do they get there?

Digging Deeper

POLLUTION IN SURFACE WATER AND GROUND WATER

Computational Models

Scientists often try to develop computational models to represent natural processes. Such models serve to make predictions, as well as to help organize thinking about a problem. Later, when there is an opportunity to make actual measurements in the field, the predictions of the model are either verified or turn out to be wrong. If they are verified, that tells the scientists that the hypotheses that went into their model are probably right. If the predictions are wrong, however, then the scientists know that they need to rethink the hypotheses, because they are either wrong or incomplete (or both!). In Part C of the investigation, you also developed a computational model in which you assumed some data and made some hypotheses.

Kinds of Pollutants

Both surface water and ground water can become polluted. Surface water generally contains a greater variety of pollutants, from a greater variety of sources. That is because there are so many ways a pollutant can get into a river or lake. Polluted ground water may contain fewer kinds of pollutants, but there may be a larger amount of each pollutant than in a typical surface water supply. It is usually much more difficult to solve problems of ground-water pollution than problems of surface-water pollution. That is because ground water moves far more slowly than surface water. It might take many human lifetimes for pollution to clear from a large aquifer by natural ground water flow.

Figure I Livestock waste can pollute water supplies.

R 189

Digging Deeper

Assign the reading for homework, along with the questions in **Check Your Understanding** if desired.

Assessment Opportunity

Reword or restructure the questions in **Check Your Understanding** for a brief quiz. Use the quiz (or a class discussion of the questions in the textbook) to assess your students' understanding of the main ideas in the reading and the activity. Three sample questions are provided below:

Question: Why is it usually more difficult to solve problems of ground-water pollution than problems of surface-water pollution?
Answer: Because ground water moves much more slowly than surface water, it might take many human lifetimes for pollution to clear from a large aquifer by natural ground-water flow.

Question: What is waste heat and why is it a problem?
Answer: Power plants use large quantities of water to cool and condense steam. This water is recycled into the environment after use. However, when it is returned to the environment, it is 5° to 10°C warmer than the water body it enters. The warmer water does not hold as much oxygen, so less is available for organisms. The warmer water can thus kill organisms and/or cause the replacement of existing organisms by warmer-water organisms.

Question: Name two nutrients that often pollute water supplies and name two sources of each.
Answer: Two common nutrient pollutants are nitrates and phosphates. Nitrates are introduced to water supplies through fertilizers and waste from organisms. Phosphates can be introduced to water supplies through fertilizers, waste from organisms, or detergents.

Teaching Tip

The photo on page R188 shows a thin film of a petroleum pollutant (gasoline) floating on top of some pond water. This photo is an opportunity to review the relationships between oil and water uncovered in **Activity 7** of **Chapter 1**, and to stimulate a discussion of potential sources for groundwater contamination like leaking underground petroleum and chemical storage tanks.

Chapter 3

Earth's Natural Resources Water Resources

Figure 2 Nutrients in fertilizers applied to fields can pollute water supplies.

Many kinds of substances can pollute water. Below is a list of the most common pollutants. Most of these fall into one of two categories: organic, or non-organic.

Organic:
    sewage
    livestock wastes
    pathogenic (disease-causing) microorganisms

Non-organic:
    nutrients in fertilizers (mainly nitrates and phosphates)
    industrial and commercial chemicals
    road salts
    agricultural pesticides
    acidic mining wastes
    waste heat
    radioactive waste

Domestic Sewage

About three-quarters of homes in the U.S. are served by municipal sewage systems. The remaining one-quarter discharge their sewage into home septic systems (or even directly into the ground!). The human wastes in sewage are not themselves especially harmful, but disease-causing microorganisms, which are common in sewage, are hazardous to health. Many illnesses, like cholera and typhoid, are caused by contact with sewage. Such illnesses are much less common in the United States than in areas of the world where people are more likely to be exposed to untreated sewage. Coliform bacteria which live in the intestinal tracts of all warm-blooded animals (including humans), are a common type of bacteria that are generally not harmful. However, their presence in sewage is commonly used as a signal of sewage contamination.

Teaching Tip

In addition to fertilizers that are applied to cropland (*Figure 2* on page R190), pesticides applied to cropland have become a major environmental issue. Pesticides can infiltrate the underlying ground water and then flow underground to join surface-water reservoirs that humans rely on for potable supplies. Factors like increased rainfall and high ground-water withdrawal can increase the movement of contaminants between ground-water and surface-water reservoirs. Discuss with students how this was illustrated in **Part B** of the investigation.

Figure 3 Water testing at Lake Cachuma, the water storage reservoir above Santa Barbara, California.

Nutrients from Fertilizers

Plants need many inorganic chemical nutrients for good growth. The most important of these are the elements nitrogen and phosphorus. Nitrogen is the most abundant gas in the Earth's atmosphere, but it cannot be used directly by plants. Certain soil bacteria convert nitrogen gas into soluble forms of nitrogen, mainly nitrate ions, which can be used by plants. Lightning bolts do the same thing, and of course man-made chemical fertilizers contain soluble nitrogen. The problem with nitrogen-containing fertilizers is that they can be "too much of a good thing": excess nitrogen promotes the growth of algae in natural surface waters, and when the algae die and decay they rob the waters of the oxygen needed by aquatic animals. Runoff from croplands is not the only culprit; fertilizer on lawns, golf courses, and home gardens often carries even more excess nitrogen than croplands.

Human wastes contain water-soluble nitrogen compounds. In areas where untreated or inadequately treated sewage finds its way into ground water supplies, nitrate levels in drinking water can be excessive. Human infants can be harmed by drinking water that has high nitrate levels because nitrate reduces the amount of oxygen carried by red blood cells. Too much nitrate can cause "blue-baby syndrome," which can lead to suffocation.

R 191

Teaching Tip

Scientists monitor the quality of water (*Figure 3* on page R191) in reservoirs. This is important, because human activities can affect the quality, distribution, and quantity of water resources.

Earth's Natural Resources Water Resources

Phosphorus is used by plants in the form of phosphate ions. The most common form of phosphorus in fertilizers is ground-up phosphate rock, a type of sedimentary rock. Deposits of a material called guano, the excrement of bats and seabirds, are also mined for phosphorus. The problem with soluble phosphorus is similar to that with nitrogen: it causes algal blooms, such as the one shown in *Figure 4*, in natural waters. Another problem with phosphorus is that the phosphates contained in household soaps and detergents can pollute ground water supplies. In many heavily populated communities that rely on ground water from water supplies, restrictions have been placed on the phosphate content of detergents.

Figure 4 Soluble phosphorus can cause algal blooms such as the one shown on the left.

Toxic Chemicals

In earlier times, the number of artificial chemicals and materials used by people in their daily lives was small. Today, however, such substances number in the tens of thousands, and the list grows rapidly. A large percentage of these are potentially or definitely a health hazard. Many of these were found to be toxic only after long use. Some of the major chemical pollutants are gasoline, chemical solvents, agricultural pesticides and herbicides, and compounds of heavy-metal elements like copper, zinc, lead, mercury, and cadmium used in industrial processes.

Unfortunately, the rate of development of new substances is far greater than the ability of government and private testing laboratories to investigate their toxicity. Inevitably, many of these toxic substances find their way into surface water supplies and ground water supplies. Partly this is from everyday use. In many places, however, toxic chemicals have been dumped, illegally, on the land

Teaching Tip

The left-hand photo in *Figure 4* on page R192 shows the surface slick of a red tide bloom that followed closely after heavy rains associated with Hurricane Floyd, which ended a very dry summer. The photo on the right shows the same area after the bloom period was over. "Red tide" is a common name for algal blooms where certain phytoplankton species contain reddish pigments and "bloom" such that the water appears to be colored red. The term "red tide" is thus a misnomer because the phenomenon is not associated with tides. The species responsible for the bloom picture in *Figure 4* (which was taken at the Bigelow Laboratory in Maine) is Prorocentrum micans.

Discuss with your students how heavy rains could induce an algal bloom by increasing runoff and increasing the concentration of soluble phosphorus in the water. The photo on the right shows the same area after the bloom period was over.

Chapter 3

Figure 5 Photograph of a toxic-waste site.

surface or stored in decaying drums at dumpsites. Cleaning up sites such as the one shown in *Figure 5*, all around the United States, is a multi-billion-dollar task.

Waste Heat

Most of the electricity in the United States is generated in power plants where burning of fossil fuels produces steam to drive turbines, which in turn drive electrical generators. Enormous quantities of water are used to cool and condense the steam back into liquid water, to be recycled in the power plant. The cooling water is put into rivers, lakes, or the ocean. It is typically 5°–10°C warmer than the water bodies that it enters. This causes three major problems. Warm water cannot hold as much dissolved oxygen as cool water, therefore less oxygen is available for aquatic animals. The higher temperatures also tend to lead to the disappearance of some species of organisms and the replacement of these species with warmer-water organisms. Sudden temperature changes can kill organisms outright, by what is called thermal shock.

Figure 6 Diversion dam for the Pit River Power Plant in central California.

Check Your Understanding

1. What are some of the health hazards of domestic sewage?

2. If coliform bacteria are generally not harmful, why is there concern about them?

3. How do chemical fertilizers cause water pollution?

4. What risks do power plants pose to the environment?

R 193

Check Your Understanding

1. Domestic sewage can contain microorganisms that are responsible for causing illnesses like cholera and typhoid.

2. Although coliform bacteria are not generally harmful, their presence in water supplies is generally a sign of sewage contamination.

3. Excesses of chemical fertilizers, like nitrates and phosphates, can promote the growth of algae in natural surface waters. When the algae die and decay, oxygen that is needed by aquatic animals is removed from the water.

4. Water that is recycled by power plants is often 5 to 10° C warmer than the water into which it is dumped. Warm water holds less oxygen than cold water, so less is available for organisms. The warmer water can kill organisms and/or cause the replacement of organisms by warmer-water organisms.

Assessment Tool

Check Your Understanding Notebook-Entry Evaluation Sheet
Use this sheet to evaluate the extent to which students understand the key concepts explored in **Activity 5** and explained in **Digging Deeper**, and to evaluate the students' clarity of expression.

Earth's Natural Resources Water Resources

Understanding and Applying What You Have Learned

1. In some northern states, harmful quantities of salt are appearing in streams, ponds, and even shallow ground water aquifers. What might be the source of the salt and how is it getting into these water sources? How might this salty water affect fresh water ecosystems?

2. The High Plains Aquifer is one of the largest ground water aquifers in the United States. It is located under Colorado, Nebraska, and parts of Texas, Oklahoma, Kansas, South Dakota, and Iowa. It is a nonrenewable water resource and has shown an increase in levels of nitrate. List the potential sources of these increases. Describe how the use of the topsoil above this aquifer may be related to its pollution.

3. Make a list of ways that citizens in your community can reduce water pollution.

4. What is the relevance of the experiment you completed in Part A of the investigation with respect to water pollution? Is your community affected by this problem? Could it be? What steps could be taken to reduce the effect?

Preparing for the Chapter Challenge

How will the construction or operation of the proposed developments pollute your town's drinking water? What factors will influence the type or amount of water pollution? What can be done to minimize water pollution from these developments? Your group should prepare a poster and present a report to your citizens' planning group addressing these questions. Prepare this poster for an audience of an intelligent group of concerned citizens who are not knowledgeable about water pollution.

Inquiring Further

1. **Local water quality**

   Research how well your local water provider has maintained the water quality in your community.

   a) Look in the provider's water quality report.
   b) Also try looking for past violations that have been reported to the United States Environmental Protection Agency (EPA). Go to the *EarthComm* web site to help you find this information. Write down the nature of any violations and the dates in which they occurred.

Understanding and Applying What You Have Learned

1. Salt enters surface water when it is used to limit ice on highways and roads. It then becomes part of the surface-water runoff system. Salty water that enters an ecosystem can kill organisms that are not adapted to dealing with saline conditions.

2. Nitrate pollution in the High Plains (Ogallala) Aquifer may be the result of synthetic fertilizer applications, the addition of soil amendment (manure) on cropland, or livestock production.

3. Answers will vary depending upon the kinds of pollutants that are present in your area.

4. If a pollutant is introduced into ground-water flow at a certain point, the pollutant travels downstream and spreads both laterally and longitudinally as it travels, creating a pollutant plume. Ground-water wells in the path of the plume become polluted. Even if the well is not located in the general direction of travel of the plume, pumping from the well can divert the plume toward the well. If the rate of pumping is sufficiently large, the pollutant plume can be drawn into the well.

Preparing for the Chapter Challenge

When they have completed **Activity 5**, students should be able to list at least five different pollutants that can be found in water, and they should be able to identify the source of these contaminants. Student presentations and reports should explain whether or not the proposed development could add pollutants to the water resources of the community, and if so, how. They should prepare this poster for an audience of intelligent—but not particularly knowledgeable—concerned citizens. The poster should be easy to see from the back of a room, and it should illustrate pollution sources that are a real threat to your community. Both surface-water and ground-water contamination should be addressed.

For assessment of this section, as suggested in the sample rubric at the start of the chapter (see **Assessment Rubric for Chapter Report on Water Resources** on pages 504 and 505), students should include a discussion on:
  • whether or not the construction and operation of these developments will pollute your town's drinking water
  • how potential for pollution can be determined

Inquiring Further

1. Local water quality
Answers to the question will vary. This is an opportunity for students to add to their knowledge of pollutants (or potential pollutants) to the local water supply. Students can consult the *EarthComm* web site for help in answering the questions.

2. Pollution from pets
Responses will vary depending upon your community.

3. Pollution in films
A Civil Action (rated PG, with some profanity) is one recent film, based on an actual case, that deals with litigation as a result of industrial pollution.

4. Thermal pollution
Responses will vary depending upon your community. The effects of increased water temperatures near some power plants have been documented to cause changes in the aquatic life present. There may be documented studies that relate to the power plant your students are investigating. If not, there are many well-documented studies of other sites that may be applicable.

NOTES

_____

_____

_____

_____

_____

_____

_____

_____

_____

_____

_____

_____

_____

_____

_____

_____

_____

_____

_____

_____

_____

Chapter 3

c) Also try searching EPA Envirofacts Warehouse— facilities information (visit the *EarthComm* web site to find this information) to research the status of industries in your county. This resource indicates if a particular industry is in compliance with federal pollution laws.

2. **Pollution from pets**

Feces and urine from cats and dogs is a major source of water pollution in some urban and suburban areas (especially those without "pooper-scooper" ordinances).

a) Try to determine the size of the cat and dog population from your community's authorities, and then, by making assumptions about the average daily output of cats and dogs, arrive at an estimate of the total weight of cat and dog feces and the volume of cat and dog urine deposited on the yards, sidewalks, streets, and other public areas of your community per year.

b) What do you think are the effects of this material on water quality in your community or in other areas surrounding your community?

3. **Pollution in films**

Water pollution has been the topic of motion pictures. As a class, view a film about water pollution. After watching, have a class discussion on the problems associated with the following issues:

a) detecting and cleaning up past industrial pollution of water sources;
b) determining whether small concentrations of industrial pollutants are harmful to human health or not;
c) whether industries that in the past used substances that are now known to be hazardous but were not thought to be hazardous at the time should be held legally liable for cleanup and for damage awards.

4. **Thermal pollution**

Locate the power plant nearest to your school, or, if there is none in your community, the power plant nearest to your community. Contact the operators of the power plant to find out their method of cooling. Some power plants use air cooling, but most use water cooling. If the plant uses water for cooling, find out the source of the water, the volume used, and the method of disposal. If the cooling water is discharged into a river or lake, try to obtain data on the temperature rise in the water body.

NOTES

ACTIVITY 6— WATER TREATMENT

Background Information

Water Treatment

Water purification occurs in nature. The processes of evaporation and condensation separate the water from substances that are dissolved in the water. Bacteria convert dissolved organic contaminants into simple and harmless compounds. Sand and gravel filter out suspended material that makes water cloudy. Rainwater and ground water that is purified by bacteria and filtered by sediment provide clean drinking water. Unfortunately, our demand for clean water is much greater than nature's ability to supply our communities, so we rely on municipal water-treatment plants.

No matter where a community's water comes from, it must be treated, because even where water is used carefully, pollutants are bound to be introduced. You can't wash clothes, take a shower, or flush the toilet without adding pollutants to the water! The goal of water treatment is to produce water that's safe for drinking and doesn't have objectionable taste, smell, or appearance.

Treatment of wastewater from municipal sewage systems involves a series of steps: screening, flocculation, filtering, and disinfecting. Depending on your local water sources and treatment needs, your community's water-treatment process may be slightly different than what is presented here. Keep in mind also that the wastewater that passes through home septic systems is not treated at all: it goes directly into the ground and tends to add to the pollutant levels in local ground-water supplies.

Screening, Flocculation, and Settling

Sewage is first passed through screens to remove large pieces of debris and other solids. Then it flows slowly through sedimentation tanks, where gritty mineral material settles to the bottom and lightweight sludge floats to the top. To aid in the settling process, alum (aluminum sulfate, $Al_2(SO_4)_3$) and slaked lime (calcium hydroxide, $Ca(OH)_2$) are sometimes added to the sewage to cause particles of sediment to coagulate (stick together) to form larger clumps, called flocs, which settle relatively fast. The sludge is sent to a separate container where the organic matter in it is digested by microorganisms.

Filtration

In the next treatment step, the water passes through sand, gravel, and activated carbon (charcoal). These filters help remove minute particles that give water a bad smell or taste. If the water contains inorganic (nonliving) components that cannot be adequately removed by filtering, a chemical exchange process—called ion exchange—can be used. This chemical process is effective in treating hard water, which is water that does not easily produce suds in the presence of soap. It can also be used to treat arsenic, chromium, excess fluoride, nitrates, radium, and uranium. As part of this stage, the wastewater is aerated (exposed to the oxygen of the atmosphere), which greatly speeds up the natural oxidation and decomposition of organic matter in the water by the activity of microorganisms.

Disinfection

Water is disinfected before it enters the distribution system. Disinfectants kill remaining dangerous microbes. Chlorine is a very effective disinfectant, and residual concentrations of chlorine guard against biological contamination in the water distribution system. Ozone is a powerful disinfectant, but it is not effective in controlling biological contaminants in the distribution pipes.

Sometimes the disinfecting process produces by-products. Disinfecting by-products (DBPs) are contaminants that form when chlorine reacts with organic matter that is in treated drinking water. Long-term exposure to some DBPs may increase the risk of cancer or have other adverse health effects. The United States Environmental Protection Agency regulates the presence of four DBPs in drinking water, and scientists are continuing to study the issue.

Levels of Treatment

Primary treatment involves only settling and perhaps chlorinating. Such treatment is better than nothing, but the water is unsatisfactory for human use. Secondary treatment involves filtering and oxidation by microorganisms, and results in wastewater of much better quality. In some places, tertiary treatment is used. Tertiary treatment involves (in addition to the processes of primary and secondary treatment) flocculation, disinfecting, and use of specific additives to remove undesirable excess compounds of iron, manganese, etc. Tertiary treatment is designed to produce high-quality domestic water that can be added to public water supplies—although the public is still not very willing to drink treated sewage, knowingly!

More Information – on the Web

Visit the *EarthComm* web site www.agiweb.org/earthcomm to access a variety of links to web sites that will help you deepen your understanding of content and prepare you to teach this activity. Many of the sites also contain images that you can download.

Chapter 3

Goals and Assessment

Clarify that the goals indicate what students should understand and be able to do as a result of the activity. Make sure students understand that Chapter Assessments are based upon these goals.

Goal	Location in Activity	Assessment Opportunity
Model and explain key processes and stages in water treatment and filtration.	**Investigate** Part A and B **Digging Deeper; Check Your Understanding** Questions 1 and 2 **Understanding and Applying What You Have Learned** Question 2	Experiments are completed correctly. Responses are based on observations, and closely match those given in Teacher's Edition.
Understand how and why water is treated before human consumption.	**Digging Deeper; Check Your Understanding** Questions 2 – 3	Responses to questions closely match those given in Teacher's Edition.
Research and describe the water treatment process used by your community.	**Investigate** Part C **Understanding and Applying What You Have Learned** Question 3	Community's water-treatment process is outlined correctly. Responses to questions are appropriate for the community.

NOTES

Chapter 3

Earth's Natural Resources Water Resources

Activity 6   Water Treatment

Goals

In this activity you will:

- Model and explain key processes and stages in water treatment and filtration.

- Understand how and why water is treated prior to human consumption.

- Research and describe the water treatment process used by your community.

Think about It

On a backpack trip high in the mountains you become separated from your group. You realize that it may be several days before you are found.

- What concerns would you have about drinking water?
- What contaminants might be found in the water?
- What precautions should you take?

What do you think? Record your ideas in your *EarthComm* notebook. Be prepared to discuss your responses with your small group and the class.

Activity Overview

Students begin the investigation by modeling the treatment of a water sample using alum and sediment. After treating the water sample with alum, students put the sample through a filtration process, using sand and gravel as a filter. Then they examine their community's water-quality report to outline the community's water-treatment process and applicable costs. **Digging Deeper** takes a close look at the steps involved in water treatment, including screening, flocculation, settling, filtration, and disinfection. The reading also examines the cost and need for water treatment to different levels: primary, secondary, and tertiary.

Preparation and Materials Needed

Parts A and B

Gather the materials in advance of the activity. You will need untreated water samples (about 2 L per group) from a lake, swamp, or stream. You can make an untreated water sample from tap water: add vanilla flavoring, a small amount of soil, 1 tbsp. of oil, and 1 to 2 drops of food coloring. Have student volunteers bring in 2-L bottles. You will need silt or clay (fine sediment) and alum (potassium aluminum sulfate)—available in the spice aisles of grocery stores or from science supply companies.

Part C

You will need to obtain a copy of your community's water-quality report. Try contacting your local municipality or check its web page.

Materials

Part A
- Water samples from an untreated lake, swamp, or stream (2 L)
- Three 2-L plastic bottles with just enough of the top part removed to make a 5-cm wide opening, OR three 1.5-L beakers
- Silt or clay
- Alum (potassium aluminum sulfate—from grocery-store spice aisle or supply catalog)
- Other materials: balance, glass stirrers, scoop or spoon, paper towels, and white notebook paper

Part B
- 2-L plastic soda bottle with the bottom 5 cm cut off
- Two 500-mL beakers
- Nylon screen or a piece of nylon stocking

- Rubber band
- Clean aquarium gravel
- Clean sand
- Dropper filled with non-chlorine bleach mixed with water

Part C
- Copy of your community's water-quality report

Think about It

Student Conceptions

Give students a few minutes to think about and respond to the questions. Hold a brief discussion. Students will likely say that they are concerned about how clean the water is, whether or not it is safe to drink, and how they will make sure they have a supply.

Answer for the Teacher Only

Possible contaminants include animal wastes, chemicals, and microorganisms (which are the most common cause of unsafe drinking water in the backcountry). Things that would indicate that the water should be treated before they would consume it include color, odor, cloudiness, its location, or evidence of animals using the water source. It is important to note that water can look and smell perfectly fine but still be unsafe for drinking. Possible precautions that can be taken if water is suspected to not be potable include boiling the water, treating of the water with iodine tablets, and filtration.

Assessment Tool

Think about It Evaluation Sheet
Use this evaluation sheet to help students understand and internalize the basic expectations for the warm-up activity.

NOTES

Chapter 3

Investigate

Part A: Modeling Water Treatment

1. Place 1 L of untreated water into a plastic bottle or beaker. Label this container "no alum" and set it aside.

2. Place 1 L of untreated water into a second plastic bottle or beaker. Label this beaker "alum."

   a) In your notebook, note any odors associated with the samples.

3. Pour the untreated water in the "alum" beaker back and forth using a third empty container. Do this 15 times.

   a) Note your observations about the appearance and odor of the water.

   b) If there is an increase in odor, explain why.

4. Add 50 g of silt or clay to each water sample. Add 20 g of alum to the "alum" beaker.

5. Stir the water in each container slowly for five minutes.

   a) What effect did the alum have on the water?

6. Allow each of the containers to sit for 15 minutes.

   a) Record your observations every five minutes. Make your observations while holding a blank sheet of white paper behind each container.

   Note: Keep the water in the "alum" beaker, undisturbed for use in Part B. While waiting you may construct the filtering device for Part B.

   Use only water samples provided by your teacher. Do not taste the water. Smell the water carefully by waving fumes toward your nose. No chemicals should be consumed. Wash hands well after the activity.

Part B: Modeling Water Filtration

1. Using a rubber band, securely attach a nylon screen or piece of stocking across the mouth of a 2-L plastic soda bottle with the bottom 5 cm cut off. Turn the bottle upside down and set it in a 500-mL beaker. Using another beaker, measure 800 mL of sand and pour it gently into the upside down bottle. Add 400 mL of gravel on top of the sand, as shown in the diagram. The sand and gravel will serve as your filter. Clean the filter by pouring 4-L of tap water slowly into it. Pour off the filtered water.

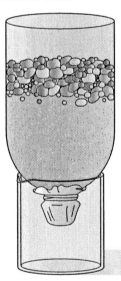

2. When the water with alum from Part A of the investigation has been undisturbed for 15 minutes, carefully pour the water into the filtering device. Try not to disturb the sediment at the bottom of the beaker.

Investigate

Part A: Modeling Water Treatment

1. Students should set aside some of the untreated water sample (labeled "no alum") for comparison at the end of the investigation.

2. In addition to odor, students can record appearance (color, anything floating in the sample, etc.). Remind them not to drink the water.

3. a) Students should observe an increase in odor.

    b) The boxing (pouring the water back and forth between containers) causes the dissolved odors to gain energy. Thus, the dissolved odors become less soluble and escape into the air.

5. a) The alum causes the fine sediment particles in the water to coagulate (clump into larger aggregates). The larger aggregates have much greater settling speeds, causing the solids to settle out rapidly and leaving the water with much less cloudiness.

6. a) Students can set up a small data table to record their observations at 5-minute intervals. They should be careful to disturb the beaker as little as possible during observation. Over time, students should observe that the alum collects fine, suspended particles in the water, which then settle to form sludge on the bottom of the beaker.

Teaching Tip

While students are waiting between observations, they can begin to construct the filtering device used in **Part B**. Remind them not to dispose of or disturb their water sample at the end of the 15-minute observation period.

Part B: Modeling Water Filtration

1. Students can use the illustration on page R197 in the text to help them construct their filters.

2. Students should pour the water sample from **Part A** slowly and carefully into the filter.

Chapter 3

Earth's Natural Resources Water Resources

3. After the water has gone through the filter, place two drops of non-chlorine bleach into the clean filtered water.

4. Compare the "no alum" sample of water with your filtered results. (Do not drink the water!)

   a) Record your observations in your notebook.

5. Dispose of the water as instructed.

Part C: Water Treatment in Your Community

1. Obtain a copy of your community's water-quality report. If your community has a population of more than 100,000, this report is required to be available to the public. Contact your local municipality or check their web page. If the information you require to complete the following steps is not in the water-quality report, have one student call the water provider and ask the question(s) for the class.

2. Examine the water-treatment process that your community uses.

   a) Outline your community's water-treatment process. Use diagrams if necessary.

   b) How much does it cost to treat water in your community?

   c) How much water is treated daily?

   d) What is the maximum capacity of the system?

3. Investigate the water supply for your community's water-treatment facility.

   a) What would happen to the water-treatment process if a great deal of soil were added to the system at once? Is there such a thing as sediment pollution?

   b) List some activities that destroy vegetation that anchors the soil.

   c) Would any of these activities occur during the construction of the proposed development?

Reflecting on the Activity and the Challenge

You have just modeled two important steps in treating water to make it safe for drinking. These two steps are probably used at your local water-treatment plant, which you investigated. Cleaning up water is not free. The more pollutants there are in the water, the more complicated and costly it is to treat it. You will need to consider the pollution the new developments will add to your community's water system.

3. This step represents the disinfection step of the water-treatment process.

4. Students should note any differences between the original, untreated water sample and the water that has passed through the filtration process. Observations will vary depending upon how contaminated the water sample was to start with.

Part C: Water Treatment in Your Community

2. Answers will vary. Students can investigate their community's water-treatment process by searching the Internet or by contacting the community's water department.

3. a) If a great deal of soil is added to the system at once, the costs and perhaps length of treatment time would increase because the sediment has to be removed. The increase in suspended particles that then settle out can clog the filters. The filtering process may slow down or even stop.

   b) Activities that destroy vegetation include farming, removal of terraces on hillsides, removal of windbreaks, which leads to wind erosion of soil in plowed fields, grading to sculpt the land surface, excessive and improper use of herbicides, and excessive water use, which lowers the water table (this also affects subsoil moisture levels).

   c) Although answers will vary depending upon your community, it is likely that some of these activities would take place during the construction of the proposed development.

Assessment Tools

EarthComm Notebook-Entry Checklist
Use this checklist as a quick guide for student self-assessment and/or an opportunity to quickly score student work. Add further criteria specific to your classroom needs or to this particular investigation.

Investigate Notebook-Entry Evaluation Sheet
Point out the criteria listed on this evaluation sheet that are relevant to this particular investigation. Encourage students to internalize the criteria by making them part of your assessment conversations as you circulate around the classroom.

Reflecting on the Activity and the Challenge

Students need to consider the real-life situation of providing water for hundreds or possibly thousands or millions of users. They modeled filtration and flocculation. It took them about 30-35 minutes to treat their small samples. They should now know which processes their community uses to treat water, and they should consider how the addition of the proposed developments will affect the treatment process.

Chapter 3

Digging Deeper

Assign the reading for homework, along with the questions in **Check Your Understanding** if desired.

Assessment Opportunity

Reword or restructure the questions in **Check Your Understanding** for a brief quiz. Use the quiz (or a class discussion of the questions in the textbook) to assess your students' understanding of the main ideas in the reading and the activity. A sample question is provided below:

Question: Label the diagram below of a typical water-treatment process. Write a short description explaining each process.
Answer:

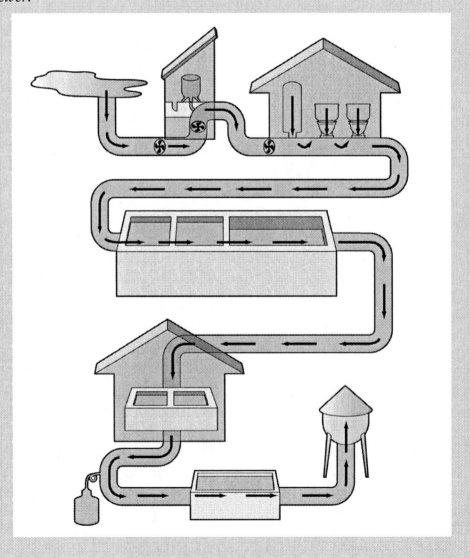

NOTES

Digging Deeper

WATER TREATMENT

Water purification occurs in nature. Evaporation and condensation separates the water from substances that are dissolved in it. Bacteria convert dissolved organic contaminants into simple and harmless compounds. Sand and gravel filter out suspended material that makes water cloudy. Rainwater and ground water that is purified by bacteria and filtered by sediment provide clean drinking water. Unfortunately, the demand for water is much greater than nature's supply, so communities rely on municipal water-treatment plants.

No matter where a community's water comes from, it must be treated in a facility similar to the one shown in *Figure 1*, because even where water is used carefully, pollutants are bound to be introduced. You cannot wash clothes, take a shower, or flush the toilet without adding pollutants to the water! The goal of water treatment is to produce water that is safe for drinking and does not have an objectionable taste, smell, or appearance. Treatment of wastewater from municipal sewage systems involves a series of steps: screening, flocculation, filtering, and disinfecting. Depending on your local water sources and treatment needs, your community's water-treatment process may be slightly different than what is presented here. Keep in mind also that the wastewater that passes through home septic systems is not treated at all: it goes directly into the ground and tends to add to the pollutant levels in local ground water supplies.

Figure 1 Water is made safe for human use at a water-treatment plant.

R 199

NOTES

Chapter 3

Earth's Natural Resources Water Resources

Screening, Flocculation, and Settling

Sewage is first passed through screens to remove large pieces of debris and other solids. Then it flows slowly through sedimentation tanks, where gritty mineral material settles to the bottom and lightweight sludge floats to the top. To aid in the settling process, alum (aluminum sulfate, $Al_2(SO_4)_3$) and slaked lime (calcium hydroxide, $Ca(OH)_2$) are sometimes added to the sewage to cause particles of sediment to coagulate (stick together) to form larger clumps called flocs, which settle relatively fast. The sludge is sent to a separate container where the organic matter in it is digested by microorganisms.

Filtration

In the next treatment step, called filtration, the water passes through sand, gravel, and activated carbon (charcoal). These filters help remove minute particles and even odors that give water a bad smell or taste. If the water contains inorganic (non-living) components that cannot be adequately removed by filtering, a chemical exchange process—called ion exchange—can be used. This chemical process is effective in treating hard water, which is water that does not easily produce suds in the presence of soap. It can also be used to treat arsenic, chromium, excess fluoride, nitrates, radium, and uranium. As part of this stage, the wastewater is aerated (exposed to the oxygen of the atmosphere), which greatly speeds up the natural oxidation and decomposition of organic matter in the water by the activity of microorganisms.

Disinfection

Water is disinfected before it enters the distribution system. Disinfectants kill remaining dangerous microbes. Chlorine is a very effective disinfectant, and residual concentrations of chlorine guard against biological contamination in the water distribution system. Ozone is a powerful disinfectant, but it is not effective in controlling biological contaminants in the distribution pipes. Sometimes the disinfecting process produces byproducts. Disinfecting byproducts (DBPs) are contaminants that form when chlorine reacts with organic matter that is in treated drinking water. Long-term exposure to some DBPs may increase the risk of cancer or have other adverse health effects. The United States Environmental Protection Agency regulates the presence of four DBPs in drinking water, and scientists are continuing to study the issue.

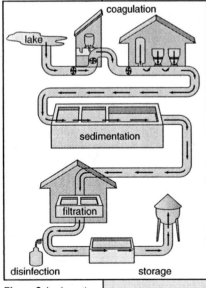

Figure 2 A schematic of a water-treatment plant.

Blackline Master Water Resources 6.1
Water-Treatment Plant

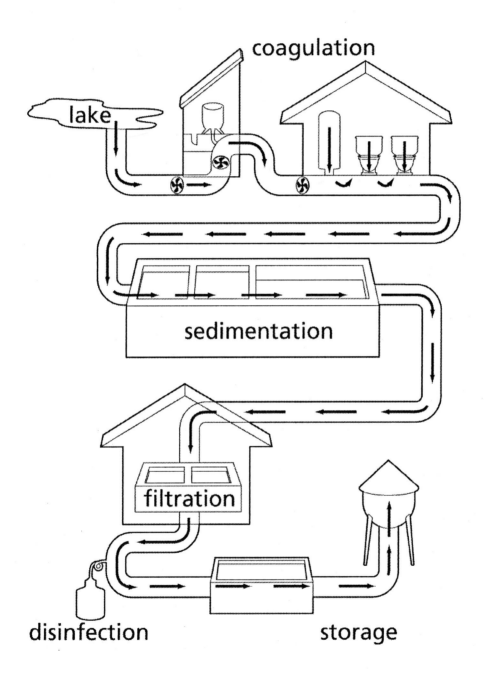

Levels of Treatment

Primary treatment involves only settling and perhaps chlorination. Such treatment is better than nothing, but the water is unsatisfactory for human use. Secondary treatment involves filtering and oxidation by microorganisms, and results in wastewater with much better quality. In some places, tertiary treatment is used. Tertiary treatment, involves flocculation, disinfecting, and use of specific additives to remove undesirable excess compounds of iron, manganese, etc., in addition to the processes of primary and secondary treatment. Tertiary treatment is designed to produce high-quality domestic water that can be added to public water supplies—although the public is still not very willing knowingly to drink treated sewage!

Treatment Costs

On a per-gallon basis, water is cheap. On average, water costs are slightly more than $2 per 1000 gallons, although the costs tend to be lower for large water systems. Treatment accounts for about 15% of that cost. Other costs are for equipment (the treatment plant and distribution system) and labor for operation and maintenance of the system.

Water Hardness

As water moves through soil, sediment, or bedrock, it dissolves some of the minerals. Water hardness reflects the concentration of dissolved solids in the water, mainly calcium and magnesium. Ground water tends to be harder than surface water, because it has had more time to be in contact with the solid materials. Hard water is also more common in areas where the bedrock is limestone, which consists of the calcium carbonate mineral calcite, which is slightly soluble in water. Some degree of hardness in drinking water is considered healthy, because calcium and magnesium are essential nutrients. Hardness also lowers the concentration of lead dissolved from the lead-based solder that is often used in copper plumbing systems. A disadvantage of hard water is that it reduces soapsuds. It also can leave hard, white deposits in teapots, on shower walls, and in water heaters and boilers. Water softeners reduce hardness by replacing the calcium and magnesium ions with sodium ions. A problem with water softeners is that high sodium levels in drinking water are undesirable for persons with high blood pressure.

Figure 3 Damaged pipes.

Check Your Understanding

1. Describe how water is purified in nature.

2. Draw a diagram of a typical water-treatment process. Label the processes and show the flow of water from source to community water mains.

3. What are the advantages and disadvantages of softening water during treatment?

Teaching Tip

Use **Blackline Master Water Resources 6.1, Water-Treatment Plant** to produce an overhead of the diagram in *Figure 2* on page R200. Incorporate this overhead into a class discussion on the process of water treatment. Ask students how the steps outlined in the diagram compare to the steps used in your own community. Consider asking students to draw a similar diagram that shows the water treatment process in your community.

Check Your Understanding

1. Natural water purification involves evaporation and condensation, which separate water from substances that are dissolved in it. Bacteria convert dissolved organic contaminants into simple and harmless compounds, and sand and gravel filter out suspended material that makes water cloudy.

2. Student diagrams should resemble the diagram on page R200 of the student text.

3. Water softeners reduce hardness by replacing the calcium and magnesium ions with sodium ions, but the high sodium levels in drinking water can cause problems for people with high blood pressure.

Assessment Tool

Check Your Understanding Notebook-Entry Evaluation Sheet
Use this sheet to evaluate the extent to which students understand the key concepts explored in **Activity 6** and explained in **Digging Deeper**, and to evaluate the students' clarity of expression.

Chapter 3

Earth's Natural Resources Water Resources

Understanding and Applying What You Have Learned

1. Assume for a moment that the new development will require that your community waterworks be expanded. Create a list of three or four concessions that developers could make that would lessen the economic impact of the proposed expansion.

2. In the investigation, you treated a water sample. What steps in the complete water treatment process were you modeling? What additional steps would be required to make your sample safe for drinking?

3. How do concerns regarding the pollution of ground water differ from those surrounding the pollution of surface water? Does your community treat these water sources differently? Why or why not?

Preparing for the Chapter Challenge

Write a short essay for inclusion in your chapter report in which you address the following questions:

1. The construction of these new developments can add sediment to local streams. How might this adversely affect your water-treatment plant? How could this negative effect be decreased?

2. How will your community's water-treatment plants handle the extra water required by these new developments? How will the additional developments affect the cost of water in your community?

Inquiring Further

1. **Softening water**

   Line three funnels with filter paper. Place sand in one, Calgon in one, and leave the third empty. Use this to design an experiment to find out whether Calgon™ or natural filtration by sand is a better method of softening water.

    Check your plan with your teacher before carrying it out. Do not drink the water. Wash hands after the activity.

2. **Recycling water in space**

   Investigate and report on current techniques involved with recycling water on long-term space ventures.

3. **Purified, distilled, spring, and mineral water**

   Investigate the differences between purified water, distilled water, spring water, and mineral water.

4. **Water analyses in your community**

   Investigate the following analyses conducted at water-treatment facilities: bacteriological analyses, general chemical and physical analyses, metal analysis, organic-compound analyses, radionuclides.

Understanding and Applying What You Have Learned

1. Concessions that developers could make include:
    *   equipping structures with low-flow appliances (shower heads, toilets, washing machines)
    *   equipping public restrooms with automatic-shutoff faucets
    *   using landscaping plants that are adapted to the local climate
    *   cutting down on area of grass lawns
    *   catching water on the roofs of buildings for landscape watering
    *   restricting industrial and commercial facilities to low-water-use applications (e.g., office parks rather than manufacturing activities)

2. Students modeled the processes of flocculation and settling in **Part A** and filtration and disinfection in **Part B** of the investigation. Several steps in the tertiary treatment were not modeled in the investigation, mainly the addition of additives to remove excess amounts of undesirable dissolved constituents like iron and manganese. Additionally, a very important step that was not modeled is the quality testing of the end-product water from the treatment process.

3. See **Activity 5, Digging Deeper**, page R189. Answers will vary depending upon your community.

Preparing for the Chapter Challenge

In **Activity 6,** students have determined two ways to improve the quality of water from a polluted resource. They have learned that it is necessary to use expensive equipment to provide the large volumes of water needed for a community. These methods also demand increases in other natural resources: for example, the energy to run the treatment facilities and the equipment that will be used. Their final report needs to address:
    *   the increase in quantities of water that will be used for domestic and commercial groups
    *   how these increases fit into their current water budget
    *   the added costs for treatment

For assessment, remind students that their essays should include a discussion on:
    *   whether or not the addition of these developments will strain the water-treatment capabilities of your town
    *   what expense will be incurred

Inquiring Further

1. Softening water
Using Calgon would be the most effective way to soften the water. Calgon contains a combination of chemical additives that are specifically designed to either sequester or precipitate out metal ions that are in solution, thus effectively softening hard water.

Chapter 3

2. Recycling Water in Space

For long-term space ventures, a supply of water is taken into space. However, there is not enough room to carry enough water to last through the entire mission. Therefore, programs have been developed that reclaim wastewater from fuel cells, from urine, from oral hygiene and hand washing, and by condensing humidity from the air. Students can learn more about how this is done by visiting the *EarthComm* web site.

3. Purified, distilled, spring, and mineral water

- Purified water is water that is treated in various ways to reduce the concentration of microorganisms and dissolved and suspended materials.
- Distilled water is extremely pure water produced by evaporation and condensation.
- Spring water is water that emerges from springs (places where the ground-water table intersects the ground surface), resulting in a flow of water from the subsurface to the land surface. Depending on the source, spring water varies greatly in quality.
- Mineral water, which varies considerably from brand to brand, is water that contains certain minerals—like calcium or magnesium—which are thought by some to be beneficial. The dissolved minerals are either contained naturally in the water or added in the bottling process.

4. Water analyses in your community

Student responses will vary depending upon the water-treatment facilities found in your community. They can visit the *EarthComm* web site to find additional information about various methods of water analyses.

NOTES

Earth Science at Work

ATMOSPHERE: *Maintenance Engineer*
Airlines employ engineers to maintain, inspect, and repair aircraft. They are responsible for maintaining wheels and brakes, flight control systems, and ice and rain protection systems.

BIOSPHERE: *Medical Technologist*
Some bacteria, viruses, and fungi are hazardous to human health. Medical-laboratory technologists are trained to detect and identify any microorganism.

CRYOSPHERE: *Figure Skater*
An adequate supply of water to "flood" an ice-skating surface is essential in maintaining good ice conditions. Figure skaters depend on a smooth ice surface to safely perform their jumps, spins, and intricate footwork.

GEOSPHERE: *Well Driller*
Many areas rely on ground water as their municipal water supply. Wells must be drilled to reach these water reservoirs.

HYDROSPHERE: *Public Works Employee*
The planning, administration, operation, and maintenance of the wastewater collection of a city require an understanding of wastewater treatment options available. This is the responsibility of a public works department.

How is each person's work related to the Earth system, and to water resources?

R 203

NOTES

Chapter 3

Water Resources and Your Community: End-of-Chapter Assessment

1. Most of the water on Earth is held in
    a) glacial ice.
    b) large lakes and streams.
    c) groundwater.
    d) the oceans.

2. Of all the fresh water on Earth (liquid and ice), most is held in
    a) glacier ice.
    b) large lakes and streams.
    c) groundwater.
    d) the atmosphere.

3. The total volume of water on Earth is
    a) almost constant but continuously in motion from one reservoir to another.
    b) increasing because water evaporated from oceans falls as precipitation on land.
    c) increasing because water held in groundwater is brought to the surface for human use.
    d) decreasing because water evaporated from the ocean gets tied up in the atmosphere.

4. Approximately what percentage of water on Earth is available for drinking water?
    a) 1%.
    b) 10%.
    c) 25%.
    d) 65%.

5. Why is the Earth's surface water not really a closed system?
    a) water is lost to the biosphere and gained from the system as rain or snow.
    b) water is lost for long periods of time in rocks and gained through volcanic eruptions.
    c) water is gained as meteorites impact the Earth.
    d) some of the water that evaporates never returns to the Earth's surface.

6. How can a small stream continue to flow when there has been no rain for weeks?
    a) Streams are closed systems.
    b) Streams take a long time to drain after it rains.
    c) Groundwater flows to the base of the stream.
    d) Plants supply water to streams through their roots.

7. Which of the following illustrates the relationship between fluxes and reservoirs in a system?
     a) the volume of water each year that evaporates from oceans.
     b) the depth to which rainwater soaks into the ground.
     c) the volume of water that a lake can hold.
     d) the number of inches of rain that fall in Kansas.

8. An aquifer with a free connection upward to the Earth's surface is known as
     a) an open aquifer.
     b) an unconfined aquifer.
     c) an aquiclude.
     d) a saturated aquifer.

Refer to the diagram below to answer questions 9 and 10.

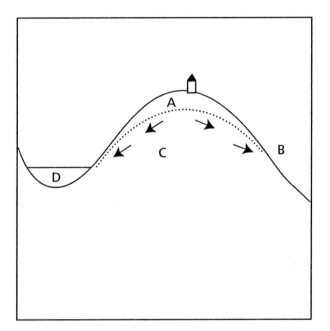

9. A farmer plans to have a well drilled near his house, which is shown in the diagram. Why should he drill at least as deep as the dashed line labeled A?
     a) The zone above the dashed line never contains water.
     b) The zone above the dashed line contains no water.
     c) The zone below the dashed line always contains water.
     d) The dashed line shows the depth at which rock or soil is currently saturated with water.
     e) Using the house for scale, the well would be at least 20 meters deep, a safe depth for producing water.

10. If the farmer walked to location B, what would you expect him to find at that spot?
   a) Dry soil.
   b) Few plants.
   c) A lake.
   d) A spring.

11. Groundwater flows
   a) from where the ground is wet to where it is dry.
   b) from where the water table is high to where it is low.
   c) from cold climates to warm climates.
   d) through lake bottoms into river channels.
   e) clockwise in the northern hemisphere.

12. The major disadvantage to using desalination to produce fresh water is that
   a) not enough cities are located near the ocean.
   b) desalination is an expensive process.
   c) desalination does not remove enough salt to protect human health.
   d) desalination plants pollute the environment.

13. The two largest uses of freshwater in the United States are:
   a) livestock and mining.
   b) power plants and irrigation.
   c) homeowners and manufacturing industries.
   d) homeowners and mining.

14. Traditional methods of irrigation are inefficient because
   a) too much water lands on the leaves of crops.
   b) water evaporates or soaks into the soil between crops.
   c) farmers plants crops too far apart.
   d) farmers plow the ground too deeply.

15. Pumping water from a well tends to
   a) lower the water table around the well.
   b) raise the water table around the well.
   c) decrease the quality of groundwater.
   d) increase the quality of groundwater.

16. Why is untreated domestic sewage a threat to the quality of water supplies?
   a) It cannot be treated in most communities.
   b) It produces phosphates that endanger humans.
   c) It contains fluoride and dihydrogen monoxide.
   d) It contains disease-causing microorganisms.

17. Why does the use of fertilizers have a negative impact on the quality of surface water?
    a) Fertilizers produce brown foam scum on surface waters.
    b) Fertilizers promote the growth of algae in surface water.
    c) Fertilizers kill most aquatic plants.
    d) Fertilizers raise the oxygen levels in surface waters, killing many fish.

18. Which of the following is NOT an example of how power plants affect water resources or aquatic ecosystems?
    a) power plants add warm carbon dioxide and dissolved oxygen to streams.
    b) the warm water released by power plants holds less dissolved oxygen.
    c) certain species of organisms that favor warm water displace species that favor cooler water.
    d) the sudden increase in the temperature of water can kill organisms.

19. What purpose does chlorine serve in the wastewater treatment process?
    a) It reduces the cloudiness of water.
    b) It kills dangerous microorganisms.
    c) It helps prevent tooth decay.
    d) It softens water, which prevents damage to plumbing.

20. Which of the following shows the correct sequence of steps in wastewater treatment?
    a) disinfection, screening, filtration, and settling.
    b) screening, filtration, disinfection, and settling.
    c) screening, settling, filtration, and disinfection.
    d) filtration, disinfection, screening, and settling.

Chapter 3

Answer Key

1. d; 2. c; 3. a; 4. a; 5. b; 6. c; 7. a; 8. b; 9. d; 10. d; 11. b; 12. b;
13. b; 14. b; 15. a; 16. d; 17. b; 18. a; 19. b; 20. c

Teacher Review

Use this section to reflect on and review the investigation. Keep in mind that your notes here are likely to be especially helpful when you teach this investigation again. Questions listed here are examples only.

Student Achievement

What evidence do you have that all students have met the science content objectives?

_____

Are there any students who need more help in reaching these objectives? If so, how can you provide this?_____

_____

What evidence do you have that all students have demonstrated their understanding of the inquiry processes?_____

_____

Which of these inquiry objectives do your students need to improve upon in future investigations? _____

_____

What evidence do the journal entries contain about what your students learned from this investigation? _____

_____

Planning

How well did this investigation fit into your class time?_____

_____

What changes can you make to improve your planning next time? _____

_____

Guiding and Facilitating Learning

How well did you focus and support inquiry while interacting with students?

_____

What changes can you make to improve classroom management for the next investigation or the next time you teach this investigation? _____

_____

How successful were you in encouraging all students to participate fully in science learning? _____

_____

How did you encourage and model the skills values, and attitudes of scientific inquiry? _____

_____

How did you nurture collaboration among students? _____

_____

Materials and Resources

What challenges did you encounter obtaining or using materials and/or resources needed for the activity? _____

_____

What changes can you make to better obtain and better manage materials and resources next time? _____

_____

Student Evaluation

Describe how you evaluated student progress. What worked well? What needs to be improved? _____

_____

How will you adapt your evaluation methods for next time? _____

_____

Describe how you guided students in self-assessment. _____

_____

Self Evaluation

How would you rate your teaching of this investigation? _____

_____

What advice would you give to a colleague who is planning to teach this investigation? _____

_____

Chapter 3

NOTES

EARTHCOMM ASSESSMENTS

Assessing the *EarthComm* Notebook

- *EarthComm* Notebook-Entry Evaluation Sheet
- *EarthComm* Notebook-Entry Checklist
- Think About It Evaluation Sheet
- Investigate Notebook-Entry Evaluation Sheet
- Check Your Understanding Notebook-Entry Evaluation Sheet

Assisting Students with Self Evaluation

- Student Evaluation of Group Participation
- Student Ratings and Self Evaluation

Assessing Student Presentations

Student Presentation Evaluation Form

- Student Presentation Evaluation Form
- Student Ratings and Self Evaluation

References:

- Doran, R., Chan, F., and Tamir, P. (1998). *Science Educator's Guide to Assessment.*
- Leonard, W.H., and Penick, J.E. (1998). *Biology – A Community Context.* South-Western Educational Publishing. Cincinnati, Ohio.

Assessment Blackline Master 1 – Journal Entry

Name: _____ Date: _____ Module: _____

Explanation: The *EarthComm* notebook is an important component of each *EarthComm* module. In using the notebook as you investigate Earth science questions, you are mirroring what scientists do. The criteria, along with others that your teacher may add, will be used to evaluate the quality of your Notebook entries. Use these criteria, along with instructions within investigations, as a guide.

Criteria

1. Entry Made

    1   2   3   4   5   6   7   8   9   10   _____

Blank   Nominal   Above average Thorough

2. Detail

    1   2   3   4   5   6   7   8   9   10   _____

Few dates   Half the time   Most days   Daily

Little detail   Some detail   Good detail   Excellent detail

3. Clarity

    1   2   3   4   5   6   7   8   9   10   _____

Vague   Becoming clearer   Clearly expressed

Disorganized   well organized

4. Data Collection/Analysis

    1   2   3   4   5   6   7   8   9   10   _____

Data collected   Data collected,   Data collected

Not analyzed   some analyzed   and analyzed

5. Originality

    1   2   3   4   5   6   7   8   9   10   _____

Little evidence   Some evidence   Strong evidence,

of originality   of originality   of originality

6. Reasoning/Higher-Order Thinking

    1   2   3   4   5   6   7   8   9   10   _____

Little evidence   Some evidence   Strong evidence

of thoughtfulness of thoughtfulness of thoughtfulness

7. Other

    1   2   3   4   5   6   7   8   9   10   _____

8. Other

    1   2   3   4   5   6   7   8   9   10   _____

EarthComm Notebook-Entry Checklist

Name: _____ Date: _____ Module: _____

Explanation: The *EarthComm* Notebook is an important component of each *EarthComm* module. In using the notebook as you investigate Earth science questions, you are mirroring what scientists do. The criteria, along with others that your teacher may add, will be used to evaluate the quality of your Notebook entries. Use these criteria, along with instructions within investigations, as a guide.

Criteria:

1. Makes entries   _____

2. Provides dates and details   _____

3. Entry is clear and organized   _____

4. Shows data collected   _____

5. Analyzes data collected   _____

6. Shows originality in presentation   _____

7. Shows evidence of higher-order thinking   _____

8. Other   _____

9. Other   _____

                                              _____

Comments:

Think About It Evaluation Sheet

Name: _____ Date: _____ Module: _____

	Strong		Fair		No Entry
Shows evidence of prior knowledge	4	3	2	1	0
Reflects discussion with classmates	4	3	2	1	0

Think About It Evaluation Sheet

Name: _____ Date: _____ Module: _____

	Strong		Fair		No Entry
Shows evidence of prior knowledge	4	3	2	1	0
Reflects discussion with classmates	4	3	2	1	0

Think About It Evaluation Sheet

Name: _____ Date: _____ Module: _____

	Strong		Fair		No Entry
Shows evidence of prior knowledge	4	3	2	1	0
Reflects discussion with classmates	4	3	2	1	0

Investigate Notebook-Entry Evaluation Sheet

Name: _____ Date: _____ Module: _____

Criteria

1. Completeness of written investigation

   1   2   3   4   5   6   7   8   9   10   _____

   Blank   Incomplete   Thorough

2. Participation in investigations

   1   2   3   4   5   6   7   8   9   10   _____

   None or little;   Needs minimal guidance,   Leads, is inquisitive
   unable to guide   sometimes helping others persistent, focused.
   self

3. Skills attained

   1   2   3   4   5   6   7   8   9   10   _____

   Few skills   Tends to use some   High degree of
   evident   appropriate skills   appropriate skills used

4. Investigation Design

   1   2   3   4   5   6   7   8   9   10   _____

   Variables not   Sometimes   Considers variables.
   considered   considers variables,   Sound rationale for
   techniques   uses logical techniques   techniques
   illogical

5. Conceptual understanding of content

   1   2   3   4   5   6   7   8   9   10   _____

   No evidence   Approaches understanding Exceeds expectations
   of understanding   of most concepts   for content attainment

6. Ability to explain/discuss inquiry

   1   2   3   4   5   6   7   8   9   10   _____

   Unable to   Some ability to   Uses scientific reasoning
   articulate   explain/discuss   to explain any
   scientific thought   the inquiry   aspect of the inquiry

7. Other

   1   2   3   4   5   6   7   8   9   10   _____

8. Other

   1   2   3   4   5   6   7   8   9   10   _____

Check Your Understanding Notebook-Entry Evaluation Sheet

Name: _____ Date: _____ Module: _____

	Strong		Fair		No Entry
Shows evidence of prior knowledge	4	3	2	1	0
Reflects discussion with classmates	4	3	2	1	0

Check Your Understanding Notebook-Entry Evaluation Sheet

Name: _____ Date: _____ Module: _____

	Strong		Fair		No Entry
Shows evidence of prior knowledge	4	3	2	1	0
Reflects discussion with classmates	4	3	2	1	0

Check Your Understanding Notebook-Entry Evaluation Sheet

Name: _____ Date: _____ Module: _____

	Strong		Fair		No Entry
Shows evidence of prior knowledge	4	3	2	1	0
Reflects discussion with classmates	4	3	2	1	0

Student Presentation Evaluation Form

Student Name_____ Date_____

Topic_____

	Excellent		Fair		Poor
Quality of ideas	4	3	2	1
Ability to answer questions	4	3	2	1
Overall comprehension	4	3	2	1

COMMENTS:

Student Presentation Evaluation Form

Student Name_____ Date_____

Topic_____

	Excellent		Fair		Poor
Quality of ideas	4	3	2	1
Ability to answer questions	4	3	2	1
Overall comprehension	4	3	2	1

COMMENTS:

Student Evaluation of Group Participation

Key:
4 = Worked on his/her part and assisted others
3 = Worked on his/her part
2 = Worked on part less than half the time
1 = Interfered with the work of others
0 = No work

My name is _____ .   I give myself a _____

The other people in my group are: I give each person:

A. _____   _____

B. _____   _____

C. _____   _____

D. _____   _____

Key:
4 = Worked on his/her part and assisted others
3 = Worked on his/her part
2 = Worked on part less than half the time
1 = Interfered with the work of others
0 = No work

My name is _____ .   I give myself a _____

The other people in my group are: I give each person:

A. _____   _   _____

B. _____   _____

C. _____   _   _____

D. _____   _____

Student Ratings and Self Evaluation

Name: _____ Date: _____ Module: _____

Key:
Highest rating _____
Lowest rating _____

1. In the chart, rate each person in your group, including yourself.

	Names of Group Members				
Quality of Work					
Quantity of Work					
Cooperativeness					
Other Comments					

2. What went well in your investigation?

3. If you could repeat the investigation, how would you change it?

NOTES

NOTES

NOTES

NOTES